DISCARD

SOURCES AND DEVELOPMENT
OF
MATHEMATICAL SOFTWARE

Wayne R. Cowell, editor

Mathematics and Computer Science Division
Argonne National Laboratory

Prentice-Hall, Inc., Englewood Cliffs, New Jersey 07632

Library of Congress Cataloging in Publication Data

Main entry under title:

Sources and development of mathematical software.

(Prentice-Hall series in computational mathematics)
Includes bibliographies and index.
1. Mathematics—Data processing. 2. Numerical
analysis—Data processing. I. Cowell, Wayne R.,
(date). II. Series.
QA76.95.S68 1984   510'.28'542   83-27012
ISBN 0-13-823501-5

Editorial/production supervision: Nancy Milnamow
Cover design: Photo Plus Art
Manufacturing buyer: Gordon Osbourne

PRENTICE-HALL SERIES IN COMPUTATIONAL MATHEMATICS
Cleve Moler, advisor

Printed in the United States of America

10 9 8 7 6 5 4 3 2 1

ISBN 0-13-823501-5

Prentice-Hall International, Inc., *London*
Prentice-Hall of Australia Pty. Limited, *Sydney*
Editora Prentice-Hall do Brasil, Ltda., *Rio de Janeiro*
Prentice-Hall Canada Inc., *Toronto*
Prentice-Hall of India Private Limited, *New Delhi*
Prentice-Hall of Japan, Inc., *Tokyo*
Prentice-Hall of Southeast Asia Pte. Ltd., *Singapore*
Whitehall Books Limited, *Wellington, New Zealand*

TABLE OF CONTENTS

10. The IMSL Library
    — *Thomas J. Aird*   264

    Introduction 264

        IMSL Company History 264
        Company Organization and Advisory Board 267
        Facts about the IMSL Library 268
        Library Contents 269
        Documentation 276
        Support 278
        Availability 279
        Subscription Fees and Policies 280

    Library Development 281

        The Research and Development Organization 283
        The Development System 284

    References 291
    Appendix A: Sample Chapter Documentation 292
    Appendix B: Sample Subroutine Documentation 297
    Appendix C: Fortran Reference Manuals 300

11. The SLATEC Common Mathematical Library
    — *Bill L. Buzbee*   302

    Introduction 302

        Motivations 302

    Organization and Administration 304

        Goals 304

    Library Development 305

        Portability 305
        Standards 306
        Code Selections 306

    Documentation 310

PREFACE

Jim Cody, in Chapter 1, characterizes the effort to produce mathematical software as the building of bridges between numerical analysts who devise algorithms and computer users who need efficient, reliable numerical software. Bridge building is an apt metaphor because it reminds us that a great deal of organization and teamwork is needed after the designer has finished. The most successful mathematical software efforts have been coalitions of software-oriented numerical analysts, programming language specialists, and those who package, deliver, and maintain the software products.

This book is a collection of essays by some who have participated in coalitions to produce mathematical software. A pair of questions was posed to each author, to be answered from his or her experience: How is high-quality mathematical software produced and delivered to the user community? What is the best mathematical software available? These themes of software construction and software critique are woven together in the belief that users will make a better selection from available offerings if they appreciate the factors affecting development, and that developers will be guided by opinions about existing software as well as by the experiences of other developers. The individual authors place different emphasis on these themes, as befit the computational areas considered and their own expertise and interests. The result is not an encyclopedia of mathematical software or software efforts, but rather a large and representative sample of current work for the guidance of students, researchers, and users of mathematical software.

Chapter 1 surveys the important events that have affected the mathematical software endeavor since the watershed Purdue conference of 1970. These events include the development of the "PACKs" — systematized collections of routines that focus on particular computational areas.

Chapters 2-5 discuss four of the PACKs in terms of algorithms, software architecture, documentation, and availability, the costs and organization of the projects, and possible future refinements.

Chapters 6-9 explore further important areas of numerical computation for which active research and development have produced a large body of mathematical software. The authors summarize the mathematical background, examine software engineering and user interface questions, and survey the available software.

Chapters 10-14 provide overviews of five publicly available libraries of mathematical software. In addition to summarizing the contents of these libraries, the authors provide a sense of the technical and organizational problems involved in creating, distributing, and maintaining a product that depends heavily on interaction with the scientific community.

The computer-based techniques used to gather and edit the chapters in this volume represent a venture into the use of a common computer-based writing/editing environment. A critical comment or two may help those who have a similar project in mind.

The chapter authors were invited to submit machine-readable manuscripts (perhaps the term "tapescripts" is better) in which the basic spacing and para-graphing instructions were embedded in the text in the form required by the Unix® processor *troff*. Because the technology is new, only about half the authors were able to comply, but we converted the remaining manuscripts.

Gail Pieper then worked her craft on each file, checking many details that guarantee uniformity among the chapters, preparing an index, and developing "macro" formatting instructions. This application of human intelligence was an essential ingredient of the computer-based production process. The authors and the editor thank Gail for all the detailed work, but most especially for making the technology friendly as well as effective.

We also thank Burton Garbow for reading every chapter and making many comments helpful to the authors and the editor.

Argonne, Illinois   *Wayne R. Cowell*
October 1983

Chapter 1

OBSERVATIONS ON THE MATHEMATICAL SOFTWARE EFFORT

W. J. Cody *
Mathematics and Computer Science Division
Argonne National Laboratory
Argonne, Illinois 60439

INTRODUCTION

John Rice coined the term "mathematical software" in 1969, and focussed attention on the subject the following year with a symposium held as part of a Special Year in Numerical Analysis at Purdue University. The movement spawned by that first meeting has been fruitful. In 1969 only a few individuals worked on what we now call mathematical software, and only a few fortunate computer sites had access to decent numerical programs. Today many talented people work in the field, large collections of good numerical software are widely available, and specialized meetings are common.

A description of the mathematical software effort is difficult because it is so broad. Its domain is that nebulous region between the discovery of numerical algorithms and the consumption of numerical software. On the one hand numerical analysts devise new computational methods, and on the other hand individuals wish to apply effective methods to their immediate problems. It is the job of the mathematical software effort to bridge the gap by packaging numerical analysts' work in software appealing to the consumer. Strictly speaking, work on mathematical software is limited to tasks related to the implementation of numerical algorithms. In practice the spectrum of activities is surprisingly wide because the process of implementation is itself worthy of study. In addition to obvious concerns with program design and testing, there are major concerns with programming practices, documentation standards, software organization and distribution methods. Other activities involve the development of programming tools to partially automate design, implementation, testing and maintenance of software, and work on the computational environment, including the design of arithmetic systems and programming languages properly supportive of good numerical software. Major contributions have been made in each of these areas by individuals who consider their primary interest to be mathematical software.

The published proceedings of the Purdue meeting contain Rice's appraisal of the mathematical software effort as it stood in 1970 [Rice, 1971], including a

---

* This survey was prepared in December 1980. The work was supported by the Applied Mathematical Sciences Research Program (KC-04-02) of the Office of Energy Research of the U.S. Department of Energy under Contract W-31-109-Eng-38.

chronological account of progress. This paper is a similar appraisal of the effort as it stands today. Instead of updating the chronological record, however, we discuss what we consider to be major milestones marking progress to this point. We then examine current problems in the field and future challenges posed by an advancing technology. This work was inspired by a panel session on the same subject that ended the week-long International Seminar on Problems and Methodologies in Mathematical Software Production held November 1980 in Sorrento, Italy, under the sponsorship of The University of Naples and the C.N.R. [Cody, 1982b]. We gratefully acknowledge the contributions of our fellow panelists, B. Ford, T. J. Dekker, M. Gentleman, J. N. Lyness and P. C. Messina, and a responsive audience. With the benefit of leisurely reflection we have reorganized and expanded some of their ideas and combined them with our own thoughts on the matter. We alone are responsible for the selection and expression of the opinions that follow, however.

The reader should be aware that the views presented below may be colored by personal bias, and that other views exist. The surveys and suggestions for research in Gear [1979], Huddleston [1979], Morris [1979], Rice [1979], and Rice *et al.* [1978] are especially recommended to the interested reader.

THE PAST

Many people associate the beginning of the mathematical software effort with Rice's 1969 call for a meeting at Purdue University. The roots go back further, however. While it would be trite to trace them to the first numerical subroutine libraries, we detect an emerging concern for software quality in the early 1960's. By then individuals at The University of Toronto, The University of Chicago, Stanford University, Bell Laboratories and Argonne National Laboratory were critically examining software and advertising their findings through technical reports and discussions at computer user group meetings. The ideas and evaluation techniques were not well enough established for publication in refereed journals, however, and efforts were hampered by poor communications. Often workers at one location were completely unaware of similar work elsewhere. Yet each of these computing centers developed outstanding program libraries by contemporary standards.

In early 1966 J. F. Traub organized SICNUM, the Special Interest Committee on Numerical Mathematics. The group grew quickly; and by midyear, when the first informal *SICNUM Newsletter* appeared, it had a membership of almost 1000. Two articles in the first *Newsletter* typify SICNUM's interests. The first announced the establishment of a working group "to investigate testing and certification techniques for numerical subroutines," and the second announced a SICNUM-sponsored evening session at the 1966 National ACM Conference.

The session included a panel discussion "in the area of machine implementation of numerical algorithms." By constantly emphasizing efforts to improve the quality of numerical software, SICNUM and its successor SIGNUM set the stage for Rice's 1969 call for a symposium.

In that call Rice [1969] defined mathematical software as "computer programs which implement widely applicable mathematical procedures." This contrasts with the definition he later included in the published proceedings, "the set of algorithms in the area of mathematics" [Rice, 1971]. These two definitions illustrate the fundamental confusion between algorithms and computer programs that plagued the early development of numerical software. The realization that an implementation is different from the underlying algorithm marks the emergence of mathematical software as a separate field of endeavor.

That difference was not widely understood in 1969. Despite early admonitions from G. Forsythe [1966] and other prominent researchers, most numerical analysts still believed that their work was finished when they had defined an algorithm. Programming was a job for programmers; numerical analysts programmed only when it was necessary for their research (and pure mathematicians never programmed). A university professor seeking advancement and tenure shied away from working on numerical software. As a result most of the early work was concentrated in government and industrial laboratories, with only a few selfless university people involved. Unfortunately, the same attitudes are still common. While work on mathematical software has gained some professional stature and there are more talented people involved in the effort today than were involved in 1969, many others still do not dare to become involved if they seek promotion. This is still especially true at many universities.

Three software projects that greatly influenced the mathematical software effort began about the time of the Purdue meeting. Each project — IMSL, NAG and NATS — resulted in a widely-used collection of high-quality numerical software. Certain software collections were publicly available before this. Computer user's groups had organized program repositories by the early 1960's, and the IBM Scientific Subroutine Package (SSP) was available on the IBM 7094, for example. Although these collections contained a few good programs, their general reputations were deservedly notorious. The IMSL, NAG and NATS collections were the first to combine quality with wide distribution.

IMSL (International Mathematical and Statistical Libraries, Inc.) was founded in 1971 by some of the people involved in the IBM SSP effort. It delivered the first purely commercial numerical subroutine library to IBM customers a year later. By mid 1973 the library had also been delivered to UNIVAC and CDC customers, and for the first time the same library of numerical programs became available on a variety of computing equipment. This enabled numerical programmers to write and distribute applications programs without worrying about the availability of a decent support library. Today IMSL supports most major computers. The success of this venture is suggested by the number of computing centers now relying on IMSL and its competitors for their core library, thus freeing local personnel to develop the specialized programs necessary for their own work.

IMSL's main competition comes from the NAG and, to a lesser extent, PORT libraries. NAG, originally Nottingham Algorithms Group but now Numerical Algorithms Group, was organized about 1970 in Great Britain as a cooperative venture between universities using ICL 1906A computers. Supported by heavy government subsidies, NAG extended its coverage to other machines and now seeks to become self-supporting. The PORT library is a product of Western Electric arising from the early library work at Bell Laboratories. It is not aggressively marketed and is therefore not as widely used as the IMSL and NAG libraries.

The NATS project, National Activity to Test Software, was conceived in 1970 and funded in 1971 by the National Science Foundation and the Atomic Energy Commission to study problems in producing, certifying, distributing and maintaining quality numerical software [Boyle et al., 1972]. This was a cooperative effort between personnel at Argonne National Laboratory, Stanford University, The University of Texas at Austin and scattered test sites to examine software production as a research problem. Intrinsic to this effort was the production of two software packages, the EISPACK collection of matrix eigensystem programs [Garbow et al., 1977; Smith et al., 1976] and the FUNPACK collection of special function programs [Cody, 1975]. The project formally ended with the distribution of extended second releases of both packages in 1976.

By any measure, the NATS project was a spectacular success. Not only did it produce superior software, but it also pioneered in organizational and technical achievements that are still being exploited. For example, the project developed an early system for automated program transformation and maintenance [Smith, Boyle and Cody, 1974] that led directly to current research on the TAMPR system [Boyle and Matz, 1977]. We believe that the NATS aids were developed before similar aids for program transformation were developed at the Jet Propulsion Laboratory [Krogh, 1977] and within the IMSL [Aird, 1977] and NAG [Du Croz, Hague and Siemieniuch, 1977] projects. They were certainly the first to be successfully used in a software project. Important as such technical achievements were, however, they were overshadowed by the organizational concepts the project developed. Machura and Sweet [1980] recently stated, "The most important lesson learned from the EISPACK project is that the development and distribution of quality software can be achieved by the joint efforts of several different organizations." Before the NATS success software was typically developed with the limited resources of one organization; since the NATS success cooperative ventures have become common.

None of this would have mattered if the NATS software had not been superior. Fortunately, the software produced by the project was well received and is still considered to be some of the best available. EISPACK, in particular, set and met high standards for performance, transportability and documentation. It has become a paradigm for thematic numerical software collections with the term "PACK" now implying all that is good in numerical software. Attesting to EISPACK's influence, the following PACKs in addition to FUNPACK either exist or are in advanced planning stages: ELLPACK [Rice, 1977], FISHPAK [Swarztrauber and Sweet, 1979], ITPACK [Grimes *et al.*, 1978], LINPACK [Dongarra *et al.*, 1979], MINPACK [Moré, Garbow and Hillstrom, 1980], PDEPACK [SCCS, 1975], QUADPACK [Piessens *et al.*, 1978], ROSEPACK [Coleman *et al.*, 1980], SPARSPAK [George and Liu, 1981], TESTPACK [Buckley, 1981] and TOOLPACK [Osterweil, 1981]. While many of these are superb packages, the use of a "PACK" name does not automatically instill quality.

It is disappointing that the NATS experience was not fully exploited. Attempts to establish a central organization for software production based on the NATS concept [Cowell and Fosdick, 1975 and 1977] failed for various political and technical reasons. This denied segments of the numerical software community access to experienced people and important resources. Many of the projects mentioned above had to rely on their own resources to coordinate production, certification and distribution of their software, duplicating similar capabilities already developed in other projects.

The first Purdue symposium was followed by two other important meetings. SIGNUM sponsored a meeting in 1971 in Ljubljana, Yugoslavia, concurrent with the 1971 IFIP Congress, that ultimately led to the establishment in late 1974 of WG 2.5, the IFIP Working Group on Numerical Software. Members of the working group now represent numerical software interests in language and hardware standardization efforts, often with detailed advice from the group as a whole. In addition the working group has organized several international workshops on software topics and has drafted and published several technical reports.

The other important meeting was the second software symposium held at Purdue in 1974. While its influence was not as great as that of the first meeting, it did lead to the establishment of the *ACM Transactions on Mathematical Software* with John Rice as editor. Since its appearance in early 1975 with papers from the Purdue meeting, *TOMS* has complemented the *SIGNUM Newsletter* by providing an outlet for refereed numerical software papers.

The second Purdue meeting was also noteworthy for the first open discussion of the BLAS, or Basic Linear Algebra Subprograms [Lawson *et al.*, 1979]. As the name implies, the BLAS are a collection of Fortran subprograms implementing low level operations, such as the dot product, from linear algebra. The project was originally organized in 1973 as a private effort to reach consensus on names, calling sequences and functional descriptions for such programs, but it quickly became a cooperative effort officially sanctioned by ACM-SIGNUM. Once conventions had been agreed on, it was possible for linear algebra programs to do fundamental operations in a uniform way. This was already a significant accomplishment, but the group also prepared efficient implementations of the BLAS for most popular computers. The project's most important contribution, however, was the concept of establishing popular "conventions" as opposed to official standards. Language designers are reluctant to augment standard languages to include something useful to only a small group. Even if that is done, years pass before the new feature is available in compilers. The establishment of private conventions outside language standards is a more reasonable approach, and the BLAS project demonstrates that it is also a practical one. As with NATS, this lead has not been fully exploited.

It is difficult to assess the importance of events in the immediate past, but we believe that the recently proposed IEEE standard for floating-point arithmetic will prove to be important. One major disappointment in numerical work has been the general lack of progress in designing clean computer arithmetic systems. High-quality software is supposed to be fail-safe and transportable; it must work properly regardless of quirks in the host arithmetic system. Software production is seriously hampered when computer arithmetic violates simple arithmetic properties such as

$$1.0 * X = X$$
$$X * Y = Y * X,$$

and

$$X + X = 2.0 * X.$$

There exist mainframes of recent design in which each of these properties fails for appropriate floating-point X and Y. Worse yet, on some machines there exist floating-point $X > 0.0$ such that

$$1.0 * X = 0.0,$$
$$X + X = 0.0,$$

or

$$[\sqrt{(x)}]^2 = overflow \ or \ underflow.$$

All these anomalies are traceable to engineering economies [Cody, 1982a].

Computer designers repeatedly ignore complaints about such mathematical atrocities, and new anomalies seem to appear with each new machine.

That may be changing, however. By 1977 technology had advanced to the point where small microprocessor manufacturers considered adding floating-point arithmetic to their chips. In an unprecedented move they turned to numerical analysts for advice. The result was the formation of a subcommittee of the IEEE Computer Society to draft a standard for binary floating-point arithmetic. The resultant draft [Stevenson, 1981] differs radically from existing arithmetic systems. Not only is it free of anomalies, but it also contains new features specifically requested and designed by numerical analysts with software experience. The first chips based on this proposed standard have now appeared [Intel, 1980], and the first microcomputers are being delivered [Klema, 1981].

These, then, are the milestones leading to where we stand today: the early work at isolated computing centers; the establishment of SICNUM; the two Purdue symposia; the establishment of commercial numerical software libraries; the NATS project and the EISPACK package; the establishment of IFIP WG 2.5 and of TOMS; the BLAS; and the drafting of a standard for floating-point arithmetic. Each of these events added something new and important to the movement. There have also been some disappointments. We mention in particular the failure to achieve full professional recognition for software work (especially at universities), the failure to fully exploit the NATS experience, and the general lack of progress in mainframe arithmetic design.

THE PRESENT

While the problems we face today are similar to those we faced ten years ago, the solutions are more complicated. We are still concerned about the production of high-quality transportable software, but we expect more from such software than we did in the past. Therefore it is more difficult to produce.

The last section pointed to many thematic numerical software packages. Some such as EISPACK and LINPACK are complete, while others such as MINPACK and QUADPACK are still under development. We believe it is significant that most of the early success involved linear algebra programs. It is true that linear algebra is a fundamental mathematical tool for other problem areas, such as optimization and partial differential equations, and that good software for these other problems was not likely to be produced until good linear algebra programs were ready. But it is also true that linear algebra had reached an algorithmic maturity that invited software production. The algorithms were well developed, well understood and backed by error analysis that clearly displayed the limitations of software implementations. Because the production of EISPACK required minimal algorithmic work, the producers could concentrate on recasting algorithms to enhance desirable software attributes. The effort thus produced a significant software package within three years of funding. In contrast, the MIN-PACK effort required about five years to produce its first small package. This lengthy development time reflects the difficulty of the task and is likely to be typical of future projects. As in many other fields, prominent researchers in optimization do not agree on the best algorithms; new methods frequently appear accompanied by confusing claims of superiority over existing methods and programs. The situation is common in a vigorous, dynamic research field, but it does not encourage the quick production of high-quality software. All the "easy" implementations may have been done already.

Despite these difficulties, we believe that some additional problem areas could be harvested for software now. We are frankly puzzled by the lack of an effort in ordinary differential equations, for example. Existing algorithms seem to be well-enough understood, but no group has emerged with the necessary dedication and support.

There is one other little-understood aspect of successful numerical software projects that we believe to be important. Part of the variation in quality in the numerous PACKs previously mentioned is due to an improper appreciation of a fundamental lesson from the NATS project. We stated above that linear algebra was in a good algorithmic position when the EISPACK work began. That does not mean the field was stagnant, however; new algorithms were being introduced. The project deliberately ignored new work because it felt that algorithms had to prove themselves before being included. Further, the project found that there is a one- to two-year delay between the completion of the first pass at software and its final release. This time is spent iteratively testing, revising and documenting to ensure that the package does what it claims. Thus there must be a one- to two-year moratorium on the introduction of new material into the package. This simple discovery has far-reaching implications. Algorithmic researchers find it almost impossible to observe such a moratorium; they are intent on wide distribution of their latest discoveries. Further, they cannot effectively polish software they feel to be inferior. Therefore, control over software projects should be vested in individuals who understand and are dedicated to software production rather than in individuals who primarily produce algorithms. Algorithm producers should be involved in software packaging, but they should not control it.

This approach has another advantage. Software packages require a uniformity of style to simplify documentation and maintenance. As EISPACK demonstrated, different programs may contain large segments of code that can be rendered almost identical, e.g., by using similar variable names and identical labels. The elements of a package also must adopt a uniform philosophy for detecting and reporting errors. The necessary surgery to produce package uniformity is best done by someone with no particular attachment to the original programs.

Aside from algorithmic development, the most difficult problem facing us today is testing. There are two fundamentally different reasons for testing, hence two fundamentally different approaches. On the one hand, algorithm creators want to show that their creations are in some way superior to existing algorithms, and they approach performance testing as a contest. The tests they design specifically highlight whatever advantage the new algorithm may have; there is usually no attempt to uncover weaknesses in the algorithm or its implementation.

On the other hand, the selection of software for general use requires complete performance evaluation. Usually some duplication of purpose is acceptable in building a library, for example, so the concern is more with eliminating unacceptable programs and in matching programs to problem characteristics than in determining the "best" program. Tests for this purpose should aggressively exercise a program in ways that will detect weaknesses, display strengths, explore robustness and probe problem-solving ability. We liken this type of testing to a physical examination. Inevitably the results of such testing will be used to compare programs, but the original intent is that a program be examined in isolation to stand or fall on its own merits.

Designing and implementing test programs is an important numerical problem that has been neglected in the rush to produce software for other purposes. Software testing locates weaknesses and leads to improvements in the next software generation. Yet, except for the ELEFUNT package of transportable Fortran test programs for the elementary functions [Cody and Waite, 1980] and collections of test programs for optimization software [Buckley, 1981; Moré, Garbow and Hillstrom, 1981], no thematic test packages exist to our knowledge. Some test materials are distributed with various PACKs mentioned earlier, but these are not intended for general use.

The trouble is that we know little about how to test most types of software. Accuracy tests, for example, are usually battery tests exercising programs on someone's haphazard collection of problems. Not only is this time-consuming, but there is little purpose behind what is done and the mass of data gathered may be incomprehensible even to those who gathered it. We must find a better way. We must back off from the problem and critically examine what we are doing; every test should have a purpose. We must find understandable and useful ways to present test results. (Note in this regard the clever use of Chernoff faces [Chernoff, 1973] to summarize evaluations of software for solving systems of nonlinear equations [Hiebert, 1981].)

There are some leads in the literature that may prove useful. J. Lyness and J. Kaganove [1976] show that numerical software falls into two broad classes. Class 1 (precision bound) programs implement methods, called "finite decision methods," that guarantee to produce results in a finite number of steps. Elementary function programs are examples of Class 1 programs. The accuracy achieved in Class 1 programs usually approaches limits imposed by the computer arithmetic system. All other programs are Class 2 (heuristic bound) programs implementing "unreliable exact arithmetic algorithms." The algorithms are such that useful results are not guaranteed in a finite number of steps even with exact arithmetic. Results that are produced are usually limited in accuracy by the algorithm and not by machine arithmetic. Quadrature and optimization programs are usually Class 2.

The importance of this classification is that while accuracy test results for Class 1 programs vary with the operating system, compiler and machine, properly structured accuracy tests for Class 2 programs produce system-independent results when the accuracy achieved is sufficiently above machine limits. Thus certain types of accuracy tests for Class 2 software need be done only once and only on one system.

But accuracy testing is just part of a complete test package; efficiency and robustness are also important. Because Class 2 programs frequently require user-supplied software with an unpredictable effect on timing, other measures of efficiency, such as the number of accesses to the user-supplied program, must be used. Where efficiency varies significantly from problem to problem, it is important to explore efficiency as a function of the problem space. In its most elegant form to date, efficiency testing has been combined with accuracy testing and parameterization of a problem space to produce "performance profiles." The prototype work on automatic quadrature programs [Lyness and Kaganove, 1977] produced curves combining probability of success and expected number of integrand evaluations as functions of requested accuracy for specific parameterized problem families. Curves for a problem family with features similar to those in a particular application should be useful in selecting a program for that application based on balancing requested accuracy and predicted cost against the probability of success. The concepts of software classification and performance profiles exemplify the abstract assault on evaluation procedures that we believe is essential to progress in this area. Except for Lyness [1979], these ideas have not been exploited beyond the work cited.

Concerns for numerical software have spawned important work in other fields as well. For example, research on the TAMPR system [Boyle and Matz, 1977] for automated program transformation and maintenance was specifically motivated by early NATS work. TAMPR is intended to accept programs in certain standard languages, map them into abstract forms, make transformations on these abstract forms, and finally recover specific realizations of the transformed programs in standard languages again. The transformations are limited conceptually only by our ability to describe what must be done. An early version of the system was used to realize all versions of the LINPACK programs from complex single-precision prototypes, for example. This application included enforcing formatting conventions and selectively implanting either calls to BLAS or inline coding with BLAS functionality, depending on the particular target computer host. Ultimately the capabilities may include automatic translation from one programming language to another by simply specifying different source and target languages in the first and last steps.

TAMPR is only one of many useful tools now under development. The TOOLPACK project is working on an extensive collection of software tools specifically designed to simplify the writing, testing, analyzing and maintaining of numerical software. The package is to combine the capabilities of TAMPR with those of formatters like POLISH [Dorrenbacher *et al.*, 1976], static analyzers like DAVE [Osterweil and Fosdick, 1976] and PFORT [Ryder, 1974], dynamic analyzers like NEWTON [Feiber, Taylor and Osterweil, 1980], and other as yet unspecified tools including text editors. Specification of the package is still incomplete, but there is agreement that the package will be portable and that package elements will be compatible in data requirements. Release of a prototype version for evaluation and comment is tentatively set for late 1982.

We earlier mentioned the work of the IEEE on standardization of binary floating-point arithmetic for microprocessors. That is only one instance of a wide concern for computer arithmetic. The IEEE has recently established a second subcommittee to draft a radix and format-independent floating-point standard that will be upward compatible with the previous effort. Although the new draft is again intended for microprocessors, its inclusion of non-binary arithmetics should interest designers of larger equipment.

The fruits of such standardization efforts will not become widely available for some time, however. In the meantime we are forced to write software for existing computers. We can improve the portability of software among such machines by explicitly including environmental dependencies in the source code. There have been several attempts to establish a fundamental set of parameters describing arithmetic systems for this purpose. IFIP WG 2.5 published one proposal [Ford, 1978] that has proven unsatisfactory in many respects and has not been widely used. A second proposal [Brown and Feldman, 1980] related to Brown's model for floating-point arithmetic [Brown, 1980] has received important support in some areas. The entire arithmetic model is embedded in ADA [Wichman, 1981], for example, much to the consternation of some numerical analysts. We return to that in a moment. Still a third proposal is being considered by the ANSI X3J3 Fortran Standards Committee for inclusion in the next

Fortran standard. This proposal defines certain parameters and reserves their names in the same way that SIN is a reserved name. The parameter names are then aliases for numerical values appropriate to the particular host environment. The difference between this approach and the ADA approach is that here only the names are specified; the numerical values provided are implementation-dependent. While the parameters are based on a model of an arithmetic system, the model is not imposed by the standard. Thus the details of the model used in a particular situation can be chosen to fit the circumstances. When portability is crucial, the model can be chosen to conservatively estimate machine parameters; when local performance is important, the model can be chosen to closely approximate the local system. Such flexibility is not available in the ADA approach, where the model specified must be conservative to be universal.

The activities and concerns just outlined are typical of the mathematical software effort today. Several large software projects are under way; others are planned. There are many ancillary activities aimed at improving the environment for software production and use. But there are also difficult problems that are not being addressed. We are not making much progress in testing methodology, for example.

THE FUTURE

Prediction of the future is always risky. Nevertheless we present a few guesses at what lies ahead. We expect that the quantity and quality of numerical software will continue to increase and that the activities just described will flourish in the future. Advancing technology and even the present success of the numerical software effort pose problems that must be overcome, however.

The most significant problem we face plagues every technical field and has been with us for a long time: communications. As we become more specialized, we lose touch with one another and especially with potential customers.

Good communications with customers is crucial. Superb software is worthless unless software consumers are persuaded to use it. It is not enough to make users aware of software existence, though that is a difficult task in itself; consumer lethargy must be overcome at the same time. Consumers are reluctant to modify running programs unless they are convinced that the software they are currently using is inferior enough to endanger their work and that the new software will remove that danger. Open literature publications have never solved this type of communications problem. The consumers we must reach are applications people who do not read numerical analysis or mathematical software literature. We must find other ways to reach them.

Several years ago both the Albuquerque and Livermore branches of Sandia Laboratories inserted library monitors in their operating systems [Bailey and Jones, 1975]. These monitors provided information on who was using which routines and on the values of certain parameters in the initial calls to those routines. This information proved valuable to both the librarians and the users. It led to improvement of frequently used programs and provision of new special purpose programs for problems previously solved with general purpose routines. It also permitted personal contact when it appeared that a program was being misused, when program bugs were found, or when better programs became available. Of course, diplomacy and tact were essential in these contacts. In a few cases users objected when they felt their privacy was being invaded or they did not appreciate proffered advice. Sandia Livermore Laboratories augmented personal contact with an advertising campaign in which new programs were featured on posters prominently placed in all terminal rooms. Such efforts are noteworthy, rare and insufficient.

Today we face a revolution in the way computers are being used. The small "personal" computer is becoming common at Argonne — and elsewhere as well, we suspect. While it is often acquired for monitoring experiments and gathering data, the temptation to use it for numerical purposes is strong. This is especially true when the cost of using a central computing facility grows and the "free" personal machine would otherwise sit idle. Such usage is not necessarily bad, because smaller machines are approaching the hardware capabilities of larger machines of only a few years ago. Software is the problem. Owners of such machines frequently write their own software or obtain it from friends. In this respect they operate as large computing centers used to twenty years ago. The software movement has completely lost whatever contact it may have had with these users, and that contact will be difficult to regain.

One possibility may be to contribute to the journals that many of these people read. *Byte*, *Personal Computing* and the like are often sources of information for such users. While some of the articles in these journals are written by highly qualified people, much of the numerical advice is amateurish, reflecting techniques that lost favor long ago. We cannot legitimately complain about this situation unless we are willing as a profession to provide the proper advice and software through these journals. We must be the ones to initiate communications with the users.

Unfortunately, we are also losing whatever communication we had with users of the larger machines. Often the original motivation for numerical software work was provided by users with applications that were endangered by poor computer programs. As our effort has matured, many of us have become more concerned with software production for the sake of production and less concerned about the real needs of users. We have tended to communicate among ourselves and to neglect the users. Perhaps that behavior pattern is typical of a new field. We hope that it will change in our field.

At the technical level we find challenges posed by new computer hardware. We have only begun to work on algorithms and software for parallel and vector machines, and now we are faced with microprocessors as well. Their coming is exciting for numerical software people. The IEEE arithmetic standard provides computational capability that was not previously available at any level. In addition to sophisticated handling of underflow and overflow, standard-conforming systems must provide square root and mod functions, among others, that are as accurate as the usual arithmetic operations. Some early implementations of the standard include square root in the hardware, where it becomes no more expensive to use than an ordinary division operation. This combination of speed and accuracy in square root coupled with other features must influence our selection of algorithms. I believe we will see dramatic changes in algorithms, software and even computer languages as these new microprocessors become common.

Overall we view the future with confidence and expectation. We will probably never satisfactorily solve the communications problem, but we expect that the quality of numerical software will continue to improve and that software production will become easier as new tools and hardware appear.

REFERENCES

Aird, T. J. [1977]. "The IMSL Fortran converter: an approach to solving portability problems." *Lecture Notes in Computer Science, Vol. 57, Portability of Numerical Software.* Ed. W. Cowell. Springer-Verlag, New York, pp. 368-388.

Bailey, C. B., and R. E. Jones [1975]. "Usage and argument monitoring of mathematical library routines." *ACM Trans. on Math. Soft.* 1:196-209.

Boyle, J. M., W. J. Cody, W. R. Cowell, B. S. Garbow, Y. Ikebe, C. B. Moler and B. T. Smith [1972]. "NATS, a collaborative effort to certify and disseminate mathematical software." *Proceedings 1972 National ACM Conference, Vol. II.* Association for Computing Machinery, New York, pp. 630-635.

Boyle, J. M., and M. Matz [1977]. "Automating multiple program realizations." *Proceedings of the M.R.I. International Symposium XXIV: Computer Software Engineering.* Polytechnic Press, Brooklyn, New York.

Brown, W. S. [1980]. *A Simple but Realistic Model of Floating-Point Computation.* Bell Laboratories Computing Science Technical Report 83.

Brown, W. S., and S. I. Feldman [1980]. "Environmental parameters and basic functions for floating-point computation." *ACM Trans. on Numer. Soft.* 6:510-523.

Buckley, A. [1981]. "A portable package for testing minimization algorithms." To appear in *Proceedings of COAL Conference on Mathematical Programming, Testing and Validating Algorithms and Software,* Boulder, Colorado, January 5-6, 1981.

Chernoff, H. [1973]. "The use of faces to represent points in k-dimensional space graphically," *J. Amer. Stat. Ass.* 68:361-368.

Cody, W. J. [1975]. "The FUNPACK package of special function subroutines," *ACM Trans. on Math. Soft.* 1:13-25.

Cody, W. J. [1982a]. "Basic Concepts for Computational Software." *Problems and Methodologies in Mathematical Software Production 142.* Ed. P. C. Messina and A. Murli. Springer-Verlag, Berlin, 1982.

Cody, W. J. [1982b]. "Panel session on the challenges for developers of mathematical software." To appear in *Proceedings of International Seminar on Problems and Methodologies in Software Production,* Sorrento, Italy, November 3-8, 1980.

Cody, W. J., and W. Waite [1980]. *Software Manual for the Elementary Functions.* Prentice Hall, Englewood Cliffs, New Jersey.

Coleman, D., P. Holland, N. Kadeen, V. Klema and S. C. Peters [1980]. "A system of subroutines for iteratively reweighted least squares computations." *ACM Trans. on Math. Soft.* 6:327-336.

Cowell, W. R., and L. D. Fosdick [1975]. *A Program for Development of High Quality Mathematical Software.* University of Colorado Department of Computer Science Report CU-CS-070-75.

Cowell, W. R., and L. D. Fosdick [1977]. "Mathematical software production." *Mathematical Software III.* Ed. J. R. Rice. Academic Press, New York, pp. 195-224.

Dongarra, J.J., J. R. Bunch, C. B. Moler and G. W. Stewart [1979]. *LINPACK User's Guide.* SIAM, Philadelphia, Pennsylvania.

Dorrenbacher, J., D. Paddock, D. Wisneski and L. D. Fosdick [1976]. *POLISH, a Program to Edit Fortran Programs.* University of Colorado Department of Computer Science Report CU-CS-050-76 (Rev.).

Du Croz, J. J., S. J. Hague and J. L. Siemieniuch [1977]. "Aids to portability within the NAG project." *Lecture Notes in Computer Science, Vol. 57, Portability of Numerical Software.* Ed. W. Cowell. Springer-Verlag, New York, pp. 390-404.

Feiber, J., R. N. Taylor and L. J. Osterweil [1980]. *NEWTON - A Dynamic Testing System for Fortran 77 Programs; Preliminary Report.* University of Colorado Department of Computer Science Technical Note.

Ford, B. [1978]. "Parameterization of the environment for transportable numerical software." *ACM Trans. on Numer. Soft.* 4:100-103.

Forsythe, G. [1966]. "Algorithms for scientific computation." *Comm. ACM* 9:255-256.

Garbow, B. S., J. M. Boyle, J. J. Dongarra and C. B. Moler [1977]. *Lecture Notes in Computer Science, Vol. 51, Matrix Eigensystem Routines - EISPACK Guide Extension.* Springer-Verlag, New York.

Gear, C. W. [1979]. *Numerical Software: Science or Alchemy?* University of Illinois Department of Computer Science Report UIUCDCS-R-79-969.

George, A., and J. W. Liu [1981]. *Computer Solution of Large Sparse Positive Definite Systems.* Prentice-Hall, Englewood Cliffs, New Jersey.

Grimes, R. G., D. R. Kincaid, W. I. MacGregor and D. M. Young [1978]. *ITPACK Report: Adaptive Iterative Algorithms Using Symmetric Sparse Storage.* Center for Numerical Analysis, University of Texas at Austin Report CNA-139.

Hiebert, K. L. [1981]. "An outline for comparison testing of mathematical software — illustrated by comparison testings of software which solves systems of nonlinear equations." To appear in *Proceedings of COAL Conference on Mathematical Programming, Testing and Validating Algorithms and Software.* Boulder, Colorado, January 5-6, 1981.

Huddleston, R. E., ed. [1979]. *Program Directions for Computational Mathematics.* Unnumbered report, Dept. of Energy, Washington, D.C.

Intel [1980]. *The 8086 Family User's Manual, Numerics Supplement.* Intel Corp., Santa Clara, California.

Klema, V. C. [1981]. Private communication.

Krogh. F. T. [1977]. "Features for Fortran portability," *Lecture Notes in Computer Science, Vol. 57, Portability of Numerical Software.* Ed. W. Cowell. Springer-Verlag, New York, pp. 361-367.

Lawson, C. L., R. J. Hanson, D. R. Kincaid and F. T. Krogh [1979]. "Basic linear algebra subprograms for Fortran usage." *ACM Trans. on Numer. Soft.* 5:308-323.

Lyness, J. N. [1979]. "A bench mark experiment for minimization algorithms." *Math. Comp.* 33:249-264.

Lyness, J. N., and J. J. Kaganove [1976]. "Comments on the nature of automatic quadrature routines." *ACM Trans. on Numer. Soft.* 2:65-81.

Lyness, J. N., and J. J. Kaganove [1977]. "A technique for comparing automatic quadrature routines." *Computer J.* 20:170-177.

Machura, M., and R. A. Sweet [1980]. "A survey of software for partial differential equations." *ACM Trans. on Numer. Soft.* 6:461-488.

Moré, J. J., B. S. Garbow and K. E. Hillstrom [1980]. *User Guide for MINPACK-1.* Argonne National Laboratory Report ANL-80-74.

Moré, J. J., B. S. Garbow and K. E. Hillstrom [1981]. "Testing unconstrained optimization software." *ACM Trans. on Numer. Soft.* 7:17-41.

Morris, A. H., Jr. [1979]. *Development of Mathematical Software and Mathematical Software Libraries.* Naval Surface Weapons Center Report NSWC TR 79-102.

Osterweil, L. J. [1981]. "TOOLPACK - an integrated system of tools for mathematical software development." To appear in *Proceedings of COAL Conference on Mathematical Programming, Testing and Validating Algorithms and Software*. Boulder, Colorado, January 5-6, 1981.

Osterweil, L. J., and L. D. Fosdick [1976]. "DAVE — a validation, error detection and documentation system for Fortran programs." *Software Practice and Experience* 6:473-486.

Piessens, R., E. De Doncker, C. W. Uberhuber and H. J. Stetter [1978]. *Detailed Test Results for Automated General Purpose Integration over Finite or Infinite Intervals*. Unnumbered Report, Applied Mathematics and Programming Division, Katholieke Universiteit Leuven, Heverlee, Belgium.

Rice, J. R. [1969]. "Announcement and call for papers, mathematical software." *SIGNUM Newsletter* 4 (3):7.

Rice, J. R., ed. [1971]. *Mathematical Software*. Academic Press, New York.

Rice, J. R. [1977]. "ELLPACK: a research tool for elliptic partial differential equations software." *Mathematical Software III*. Ed. J. R. Rice. Academic Press, New York, pp. 319-341.

Rice, J. R. [1979]. "Software for numerical computation." *Research Directions in Software Technology*. Ed. P. Wegner. MIT Press, Cambridge, Massachusetts, pp. 688-708.

Rice, J. R., C. W. Gear, J. M. Ortega, B. N. Parlett, M. Schultz, L. F. Shampine and P. Wolfe [1978]. *Numerical Computation, Panel Report for the COSERS Project*. Special issue *SIGNUM Newsletter*.

Ryder, B. G. [1974]. "The PFORT verifier." *Software Practice and Experience* 4:359-377.

SCCS [1975]. *PDEPACK: Partial Differential Equations Package User's Guide*. Scientific Computing Consulting Services, Manhattan, Kansas.

Smith, B. T., J. M. Boyle and W. J. Cody [1974]. "The NATS approach to quality software." *Software for Numerical Mathematics*. Ed. D. J. Evans. Academic Press, New York, pp. 393-405.

Smith, B. T., J. M. Boyle, J. J. Dongarra, B. S. Garbow, Y. Ikebe, V. C. Klema and C. B. Moler [1976]. *Lecture Notes in Computer Science, Vol. 6, Second Edition, Matrix Eigensystem Routines — EISPACK Guide.* Springer-Verlag, New York.

Stevenson, D., Chairman IEEE P754 [1981]. "A proposed standard for binary floating-point arithmetic, Draft 8.0 of IEEE Task P754." *Computer* 14 (3):51-62.

Swarztrauber, P., and R. Sweet [1979]. "Efficient FORTRAN subroutines for the solution of separable elliptic equations. Algorithm 541." *ACM Trans. on Math. Soft.* 5:352-364

Wichman, B. A. [1981]. *Tutorial Material on the Real Data-Types in ADA.* Final Technical Report, U.S. Army European Research Office, London.

Chapter 2

LINPACK — A PACKAGE FOR SOLVING LINEAR SYSTEMS

J. J. Dongarra*
Mathematics and Computer Science Division
Argonne National Laboratory
Argonne, Illinois 60439

G. W. Stewart†
Department of Computer Science
University of Maryland
College Park, Maryland 20742

INTRODUCTION

LINPACK is a collection of Fortran subroutines that analyze and solve linear equations and linear least squares problems. The package solves linear systems whose matrices are general, banded, symmetric indefinite, symmetric positive definite, triangular, and tridiagonal. In addition, the package computes the QR and singular value decompositions of rectangular matrices and applies them to least squares problems.

The software for LINPACK can be obtained from either

National Energy Software Center (NESC)
Argonne National Laboratory
9700 South Cass Avenue
Argonne, Illinois 60439
Phone: 312-972-7250
Cost: Determined by NESC policy

or

IMSL
Sixth Floor, NBC Building
7500 Bellaire Boulevard
Houston, Texas 77036
Phone: 713-772-1927
Cost: $75.00 (Tape included).

Requestors in European Organization for Economic Cooperation and Development countries may obtain the software by writing to

---

* Work supported in part by the Applied Mathematical Sciences Research Program (KC-04-02) of the Office of Energy Research of the U.S. Department of Energy under Contract W-31-109-Eng-38.
† Work supported in part by the Computer Science Section of the National Science Foundation under Contract MCS 7603297.

NEA Data Bank
B.P. No. 9 (Bat. 45)
F-91191 Gif-sur-Yvette
France
Cost: Free.

The documentation for the codes is contained in the following book:

J.J. Dongarra, J.R. Bunch, C.B. Moler, G.W. Stewart,
LINPACK Users' Guide
Society for Industrial and Applied Mathematics (1979)
Cost: $17.00.

HISTORY

In June 1974 Jim Pool, then director of the Research Section of the Applied Mathematics Division (AMD) at Argonne National Laboratory, initiated informal meetings to consider producing a package of high-quality software for the solution of linear systems and related problems. The participants included members of the AMD staff, visiting scientists, and various consultants and speakers at AMD colloquia. It was decided that such a package was needed and that a secure technological basis was available for its production.

In February 1975 the participants met at Argonne to lay the groundwork for the project and hammer out what was and was not to be included in the package. A proposal was submitted to the National Science Foundation (NSF) in August 1975. NSF agreed to fund such a project for three years beginning January 1976; the Department of Energy also provided support at Argonne.

The package was developed by four participants working from their respective institutions:

J.J. Dongarra	Argonne National Laboratory
J.R. Bunch	University of California at San Diego
C.B. Moler	University of New Mexico
G.W. Stewart	University of Maryland

Argonne served as a center for the project. The participants met there summers to coordinate their work. In addition Argonne provided administrative and technical support; it was responsible for collecting and editing the programs as well as distributing them to test sites.

In the summer of 1976 the members of the project visited Argonne for a month. They brought with them a first draft of the software and documentation. It became clear that more uniformity was needed to create a coherent package. After much discussion and agonizing, formats and conventions for the package were established. By the end of that summer an early version of the codes emerged, along with rough documentation.

The fall of 1976 and winter of 1977 saw the further development of the package. The participants worked on their respective parts at their home institutions and met once during the winter to discuss progress.

During the summer of 1977 the participants developed a set of test programs to support the codes and documentation that were to become LINPACK. This set was sent to 26 test sites in the fall of 1977. The test sites included universities, government laboratories, and private industry. In addition to running the test programs on their local computers and reporting any problems that occurred, the sites also installed the package at their computer centers and announced it to their user communities. Thus, the package received real user testing on a wide variety of systems.

As a result of the testing, in mid 1978 some changes were incorporated into the package, and a second test version was sent to the test sites. The programs were retested and timing information returned for the LINPACK routines on various systems.

By the end of 1978 the codes were sent to NESC and IMSL for distribution, and the users' guide was completed and sent to SIAM for printing. At the beginning of 1979 both the programs and the documentation were available to the public.

LINPACK AND MATRIX DECOMPOSITIONS

LINPACK is based on a *decompositional* approach to numerical linear algebra. The general idea is the following. Given a problem involving a matrix A, one factors or decomposes A into a product of simpler matrices from which the problem can easily be solved. This divides the computational problem into two parts: first the computation of an appropriate decomposition, then its use in solving the problem at hand. Since LINPACK is organized around matrix decompositions, it is appropriate to begin with a general discussion of the decompositional approach to numerical linear algebra.

Consider the problem of solving the linear system

$$Ax = b, \tag{1}$$

where A is a nonsingular matrix of order n. In older textbooks this problem is treated by writing (1) as a system of scalar equations and eliminating unknowns in such a way that the system becomes upper triangular (Gaussian elimination) or even diagonal (Gauss-Jordan elimination). This approach has the advantage that it is easy to understand and that it leads to pretty computational tableaus suitable for hand calculation. However, it has the drawback that the level of detail obscures the very broad applicability of the method.

In contrast, the decompositional approach begins with the observation that it is possible to factor A in the form

$$A = LU, \tag{2}$$

where L is a lower triangular matrix with ones on its diagonal and U is upper triangular. * The solution to (1) can then be written in the form

$$x = A^{-1}b = U^{-1}L^{-1}b = U^{-1}c,$$

where $c = L^{-1}b$. This suggests the following algorithm for solving (1):

1: *Factor A in the form* (2);

2: *Solve the system* $Lc = b$;   (3)

3: *Solve the system* $Ux = c$.

Since both L and U are triangular, steps 2 and 3 of the above algorithm are easily done.

The approach to matrix computations through decompositions has turned out to be quite fruitful. Here are some of the advantages. First, the approach separates the computation into two stages: the computation of a decomposition and the use of the decomposition to solve the problem at hand. These stages are exemplified by the contrast between statement 1 and statements 2 and 3 of (3). In particular, it means that the decomposition can be used repeatedly to solve new problems. For example, if (1) must be solved for many right-hand sides b, it is necessary to factor A only once. This may represent an enormous savings, since the factorization of A is an $O(n^3)$ process, whereas steps 2 and 3 of (3) require only $O(n^2)$ operations.

Second, the approach suggests ways of avoiding the explicit computation of matrix inverses or generalized inverses. This is important because the first thing a computationally naive person thinks of when faced with a formula like $x = A^{-1}b$ is to invert and multiply; and such a procedure is always computationally expensive and numerically risky.

---

* This is not strictly true. It may be necessary to permute the rows of A (a process called pivoting) in order to ensure the existence of the factorization (2). In finite precision arithmetic, pivoting *must* be done to ensure numerical stability.

Third, a decomposition practically begs for new jobs to do. For example, from (2) and the fact that $det(L) = 1$, it follows that

$$det(A) = det(L) det(U) = det(U).$$

Since U is triangular, $det(A)$ is just the product of the diagonal elements of U. As another example, consider the problem of solving the transposed system $A^T x = b$. Since $x = A^{-T}b = (LU)^{-T}b = L^{-T}U^{-T}b$, this system may be solved by replacing statements 2 and 3 in (3) with

2′: *Solve* $U^T c = b$;

3′: *Solve* $L^T x = c$.

Note that it is not at all trivial to see how row elimination as it is usually presented can be adapted to solve transposed systems.

Fourth, the decompositional approach introduces flexibility into matrix computations. There are many decompositions, and a knowledgeable person can select the one best suited to his application.

Fifth, if one is given a decomposition of a matrix A and a simple change is made in A (e.g., the alteration of a row or column), one frequently can compute the decomposition of the altered matrix from the original decomposition at far less cost than the *ab initio* computation of the decomposition. This general idea of *updating* a decomposition has been an important theme during the past decade of numerical linear algebra research.

Finally, the decompositional approach provides theoretical simplification and unification. This is true both inside and outside of numerical analysis. For example, the realization that the Crout, Doolittle, and square root methods all compute LU decompositions enables one to recognize that they are all variants of Gaussian elimination. Outside of numerical analysis, the spectral decomposition has long been used by statisticians as a canonical form for multivariate models.

LINPACK is organized around four matrix decompositions: the LU decomposition, the (pivoted) Cholesky decomposition, the QR decomposition, and the singular value decomposition. The term LU decomposition is used here in a very general sense to mean the factorization of a square matrix into a lower triangular part and an upper triangular part, perhaps with some pivoting. These decompositions will be treated at greater length later, when the actual LINPACK subroutines are discussed. But first a digression on nomenclature and organization is necessary.

NOMENCLATURE AND CONVENTIONS

The name of a LINPACK subroutine is divided into a prefix, an infix, and a suffix as follows:

$$TXXYY$$

The prefix **T** signifies the type of arithmetic and takes the following values:

S	single precision
D	double precision
C	complex

Where it is supported, a prefix of **Z,** signifying double precision complex arithmetic, is permitted.

The infix **XX** is used in two different ways, which reflects a fundamental division in the LINPACK subroutines. The first group of codes is concerned principally with solving the system $Ax = b$ for a square matrix A, and incidentally with computing condition numbers, determinants, and inverses. Although all of these codes use variations of the LU decomposition, the user is rarely interested in the decomposition itself. However, the structural properties of the matrix, such as symmetry and bandedness, make a great deal of difference in arithmetic and storage costs. Accordingly, the infix **XX** in this part of LINPACK* is used to designate the structure of the matrix.

In the square part the infix can have the following values:

GE	General matrix — no assumptions about the structure
GB	General banded matrix
PO	Symmetric positive definite matrix
PP	Symmetric positive definite matrix in packed storage format
PB	Symmetric positive definite banded matrix
SI	Symmetric indefinite matrix
SP	Symmetric indefinite matrix in packed storage format
HI	Hermitian indefinite matrix (with prefix **C** or **Z** only)
HP	Hermitian indefinite matrix in packed storage format (with prefix **C** or **Z** only)
GT	General tridiagonal matrix
PT	Positive definite tridiagonal matrix

---

* Because this part deals exclusively with square matrices, it will be called the *square part.*

The second part of LINPACK is called the least squares part because one of its chief uses is to solve linear least squares problems. It is built around subroutines to compute the Cholesky decomposition, the QR decomposition, and the singular value decomposition. Here nothing special is assumed about the form of the matrix, except symmetry for the Cholesky decomposition; however, it is now the decomposition itself that the user is interested in. Accordingly, in the least squares part of LINPACK the infix is used to designate the decomposition. There are three options:

CH	Cholesky decomposition
QR	QR decomposition
SV	Singular value decomposition

The suffix **YY** specifies the task the subroutine is to perform. The possibilities are

FA[*]	Compute an LU factorization.
CO[*]	Compute an LU factorization and estimate the condition number.
DC[†]	Compute a decomposition.
SL	Apply the results of **FA, CO,** or **DC** to solve a problem.
DI[*]	Compute the determinant or inverse.
UD	Update a Cholesky decomposition.
DD	Downdate a Cholesky decomposition.
EX	Update a Cholesky decomposition after a permutation (exchange).

One small corner of LINPACK, devoted to triangular systems, is not decompositional, since a triangular matrix needs no reduction. Codes in this part are designated by an infix of **TR,** and the only two suffixes are **SL** and **CO.**

In addition to the uniform manner of treating subroutine names, there are a number of other uniformities of nomenclature in LINPACK. A square matrix is always designated by **A** (**AP** and **ABD** in the packed and banded cases), and the dimension is always **N**. Rectangular input matrices are denoted by **X**, and they are always **N** × **P**. The use of **P** as an integer is the sole deviation in the calling sequences from the Fortran implicit typing convention; it was done to keep LINPACK in conformity with standard statistical notation.

Since Fortran associates no dope vectors with arrays, it is necessary to pass the first dimension of an array to the subroutine. This number is always denoted by **LDA** or **LDX,** depending on the array name. A frequent, though unavoidable, source of errors in the use of LINPACK is the confusion of **LDA** (the leading dimension of the *array* A) with **N** (the order of the *matrix* A), which may be smaller than **LDA**.

Since many LINPACK subroutines perform more than one task, it is necessary to have a parameter to say which tasks are to be done. This parameter

---

[*] Square part only.
[†] Least squares part only.

is always called **JOB** in LINPACK, although the method of encoding options varies from subroutine to subroutine. (The **JOB** parameter in **SQRSL,** which has a lot to do, is of Byzantine complexity, although it is very easy to use once the trick is known.)

The status of a computation on return from a LINPACK subroutine is always signaled by **INFO.** The name was deliberately chosen to have neutral connotations, since it is not necessarily an error flag. As with **JOB**, the exact meaning of **INFO** varies from subroutine to subroutine.

THE SQUARE PART OF LINPACK

The square part of LINPACK is well illustrated by the codes with the infix **GE,** i.e., those codes dealing with a general square matrix. The basic decomposition used by the **GE** codes is of the form

$$PA = LU,$$

where L is a unit lower triangular matrix and U is upper triangular. The matrix P is a permutation matrix that represents row interchanges made to ensure numerical stability. The algorithm used to determine the interchanges is called *partial pivoting* (cf. Forsythe and Moler [1967] or Stewart [1974]). The decomposition is computed by the subroutine **SGEFA,*** which uses Gaussian elimination and overwrites A with L and U. (This is typical of the LINPACK codes; the original matrix is always overwritten by its decomposition.) Information on the interchanges is returned in **IPVT.**

There are two **GE** subroutines to manipulate the decomposition computed by **SGEFA.** **SGESL** solves the system $Ax = b$ or the system $A^T x = b$, as specified by the parameter **JOB.** The solution x overwrites the right-hand side b. The subroutine **SGEDI** computes the determinant and inverse of A. Because the value of a determinant can easily underflow or overflow (if A is 50×50, $det(10*A) = 10^{50} det(A)$), the determinant is coded into an array **DET** of length two in the form

$$det(A) = \textbf{DET} (1)* 10** \textbf{DET} (2).$$

As was indicated earlier, an explicit matrix inverse is seldom needed for most problems. Consequently, **SGEDI** should be used to compute the inverse only when there is no alternative. The inverse overwrites the LU decomposition, so that the array A cannot be used in subsequent calls to **SGESL** and **SGEDI.** It is technically possible to recover A by factoring A^{-1} (**SGEFA**) and inverting it (**SGEDI**); but, owing to rounding errors, the matrix obtained in this way may differ from the original.

---

* For definiteness the prefix **S** will be used for all LINPACK codes, it being understood that **D, C,** and **Z** are also options.

A perennial question that arises when linear systems are solved is how accurate the computed solution is. The answer to this question is usually cast in terms of the *condition number*, $\kappa(A)$, defined by

$$\kappa(A) = ||A||||A^{-1}||, \tag{4}$$

where $||\cdot||$ is a suitable matrix norm. For example, if the system $Ax = b$ is solved by **SGEFA** and **SGESL** in t-digit decimal arithmetic and $\bar{x}$ is the computed solution, then

$$\frac{||x - \bar{x}||}{||\bar{x}||} \leqslant f(n)\kappa(A)10^{-t}, \tag{5}$$

where $f(n)$ is a slowly growing function of the order n of A [Wilkinson, 1963].

The LINPACK subroutine **SGECO,** in addition to factoring A, returns an estimate of $\kappa(A)$. Because $\kappa(A)$ can become arbitrarily large, **SGECO** actually returns the reciprocal of $\kappa(A)$ in the parameter **RCOND.** The user can inspect this number to determine the accuracy of the purported solution. However, it must be borne in mind that the interpretation of the condition number depends on how the problem has been scaled, a point that will be discussed further in the section on Numerical Properties.

Most of the remaining square part of LINPACK can be regarded as adaptations of the **GE** routines to special forms of square matrices. One of the most frequently occurring forms is the positive definite matrix, where the matrix A is symmetric ($A^T = A$) and satisfies $x^T A x > 0$ whenever $x \neq 0$. In this case the LU factorization can be written in the form

$$A = R^T R, \tag{6}$$

where R is upper triangular.* This *Cholesky factorization* is what the subroutine **SPOFA** computes. The advantage of **SPOFA** over **SGEFA** is that it requires half the storage and half the work. Specifically, since A is symmetric, only its upper half need be stored. Likewise, the upper triangular factor R in (6) can overwrite the upper half of A. The lower part of the array containing A is not referenced by **SPOFA** and hence can be used to store other information. The **PO** routines include an **SL** subroutine to solve linear systems and a **DI** subroutine to compute the determinant or the inverse.

One conventional way of storing a symmetric matrix is to pack either its lower or upper part into a linear array. The LINPACK subroutine **SPPFA** computes the Cholesky factorization of a positive definite matrix with its upper part packed in the order indicated below:

---

* This decomposition is often written $A = L^T L$, where L is lower triangular. The upper triangular form was chosen for its consistency with the QR decomposition.

$$
\begin{array}{ccccc}
1 & 2 & 4 & 7 & 11 \\
  & 3 & 5 & 8 & 12 \\
  &   & 6 & 9 & 13 \\
  &   &   & 10 & 14 \\
  &   &   &   & 15
\end{array}
$$

The Cholesky factor R is returned in the same order. There are corresponding **SPPCO, SPPSL,** and **SPPDI** subroutines.

When a symmetric matrix is indefinite (i.e., there are vectors x and y such that $y^T A y < 0 < x^T A x$), it has no Cholesky factorization. However, there is a permutation matrix P such that $P^T A P$ can be decomposed stably in the form

$$P^T A P = U D U^T, \tag{7}$$

where U is triangular and D is block diagonal with only 1×1 or 2×2 blocks. The subroutines **SSIFA** and **SSICO** compute this factorization, and **SSISL** and **SSIDI** apply it, as usual, to solve systems or compute determinants and inverses. **SSIDI** also computes the inertia of A. There are corresponding packed-storage subroutines.

There are two exceptional aspects of symmetric indefinite programs. First, the matrix U in (7) is *upper* triangular, giving the factorization the appearance

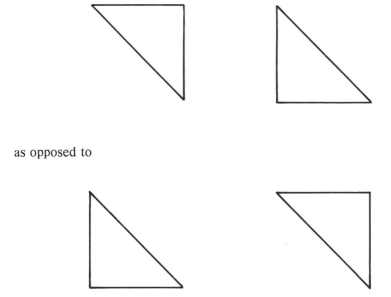

as opposed to

for the classical LU factorization (2). This unusual factorization is required so that the algorithm can both work with the upper half of A (which keeps it consonant with the **PO** routines) and also remain column oriented for efficiency (more on this later).

The other aspect concerns the passage from real to complex arithmetic. A complex positive definite matrix is generally required to be Hermitian (i.e., equal to its conjugate transpose); hence there is no ambiguity in the properties of A for the **CPOYY** routines. In the indefinite case, a complex matrix may be either symmetric $(A^T = A)$ or Hermitian. This point was resolved by letting the **CSIYY** routines handle complex symmetric matrices and devising a new infix **HI** for Hermitian matrices.

In many applications the nonzero elements of A are clustered around the diagonal of A. Such a matrix is called a *band matrix*. The distance m_l to the left along a row from the diagonal to the farthest off-diagonal element is the *lower band width*; the distance m_u to the right from the diagonal to the farthest element is the *upper band width*. The structure of an 8×8 matrix with lower band width one and upper band width two is illustrated below:

$$
\begin{array}{cccccccc}
X & X & X & O & O & O & O & O \\
X & X & X & X & O & O & O & O \\
O & X & X & X & X & O & O & O \\
O & O & X & X & X & X & O & O \\
O & O & O & X & X & X & X & O \\
O & O & O & O & X & X & X & X \\
O & O & O & O & O & X & X & X \\
O & O & O & O & O & O & X & X \\
\end{array}
$$

By storing only the nonzero diagonals of a band matrix A, the matrix may be placed in an array of dimensions $n \times (m_l + m_u + 1)$, where n is the order of A. Thus, if m_l and m_u are fixed, the storage requirements grow only linearly with increasing n, and it is possible to represent very large systems in a modest amount of memory. The amount of work required to solve band systems also grows linearly with n.

LINPACK provides routines **SGBFA, SGBCO, SGBSL,** and **SGBDI** to manipulate band matrices. Because pivoting is required for numerical stability, the user must arrange the nonzero elements in an $n \times (2m_l + m_u + 1)$ array. Several possible schemes for storing band matrices were considered for LINPACK. The final choice was dictated by a decision to have the **GB** routines reflect the loop structure and arithmetic properties of the **GE** routines. For matrix elements within the band structure, the two sets of routines perform the same arithmetic operations in the same order. The computed solution to a band system obtained by **SGBSL** and the solution to the same system stored as a full matrix and computed by **SGESL** should agree to the last bit.

Some users of LINPACK have found that its approach to storing band matrices is complicated and unnatural. However, a simple program, listed in the *LINPACK Users' Guide,* will automatically place the elements where they belong. The subroutine **SGBDI** computes only the determinant, since the inverse of a band matrix is not itself a band matrix.

There are corresponding routines with infix **PB** for banded positive definite matrices. Since these are symmetric and require no pivoting, the storage requirement is reduced to $n \times (m_u + 1)$, and the work is correspondingly reduced. For the important case of tridiagonal matrices $(m_l = m_\mu = 1)$, two special subroutines, **SGTSL** and **SPTSL,** are provided. The latter subroutine employs the "Millay" algorithm, which simultaneously factors the matrix from each end to save looping overhead.*

THE LEAST SQUARES PART OF LINPACK

Although the least squares part of LINPACK has many and varied applications, it will unify the exposition if we concentrate on the linear least squares problem. Here we are given an $n \times p$ matrix X and an n-vector y and wish to determine a p-vector b such that

$$\|y - Xb\| = \min, \tag{8}$$

where $\|\cdot\|$ denotes the usual Euclidean norm. It can be shown that any solution of (8) must satisfy the *normal equations*

$$Ab = c, \tag{9}$$

where

$$A = X^T X, \quad c = X^T y. \tag{10}$$

At any solution the vector Xb is the projection of y onto the column space of X, and the residual vector $r = y - Xb$ is the projection of y onto the orthogonal complement of the column space of X.

If the columns of X are independent, the normal equations (9) are positive definite. Consequently, the solution b can be obtained by first calling **SPOFA** to compute the Cholesky factorization of A and then calling **SPOSL** to solve (9). In fact, any least squares problem involving an initial set of columns of X can be solved in this manner. To see this, partition X in the form $X = (X_1 \ X_2)$, where X_1 is $n \times k$, and consider the problem

$$\|y - X_1 \bar{b}\| = \min. \tag{11}$$

Then the normal equations assume the form

$$A_{11} \bar{b} = c_1,$$

---

* *My candle burns at both ends;/ It will not last the night;/ But, ah my foes, and, oh, my friends —/ It makes a lovely light.* Edna St. Vincent Millay, 1892-1950.

where $A_{11} = X_1^T X_1$. Moreover, if the Cholesky factor R of A is partitioned in the form

$$R = \begin{bmatrix} R_{11} & R_{12} \\ 0 & R_{22} \end{bmatrix},$$ (12)

where R_{11} is $k \times k$, then

$$A_{11} = R_{11}^T R_{11}.$$

Thus, R_{11} is the Cholesky factor of A_{11}. From the point of view of LINPACK, this means that once the Cholesky factor of A has been obtained from **SPOFA,** the reduced problem (11) can be solved by calling **SPOSL** with k replacing p as the order of the matrix.

It frequently happens that the columns of the least squares matrix are linearly dependent, or nearly so. In this case it is necessary to stabilize the least squares solution in some way. The subroutine **SCHDC** provides one way by effectively moving the dependent columns to the end of X. Specifically, when **SCHDC** is applied to A, it produces a permutation matrix P and a Cholesky factorization

$$P^T A P = R^T R$$ (13)

that satisfies

$$r_{kk}^2 \geqslant \sum_{i=k}^{j} r_{ij}^2 \quad (j = k, k+1, \cdots, p).$$

Thus in the partition (12), if the leading element of R_{22} is small, all the elements are small, and the last columns of XP are nearly dependent on the first ones. These may then be discarded, and a least squares solution involving the initial columns of XP may be obtained by calling **SPOSL** as described above. Incidentally, the ratio $(r_{11}/r_{pp})^2$ from the pivoted Cholesky factorization is a reliable estimator of the condition number of A, and its reciprocal may be used in place of the number **RCOND** produced by **SPOCO** [Stewart, 1980].

In some applications, it is necessary to add or delete rows from a least squares fit. For the addition of a row x^T, this amounts to computing the Cholesky factorization of

$$\tilde{A} = A + xx^T,$$

where A is defined in (10). The subroutine **SCHUD** (**UD** = update) provides a way of computing the Cholesky factor $\tilde{R}$ of $\tilde{A}$ from that of A. This procedure is cheaper $[O(p^2)]$ than computing $\tilde{A}$ from A and factoring with **SPOFA** $[O(p^3)]$. **SCHUD** may also be used to solve least squares problems for which n is too large to allow X to fit into high-speed memory. The trick is to bring in X one row at a time and use repeated calls to **SCHUD** to incorporate the rows into R.

The deletion of a row amounts to computing the Cholesky decomposition of $\tilde{A} = A - xx^T$ from that of A, a process that is sometimes called *downdating*. The subroutine **SCHDD** accomplishes this; however, it is important to realize that while updating is a very stable numerical procedure, downdating is not, and the uncritical use of **SCHDD** can result in anomalous output.

In data analysis and model building it is often necessary to compare the least squares approximations corresponding to different subsets of columns of X. We have seen that this can be done if the subset in question can be moved to the beginning of X, that is, if the Cholesky decomposition of $P^T AP$ can be computed from that of A, where P is a permutation matrix such that XP has the selected columns at the beginning. The subroutine **SCHEX** (**EX** = exchange) does this for a class of permutations from which all others can be built up.

It is a commonplace in numerical linear algebra that whenever possible one should avoid using the normal equations to solve linear least squares problems. One way of accomplishing this is by the row-wise formation of R described above. Another way is to use the LINPACK subroutines **SQRDC** and **SQRSL** to compute and manipulate the QR decomposition of X. This decomposition has the form

$$Q^T X = \begin{bmatrix} R \\ 0 \end{bmatrix},$$

where Q is an orthogonal matrix and R is upper triangular. From the orthogonality of Q it follows that

$$R^T R = XQQ^T X = X^T X,$$

which implies that the R factor in the QR decomposition of X is just the Cholesky factor of $X^T X$. It can further be shown that if Q is partitioned in the form

$$\begin{array}{cc} p & n-p \end{array}$$
$$Q = (\begin{array}{cc} Q_X & Q_\perp \end{array}),$$

then the least squares solution b satisfies

$$Rb = z,$$

where $z = Q_X^T y$. Moreover,

$$Xb = Q_X z, \qquad r = y - Xb = Q_\perp s,$$

where $s = Q_\perp^T y$. Thus the least squares approximation to y and its residual vector can be obtained from the QR decomposition.

Since Q is an $n \times n$ matrix, it will be impossible to store it explicitly, even when the $n \times p$ matrix X can be stored. To circumvent this difficulty, **SQRDC** computes Q in the form

$$Q = H_1 H_2 \cdots H_p,$$

where each H_j is a Householder transformation requiring only $n-j+1$ words of storage for its representation. Thus the entire QR decomposition, consisting of R and the factored form of Q, can be stored in X and an auxiliary array of length p.

SQRSL manipulates the QR decomposition computed by **SQRDC.** Under the control of the **JOB** parameter, **SQRSL** can return the vectors $Q^T y$, Qy, b, Xb, and r. Moreover, it can return the analogous quantities corresponding to the first k columns of X, in the same way as can be done with the Cholesky decomposition; cf. (10) and the following discussion.

SQRDC also has a pivoting option, which results in the computation of the QR decomposition of the permuted matrix XP. In the absence of rounding errors, the permutation matrix P is the same as the one produced by the pivoting Cholesky decomposition of $X^T X$; cf. (13). Thus the pivoting option in **SQRDC** can be used to estimate condition numbers and detect near-degeneracies in rank. When $n < p$, the pivoting option can be used to collect a well-conditioned $n \times n$ submatrix of X, providing one exists.

The final decomposition computed by LINPACK is the singular value decomposition. As above, let X be an $n \times p$ matrix where, for definiteness, $n \geqslant p$. Then there are orthogonal matrices U and V such that

$$U^T X V = \left[\begin{array}{c} \Sigma \\ 0 \end{array} \right],$$

where $\Sigma = diag(\sigma_1, \cdots, \sigma_p)$ with

$$\sigma_1 \geqslant \sigma_2 \geqslant \cdots \geqslant \sigma_p \geqslant 0.$$

This decomposition is a supple theoretical tool that is widely used in analysis involving matrices. If Σ is known, it can be used to detect degeneracies and compute the condition number, which is σ_1/σ_p in the 2-norm. The decomposition is also required in many kinds of statistical computations — ridge regression, canonical correlations, and cross validation, to name just three.

The singular value decomposition is computed by **SSVDC**, which implements the only iterative algorithm in LINPACK. In addition to Σ, **SSVDC** will return at the user's request any of V, U, or the first p columns of U.

NUMERICAL PROPERTIES

The previous four sections addressed the question "What does LINPACK do?" Here we shall consider the related question "How well does it do it?" In other words, how do LINPACK codes behave on actual computers? It will be convenient to divide this question into two parts and ask, first, how LINPACK codes perform in the presence of inexact arithmetic and, second, how efficient the LINPACK codes are. The first question will be treated in this section and the second in the next.

For the nonexpert there is perhaps no more bewildering subject than rounding-error analysis. The attitudes about rounding error generally range from exaggerated fears that it hopelessly contaminates everything to unjustified confidence that it is never really important. This is not the place to enter into a detailed discussion of rounding-error analysis, which has been exhaustively treated by Wilkinson [1963, 1965] and others. However, it is impossible to describe how rounding-error affects LINPACK codes without presenting some technical background.

The natural question to ask about a computed solution to a problem is how accurate it is. In numerical linear algebra, however, the question is best asked in two stages. The first stage is to ask if the solution is *stable*. The term stable is used here in a very specific sense, which can be illustrated by considering the problem of solving the system $Ax = b$. Suppose that this system is solved in t-digit decimal floating-point arithmetic to give a computed solution $\bar{x}$. Then $\bar{x}$ is said to be stable if there is a small matrix E such that

$$(A + E)\bar{x} = b, \tag{14}$$

i.e., $\bar{x}$ satisfies the slightly perturbed system (14). The matrix E is required to be small in the sense that if $\alpha = \max(|a_{ij}|)$ and $\epsilon = \max(|e_{ij}|)$, then $\epsilon/\alpha = O(10^{-t})$. Thus if eight digits are carried in the computation and A is scaled so that its largest element is one, then the largest element of E must not exceed about 10^{-8}.

The notion of stability is useful because it frequently makes the problem of accuracy moot. Suppose, for example, that a user of LINPACK must solve a linear system $A_T x_T = b$, where A_T is the true value of the matrix and x_T is the true solution. Suppose, further — as often happens in practice — that owing to measurement or computational errors the matrix actually given to the LINPACK code is

$$A = A_T + F. \tag{15}$$

Then there are three "solutions" floating around: the true solution x_T; the exact solution x of $Ax = b$; and finally the computed solution $\bar{x}$ of $Ax = b$, which satisfies (14). Now often E in (14) will be smaller than F in (15), and consequently $\bar{x}$ and x will be nearer to each other than either is to x_T. In this case, the noise in the true matrix has already altered the solution more than the subsequent errors made by the LINPACK codes. If the user is unhappy with the answer, further computational refinements will not relieve the situation; the user must go back and get more accurate data, i.e., a better approximation to A_T.

All the decompositions computed by LINPACK are stable. For example, if $\bar{R}$ denotes the computed Cholesky factor of the positive definite matrix A, then $\bar{R}^T\bar{R}$ is equal to $A + E$ for some small matrix E. Again, the output of **SQRDC** is the numbers that would be obtained by performing exact computations on $X + E$, where E is small.

It is not to be expected that all things computed by LINPACK are done stably, although most of them are. The solutions of linear systems and linear least squares problems are stable, as are determinants. The updating routines are stable, with the exception of **SCHDD**. The most widespread unstable computation is that of the inverse.

Turning now from the question of stability to that of accuracy, we note that the inequality (5) already provides an answer in terms of the condition number $\kappa(A)$. Problems with large values of $\kappa(A)$ are said to be *ill conditioned*, and their solutions are very sensitive to perturbations in the matrix A. For the LINPACK routines, a rule of thumb is that if $\kappa(A) = 10^k$, one can expect to lose about k decimal digits of accuracy in the solution. It is important to remember that this rule accounts only for rounding errors made by LINPACK itself, and that it compares x and $\bar{x}$ defined as above; cf. (14). The relation between x and the true value x_T can be assessed only if the user is willing to provide additional information about the error F in (15).

The condition estimate provided by the LINPACK routines with suffix **CO** is generally reliable, although it can be fooled by highly contrived examples [O'Leary, 1980]. For large systems it is cheap to use, requiring only $O(n^2)$ operations as opposed to $O(n^3)$ for the factorization of A. The condition estimate obtained from the pivoted Cholesky decomposition is about as reliable [Stewart, 1980]. The singular value decomposition provides a completely reliable computation of $\kappa(A)$ in the spectral matrix norm.

We close this discussion of stability and condition estimates with a caveat. Bounds like (5) attempt to summarize the behavior of computed solutions to linear systems with a few numbers, and it is not surprising that something can be lost in this compression of the data. In particular, neither the condition number (4) nor the interpretation of the bound (5) is invariant under the scaling of the rows and the columns of A. Exactly what is the proper scaling is a complicated matter that at present is imperfectly understood. The *LINPACK Users' Guide* offers some advice, which must, however, be regarded as tentative.

The focus of the discussion up to this point has been on the effects of finite precision arithmetic on the LINPACK routines. However, something must also be said about how LINPACK deals with the finite range of the arithmetic, i.e., with underflow and overflow. The ideal in this regard would be to produce programs that succeed when both the problem and its answer are representable in the computer. Although the LINPACK programs do not attain this austere goal, they come near it by scaling strategic computations in such a way that overflows do not occur and underflows may be set to zero without affecting the results. Unfortunately this is not true everywhere, and some LINPACK programs can be made to fail by giving them data very near the underflow and overflow points. However, it is safe to say that LINPACK goes far in freeing the user from many of the difficulties associated with scaling.

EFFICIENCY

The efficiency of programs for manipulating matrices is not easy to discuss, since features that speed up a program on one system may slow it down on another. In this section we shall discuss in some detail the effects of two aspects of LINPACK on efficiency: the column orientation of the algorithms and the use of Basic Linear Algebra Subprograms (BLAS).

There was a time when one had to go out of one's way to code a matrix routine that would not run at nearly top efficiency on any system with an optimizing compiler. Owing to the proliferation of exotic computer architectures, this situation is no longer true. However, one of the new features of many modern computers — namely, hierarchical memory organization — can be exploited by some algorithmic ingenuity.

Typically, a hierarchical memory structure involves a sequence of computer memories, ranging from a small but very fast memory at the bottom to a capacious but slow memory at the top. Since a particular memory in the hierarchy (call it M) is not as big as the memory at the next higher level (M'), only part of the information in M' will be contained in M. If a reference is made to information that is in M, then it is retrieved as usual. However, if the information is not in M, then it must be retrieved from M', with a loss of time. In order to avoid repeated retrieval, information is transferred from M' to M in blocks, the supposition being that if a program references an item in a particular block, the next reference is likely to be in the same block. Programs having this property are said to have *locality of reference.*

LINPACK uses column-oriented algorithms to preserve locality of reference. By column orientation is meant that the LINPACK codes always reference arrays down columns, not across rows. This works because Fortran stores arrays in column order. Thus, as one proceeds down a column of an array, the memory references proceed sequentially. On the other hand, as one proceeds across a row the memory references jump, the length of the jump being proportional to the length of a column. The effects of column orientation are quite dramatic; on systems with virtual or cache memories, the LINPACK codes will significantly outperform comparable codes that are not column oriented.

There are two comments to be made about column orientation. First, the textbook examples of matrix algorithms are seldom column-oriented. For example, the classical recursive algorithm for solving the lower triangular system $Lx = b$ is the following:

> *for* **i** := **1** *to* **n** *loop*
> $x_i := b_i$;
>   *for* **j** := **i+1** *to* **n** *loop*
> $x_i := x_i - l_{ij}x_j$;
>   *end loop*;
> $x_i := x_i/l_{ii}$;
> *end loop*;

On the other hand, the column-oriented algorithm is the following:

> $x_j := b_j$ *(j:=1,2,...,n)*;
> *for* **j** := **1** *to* **n** *loop*
> $x_j := x_j / l_{jj}$;
>   *for* **i** := **j+1** *to* **n** *loop*
> $x_i := x_i - l_{ij}x_j$;
>   *end loop*;
> *end loop*;

From this example it is seen that translating from a row-oriented algorithm to a column-oriented one is not a mechanical procedure. An even more extreme example, cited in the discussion of matrix decompositions, is the algorithm for symmetric indefinite matrices, where column orientation requires that the underlying decomposition be modified.

The second comment is that LINPACK codes should not be translated into languages such as PL/I or PASCAL, where arrays are stored by rows. Instead, row-oriented subroutines with the same calling sequences should be produced. Unfortunately, doing this for all of LINPACK is a formidable task.

Another important factor affecting the efficiency of LINPACK is the use of the BLAS [Lawson *et al.*, 1979]. This set of subprograms performs basic operations of linear algebra, such as computing an inner product or adding a multiple of one vector to another. In LINPACK the great majority of floating-point calculations are done within the BLAS. The reasons why the BLAS were used in LINPACK will be discussed in the next section. Here we are concerned with the effects of the BLAS on the efficiency of the programs.

The BLAS affect efficiency in three ways. First, the overhead entailed in calling the BLAS reduces the efficiency of the code. This reduction is negligible for large matrices, but it can be quite significant for small matrices. The point at which it becomes unimportant varies from system to system; for square matrices it is typically between $n = 25$ and $n = 100$. If this should seem like an unacceptably large overhead, remember that on many modern systems the solution of a system of order twenty five or less is itself a negligible calculation. Nonetheless, it cannot be denied that a person whose programs depend critically on solving small matrix problems in inner loops will be better off with BLAS-less versions of the LINPACK codes. Fortunately, the BLAS can be removed from the smaller, more frequently used programs in a short editing session.

The BLAS improve the efficiency of programs when they are run on nonoptimizing compilers. This is because doubly subscripted array references in the inner loop of the algorithm are replaced by singly subscripted array references in the appropriate BLAS. The effect can be seen in matrices of quite small order, and for large orders the savings are quite large.

Finally, improved efficiency can be achieved by coding a set of BLAS to take advantage of the special features of the computers on which LINPACK is being run. For most computers, this simply means producing machine-language versions. However, the code can also take advantage of more exotic architectural features, such as vector operations.

An important conclusion to be drawn from the foregoing discussion is that on today's computers efficiency is not portable. The modifications that make a program efficient on one system may make it inefficient on another, and would-be designers of portable software must be prepared to draw flak no matter what they do.

DESIGN AND IMPLEMENTATION

One of the aims of LINPACK was to provide easy-to-use software for solving linear equations and least squares problems. Although such software existed before LINPACK, the best codes were often difficult to obtain and not readily transportable across machines. The algorithms in LINPACK are built around four standard decompositions of a matrix and are not new. The contribution of LINPACK has been in the directions of uniformity, portability, and efficiency.

The software in LINPACK owes its form to a set of decisions made early in the course of the project. Some of the decisions were determined by the nature of the package, but others were arbitrary in the sense that other ways of proceeding would have worked equally well.

Two of the major decisions were already discussed in the section on efficiency: namely, the use of column-oriented algorithms and the use of the BLAS. Against the former, one can argue that, in some algorithms, it prevents the user from obtaining greater accuracy by accumulating inner products in double precision. We felt that the sacrifice of this feature, which is not portable, was a small price to pay for the superior performance of LINPACK on systems with hierarchical memories. The decision to use the BLAS was more problematical. It was made in the absence of complete information, and the timings collected subsequently can be used to argue pro or con, depending on one's application. If considerations of efficiency are dropped, then the BLAS are a clear plus, since they reduce the amount of code while they improve clarity.

Two other major decisions are rather controversial. The first concerns the absence of driver programs to coordinate the LINPACK subroutines. Most problems will require the use of two LINPACK subroutines, one to process the coefficient matrix and one to process a particular right-hand side. This modularity results in significant savings in computer time when there is a sequence of problems involving the same matrix but different right-hand sides. Such a situation is so common and the savings so important that no provision has been made for solving a single system with just one subroutine. Actually, this should cause few problems for the user, since there is nothing in LINPACK as complicated as the EISPACK [Smith *et al.*, 1976; Garbow *et al.*, 1977] codes, where many routines are needed to solve a given problem. Another reason for not providing driver programs is to keep the size of the package manageable. Given the way the package is structured, an extra driver for each matrix structure and decomposition would have significantly increased the number of routines.

The second controversial decision concerns error checking. No checks are made on quantities such as the order of the matrix or the leading dimension of the array. The reason is that, except in the simplest programs, the number of things to check is very large and the checking would add significantly to the length of the code. Moreover, an elaborate data structure would have to be devised to report errors, so elaborate as to be beyond most casual users. Although the LINPACK programs will inform the user if a computed decomposition is exactly singular, no attempt is made to check for near singularities, much less to recover from them. The reason is that what constitutes a near singularity depends on how the problem has been scaled, a process that is imperfectly understood at this time. However, the user may monitor the condition number and do something if it is too large.

The naming conventions and data structures for the package have already been discussed. By and large, the decisions were easy, either because they were necessary or because it did not make much difference as long as something was decided.

Each of the LINPACK routines has a standard prologue:

Subroutine statement
Declaration of subroutine parameters
Brief description of the routine
List of the input arguments, their types and dimensions,
    and role in the algorithm
List of the output arguments, their types and dimensions
Date and author of the routine
External user routines needed, such as the BLAS
Fortran functions used
Declaration of variables for internal subroutine usage

The subroutine arguments are arranged in a specific order. The first three parameters are usually the matrix, **A**; the leading dimension of the matrix, **LDA**; and the order of the matrix, **N**. As has been pointed out, **LDA** and **N** should not be confused. Information about additional matrices, if any, is entered in the same way. The last two parameters are the **JOB** parameter, which tells the subroutine what to do, and the **INFO** parameter, which tells the user what the subroutine has done.

The parameters for a subroutine are declared in the order **INTEGER, LOGICAL, REAL, DOUBLE PRECISION, COMPLEX,** and **COMPLEX*16**. They are arranged in order of appearance within each classification.

After a description of the subroutine's purpose there follows a description of the parameters passed to and from each subroutine. This section is divided into two parts: information the routine needs for execution, and information generated and returned from the routine. These two parts are headed ON ENTRY and ON RETURN. The parameters listed are typed; dimensioning information is supplied, as well as a brief description of the function of the parameters. This way of specifying the input and output works well when the calling sequence is short. However, when the parameter list is long and complicated, as in the singular value routine, the convention is not so clean.

After the parameter information, the author's name appears with a date. This date shows when the routine was last updated and is very important in maintaining the package.

Next is a list of routines used by the documented routine. This list has the following order: LINPACK routines used, BLAS used, and Fortran functions used. Finally, the declarations for the internal variables are given.

Four versions of LINPACK exist, corresponding to the data types single precision, double precision, complex, and double precision complex. Because the algorithms are essentially the same for all data types, we decided to write code for the complex version only. This complex master version was then processed by an automatic programming-editing system called TAMPR [Boyle, 1980; Boyle and Dritz, 1974] to produce the other three types of programs. This automatic processing of the complex version reduced the coding and debugging time while contributing to the integrity of the package as a whole.

It was decided that the LINPACK programs should follow the canons of the structured programming school; that is, they should use only simple control structures, such as *if—then—else*. Since these structures do not exist in Fortran IV, it was necessary to simulate them. This process was greatly aided by the TAMPR system, which has the ability to recognize structure and generate appropriately indented code.*

The programs in LINPACK conform to the ANSI Fortran 66 Standard [ANSI, 1966]. We adopted Fortran because it is the language most widely used in scientific computations. Actually, the codes are restricted to a subset of the Standard that excludes **COMMON**, **EXTERNAL**, **EQUIVALENCE**, and any input or output statements. While the use of these excluded forms does not necessarily lead to unreadable, nonportable programs, it was felt that they should be avoided if possible [Smith, 1976].

The documentation of LINPACK is designed to serve both the casual user and the person who must know the technical details of the programs. The *LINPACK Users' Guide* treats, by chapter, the various types of matrix problems solved by LINPACK. In addition, appendices give related information such as the BLAS, timing data, and program listings.

Each chapter is self-contained and usually consists of seven sections:

> Overview
> Usage
> Examples
> Algorithmic Details
> Programming Details

---

* At one point the participants considered publishing the coding conventions by which the structures were implemented, but the appearance of Fortran 77 made it pointless to do so.

Performance
Notes and References

The first three sections of each chapter contain the user-oriented material. Basic information on how to use the routines and some examples of common usage are presented. The technical material is contained in the remaining sections. These give detailed descriptions of the algorithms and their implementations, as well as operation counts and a discussion of the effects of rounding error. The final section gives historical information and references for further reading.

TESTING

Only too often software is produced with little testing and evaluation. In the development of LINPACK, considerable time and effort were spent in designing and implementing a test package. In some cases, the test programs were harder to design than the programs they tested. The chief goal was to ensure the numerical stability and the portability of the programs.

We have already observed that no program can be expected to solve ill-conditioned problems accurately. Yet ill-conditioned problems must be included in any test package, since these problems often cause an algorithm to fail catastrophically. This creates the problem of how to judge the solution of an ill-conditioned problem. The answer adopted for the LINPACK tests was to demand that the solution be stable — that is, that it be the exact solution of a slightly perturbed problem. For example, in testing the program for computing the QR factorization of a matrix X, it was required of the computed Q and R that QR reproduce X to within a modest multiple of the rounding unit. Similarly, the computed solution $\bar{x}$ of $b - Ax$ was required to satisfy

$$\frac{\|b - A\bar{x}\|}{\|A\| \, \|\bar{x}\|} \leqslant \epsilon,$$

where ϵ is near the rounding unit. This is equivalent to testing for stability.

In problems such as matrix inversion, which are not done stably, the LINPACK programs were required at least to produce accurate solutions to well-conditioned problems.

The programs were also tested at "edge of the machine"; that is, the programs were given problems with entries near the underflow point or near the overflow point of the machine. Our purpose was to test how close LINPACK came to being free of overflow and underflow problems. An interesting fact that emerged from these tests is that a 64-bit floating-point word with an 8-bit binary exponent is a handicap in serious computations.

An important, though technically unachievable, LINPACK goal was to produce a completely portable package. Completely portable means that *no* changes need to be made to the software to run on any system. The testing program was critical in approximating this goal. As the results came back, it was found that this or that system had unexpected, perverse features. As each of these problems was circumvented, LINPACK came nearer and nearer to complete portability.

The test programs that were used are distributed with the package. They will help spot major flaws in the installation of the LINPACK routines, although they are not designed to test the codes exhaustively. It must be admitted that the quality of the test programs is not as high as LINPACK itself, although every effort was made to ensure their portability.

Timing information was collected on a wide range of computer-compiler combinations. In the multiprogramming environment of modern computers, it is often very difficult to measure the execution time of a program reliably. Significant variations can occur, depending on the load of the machine, the amount of I/O interference, and the resolution of the timing program. The timing data were gathered by a number of people in quite different environments. Our experience with gathering timing data indicates that we can expect a variation of 10 to 15 per cent if the timings are repeated.

Execution times also vary widely from computer to computer. It was found that this variability can be reduced by dividing the raw time for a process by the raw time for another process that involves the same amount of work. For example, the time required to execute **SGESL** might be compared to the time required to compute $A*x$. By the use of this scaling technique the authors of LINPACK were able to extract meaningful results from the mass of raw timing data collected during the testing. These are reported in the *LINPACK Users' Guide*.

LINPACK was tested by various people on many different compilers and operating systems. The authors of LINPACK performed the initial testing on their own computers to ensure that the programs were working correctly. Then, the package was sent out to the test sites for further testing. To say that LINPACK would have been impossible without the help of the test sites is not an exaggeration. They twice nursed our test programs through their systems, made the routines available to their user communities, commented on the documentation, and reported results. Listed below are the machines used in the testing:

Amdahl 470/V6
Burroughs 6700
CDC Cyber 175
CDC 6600
CDC 7600
CRAY-1
DEC KL-20
DEC KA-10
Data General Eclipse C330

Honeywell 6030
IBM 360/91
IBM 370/158
IBM 360/165
IBM 370/168
IBM 370/195
Itel AS/5
Univac 1110

As it turned out, few errors were revealed by the testing; on the contrary most failures were due to errors in compilers and operating systems. The test package is being used by CRAY as one of their Fortran compiler tests. While the package does not exercise all of the Fortran language, it does provide a good check of the numerical environment.

As a result of the extensive testing and our efforts in writing the codes, LINPACK comes close to being a fully portable package. There are no machine-dependent parameters or constants. The routines run, without modification, on all Fortran-based systems we know of.

CONCLUSIONS

In this final section we should like to offer some subjective conclusions on the LINPACK project and its implementation. The reader should keep in mind that these conclusions are the opinions of the two authors of this paper and do not necessarily reflect those of the other LINPACK participants.

LINPACK was not funded as a software development project; rather the National Science Foundation regarded it as research into methods for producing mathematical software. Although many useful ideas emerged from the project, it is safe to say the authors were more interested in development of the package than in software research. The reason for the curious rationale is that the National Science Foundation is constrained to help "research," which excludes software development. We feel that this constraint is detrimental to scientific endeavors in all fields and should somehow be removed. At the very least, it could be recognized that software development, by its very nature, involves a great deal of unstructured research, and this is sufficient justification for supporting such projects.

The fact that the LINPACK authors were scattered across the country did not impede the project. This is important because such an arrangement is a way of getting senior people from universities and other institutions involved in software development. In these days of computer networks, communications are not a problem. But the arrangement does require that the participants set aside two or three weeks a year to meet at a fixed location. There is no substitute for face-to-face contact.

The major difficulty with the way the project was organized was that there was no senior member with final authority to decide hard cases. Instead, each participant was responsible for a specific part of the package, and common matters — such as nomenclature, programming conventions, and documentation — were decided by consensus. As might be expected, the process frequently degenerated into bickering, usually about matters that did not seem very important a few months later. Agreements were always reached, and we do not feel that LINPACK suffered from the compromises. But we would advise anyone embarking on a project of this sort to set up a court of last resort, especially if more than three or four people are involved.

It goes without saying that LINPACK could not have succeeded without the support of the Applied Mathematics Division at Argonne National Laboratory. Not only did they provide tangible support in the form of offices, computer time, and secretarial assistance, but they also handled the many administrative details associated with the project. Most important of all, the members of the division treated the project participants with warm hospitality.

Only recently have people become aware of how greatly the development of mathematical software is aided by appropriate computer tools. We were fortunate in having the TAMPR system available to generate code from master complex programs and format it according to its structure. We also used the PFORT verifier to check our programs for portability. Although, strictly speaking, it is not a software tool, we found the WATFIV system with its extensive error checking useful in debugging our programs. We regret that we did not have one of the mathematical typesetting systems that are now appearing; if we had, we would undoubtedly have prepared the *LINPACK Users' Guide* on it.

At the beginning of the project, we decided to get as much advice from others as possible. To let people know what we were doing, we distributed informal reports under the title "LINPACK Working Notes." We also made early versions of the programs available to those who requested them. Although we had to spend a great deal of time justifying specific decisions to people who would have done things otherwise, the valuable suggestions we got more than compensated for the trouble.

We learned not to expect that tests, however extensive, would uncover all program bugs. By the time a well-written piece of mathematical software has been run on two or three systems, most of the obvious errors have been detected, and the remaining errors are quite subtle. For example, the shift of origin in the singular value routine is calculated incorrectly. This was not discovered during testing because it had no dramatic effect on the convergence of the algorithm and no effect at all on the stability of the result. By no means do we intend to imply that extensive testing is pointless; we have already noted that the tests helped improve the portability of LINPACK. But we found that there is considerable truth to the inverse of Murphy's law: If something *must* go wrong, it won't.

LINPACK is by no means perfect, and each of the participants has his particular regrets. The condition estimator would have been more useful if it had been a 2-norm and null vector estimator. The package should contain programs for packed triangular matrices and for updating the *QR* decomposition. We could have spent more time polishing the test drivers.

But perfection is an elusive thing. We are convinced that LINPACK is a good package and do not regret the time we spent producing it. We are reminded of Darwin's advice to the would-be world traveler:

But I have too deeply enjoyed the voyage, not to recommend any naturalist ... to start. He may feel assured, he will meet with no difficulties or dangers, excepting in rare cases, nearly so bad as he beforehand anticipates. In a moral point of view, the effect ought to be, to teach him good-humoured patience, freedom from selfishness, the habit of acting for himself, and of making the best of every occurrence. ... Travelling ought also to teach him distrust; but at the same time he will discover, how many truly kindhearted people there are, with whom he never before had, or ever again will have any further communication, who yet are ready to offer him the most disinterested assistance.

REFERENCES

ANSI [1966]. *FORTRAN.* ANS X3.9-1966, American National Standards Institute, New York.

Boyle, J. [1980]. "Software Adaptability and Program Transformation." *Software Engineering.* Eds. W. Freeman and P. M. Lewis. Academic Press, New York, pp. 75-90.

Boyle, J., and K. W. Dritz [1974]. "An Automated Programming System to Aid the Development of Quality Mathematical Software." *IFIP Proceedings,* North-Holland, Amsterdam, pp. 542-546.

Dongarra, J. J., J. R. Bunch, C. B. Moler, and G. W. Stewart [1979]. *LINPACK Users' Guide*. SIAM Publications, Philadelphia.

Forsythe and Moler [1967]. *Computer Solution of Linear Algebra Systems*. Prentice-Hall, Englewood Cliffs, New Jersey.

Garbow, B. S., *et al.* [1977]. *Matrix Eigensystem Routines — EISPACK Guide Extension*. Lecture Notes in Computer Science, Vol. 51. Springer-Verlag, Berlin.

Lawson, C., R. Hanson, D. Kincaid, and F. Krogh [1979]. "Basic linear algebra subprograms for Fortran usage." *ACM Trans. on Math. Soft.*, 5:308-323.

O'Leary [1980]. "Estimating matrix condition numbers." *SIAM Scientific and Statistical Computing,* 2:205-209.

Smith, B. T. [1976]. *Fortran Poisoning and Antidotes*. In Lecture Notes in Computer Science, Vol. 57. *Portability of Numerical Software*. Ed. W. Cowell, Springer-Verlag, Berlin.

Smith, B. T. *et al.* [1976]. *Matrix Eigensystem Routines — EISPACK Guide*. Lecture Notes in Computer Science, Vol. 6. 2nd ed. Springer-Verlag, Berlin.

Stewart, G. W. [1974]. *Introduction to Matrix Computations*. Academic Press, New York.

Stewart, G. W. [1980]. "The efficient generation of random orthogonal matrices with an application to condition estimators." *SIAM Numer. Anal.,* 17:403-409.

Wilkinson, J. H. [1963]. *Rounding Errors in Algebraic Processes*. Prentice-Hall, Englewood Cliffs, New Jersey.

Wilkinson, J. H. [1965]. *The Algebraic Eigenvalue Problem*. Oxford University Press, London.

Chapter 3

FUNPACK — A PACKAGE OF SPECIAL FUNCTION ROUTINES

W. J. Cody *
Mathematics and Computer Science Division
Argonne National Laboratory
Argonne, Illinois 60439

INTRODUCTION

FUNPACK is a modest systematized collection of machine-specific special function Fortran programs prepared under the NATS (National Activity to Test Software) project. Versions exist only for large-scale IBM, CDC and Univac computing systems. The package includes subroutines to evaluate certain Bessel functions, complete elliptic integrals, exponential integrals, Dawson's integral, and the psi function with secondary entries for exponential scaling or alternative arguments where appropriate. The package also includes a versatile error-monitoring routine.

Software for FUNPACK may be obtained from

> National Energy Software Center (NESC)
> Argonne National Laboratory
> 9700 South Cass Avenue
> Argonne, Illinois 60439
> (Phone: 312-972-7250)
> Cost: Determined by NESC policy

After its public release the package was modified by International Mathematical and Statistical Libraries, Inc. (IMSL), a commercial software house, and inserted in their library. This unofficial version is available on all machines serviced by IMSL.

Much has been written elsewhere about the software philosophy and production techniques developed by NATS (see, e.g., Boyle *et al.* [1972]; Smith, Boyle and Cody [1974]), but little has been written specifically about the production of FUNPACK. This paper reconstructs the events and decisions leading to FUNPACK as best we can recall them.

---

* This work was supported by the Applied Mathematical Sciences Research Program (KC-04-02) of the Office of Energy Research of the U.S. Department of Energy under Contract W-31-109-Eng-38.

HISTORY

The NATS project was conceived in 1970 and funded in 1971 by the National Science Foundation and the Atomic Energy Commission. This was a three-year cooperative effort among personnel at Argonne National Laboratory, Stanford University, The University of Texas at Austin and scattered test sites to study production, certification, distribution and maintenance of high-quality numerical software as a research problem. Although software production was intrinsic to the study, the particular software produced was of secondary importance provided it was useful and of high quality.

That is not to say that the software to be implemented was selected capriciously; there were requirements. The project naturally hoped that whatever software was produced would have a useful lifetime long enough to prove its quality, and that such software would be used often enough to attract attention to the project and its production methods. This meant that the underlying algorithms had to be drawn from an algorithmically mature, yet not stagnant, field rather than from a rapidly evolving research field. Ideally, the algorithms should be recent, widely accepted by leading researchers in the field, and backed by solid numerical and error analysis.

The project developers decided early to build on existing software efforts at Argonne rather than to start new efforts. There were several reasons for this decision. Argonne had a reputation for producing good programs for its own library; hence a reservoir of high-quality programs already existed there. Existing numerical expertise could be tapped, thus freeing the principal investigators to concentrate on production problems rather than algorithmic problems. Finally, two contemporary Argonne efforts posed contrasting production problems while meeting all other project requirements, thus giving the project variety and a running start at the same time. Mrs. V. C. Klema had recently initiated a translation of well-regarded matrix eigensystem codes [Wilkinson and Reinsch, 1971] from Algol into Fortran IV, and there also existed a long-standing effort to produce programs evaluating special functions. One effort blossomed into the highly transportable EISPACK package [Smith *et al.*, 1976; Garbow *et al.*, 1977] discussed in Chapter 4 of this book, while the other resulted in the highly machine-dependent FUNPACK package [Cody, 1975].

Of the two packages produced, EISPACK was clearly more important. It attracted more attention because matrix eigenanalysis software had a larger potential user community than special function software, and the production problems posed by EISPACK were more nearly typical of other software areas. The project therefore staked its reputation on EISPACK. Although the

FUNPACK effort was small and lagged behind the EISPACK effort, it profited from the understanding, techniques and tools forged and honed for EISPACK.

The typical work scenario for NATS software was as follows [Boyle *et al.*, 1972]. Prototype code, testing aids and preliminary documentation were written, systematized and debugged on IBM equipment at Argonne. Material for a different machine was then produced from the prototype and sent to one or more specific test sites for preliminary checking. This material was refined by iteration between Argonne and the test site until both were satisfied. When the package had been assembled, it was sent to cooperating test sites for extensive field testing under a variety of operating systems and compilers. The package iterated between the field test sites and Argonne until everyone was satisfied that the programs performed well and that the documentation was complete and correct. At that point the package was judged ready for release and was turned over to the software distribution agencies NESC (then called the Argonne Code Center) and IMSL for public distribution.

The first small version of FUNPACK went to test sites in the summer of 1972. Field testing uncovered a design flaw that seriously hampered what little transportability the package had. This problem was solved by repackaging the software, replacing the traditional FUNCTION subprograms with "packets" of related subprograms. An enlarged, redesigned version passed field testing in early 1973 and was delivered to the Argonne Code Center for public distribution later that year coincident with the end of NATS funding. FUNPACK I, publicly announced on September 23, 1973, contained only four different monovariate function packets (six for IBM machines) and an error monitor.

In 1974 NATS II was funded to continue the NATS effort with minor changes for another three-year period. Among the changes, personnel from Jet Propulsion Laboratory and the University of Kentucky joined the FUNPACK effort, and the work was extended to include algorithm development and implementation for multivariate functions, introducing new production difficulties into the study. The extended project now examined all aspects of software production beginning with the numerical research leading to algorithms and ending with the distribution of certified software implementations. Work on FUNPACK II proceeded on schedule with few difficulties. The complete package of twelve function packets and the error monitor (see Table I) went to test sites for final review in April 1976 and to the Argonne Code Center that September. The NATS project ended with the formal announcement of availability of FUNPACK II a month later.

Table I

Elements of FUNPACK

Category	Packet	Entry	Computation [Abramowitz and Stegun, 1964]
Bessel functions	NATSJ0	BESJ0(X)	$J_0(x)$
	NATSJ1	BESJ1(X)	$J_1(x)$
	NATSI0	BESI0(X)	$I_0(x)$
		BESEI0(X)	$e^{-|x|}I_0(x)$
	NATSI1	BESI1(X)	$I_1(x)$
		BESEI1(X)	$e^{-|x|}I_1(x)$
	NATSK0	BESK0(X)	$K_0(x)$
		BESEK0(X)	$e^x K_0(x)$
	NATSK1	BESK1(X)	$K_1(x)$
		BESEK1(X)	$e^x K_1(x)$
	NATSBESY	[D]YNU(X,V)	$Y_\nu(x)$
Complete elliptic integrals	DELIPK	[D]ELIPK(EM)	$K(m)$
		[D]ELIK1(CAY)	$K(m),\ m=k^2$
		[D]ELIKM(ETA)	$K(m),\ m=1-\eta$
	DELIPE	[D]ELIPE(EM)	$E(m)$
		[D]ELIE1(CAY)	$E(m),\ m=k^2$
		[D]ELIEM(ETA)	$E(m),\ m=1-\eta$
Exponential integrals	DEI	[D]EI(X)	$Ei(x)$
		[D]EXPEI(X)	$e^{-x}Ei(x)$
		[DP]EONE(X)	$E_1(x)$
Dawson's integral	DDAW	[D]DAW(X)	$e^{-x^2}\int_0^x e^{t^2}dt$
Psi function	NATSPSI	[D]PSI(X)	$\psi(x)=\Gamma'(x)/\Gamma(x)$
Error monitor	MONERR	MONERR	report or set error flags

DESIGN

The first version of FUNPACK harvested past research work on function approximation. Over a period of years a small group at Argonne had produced minimax rational approximations that were used to prepare a large collection of special function programs. This collection was written for optimal performance on local computers but with little thought about how it might be moved to machines with different architecture. The decision to select candidates for FUN-PACK I from this collection dictated many of the decisions that followed.

The list of candidates included software for a few important functions such as elliptic integrals, exponential integrals and Dawson's integral, which were not often found in computer libraries at that time. By choosing these functions for FUNPACK I and ignoring more common functions such as the error function, gamma function and Bessel functions, we hoped that the collection would not duplicate capabilities available elsewhere, and would therefore be used.

Certain ideals for performance that emerged from the previous work on function software and on EISPACK were immediately adopted as goals for FUN-PACK. Within reason, we wanted subroutines to return accurate function values whenever both the argument and function value exist and are representable in the host arithmetic system, and to return an error in all other cases. Thus accuracy and argument domains were to approach limits naturally imposed by machine architecture whenever that could be achieved with reasonable effort. The computations were to be completely free of underflow and overflow, with unavoidable problems properly recorded and reported to the user. We insisted that the programs detect and report every error condition themselves, avoiding inconsistent, unreliable and sometimes misleading system-generated error returns. Finally, the software was to be as efficient as possible within the constraints imposed by accuracy and dependability. These and similar design goals for EISPACK led to the software concepts of reliability and robustness discussed at length by Smith, Boyle and Cody [1974].

Portability was never an important issue with FUNPACK. Indeed, the lack of portability deliberately built into FUNPACK directly contrasted with the portability of EISPACK and broadened the NATS experience. High-performance function programs are intrinsically more machine-dependent than linear algebra programs, although some special function algorithms are less machine-dependent than others. We will see later that considerations independent of the underlying algorithm require major changes in function programs as they move from one machine to another, thus further reducing portability.

Even if this were not the case, only the complete elliptic integrals among the initial FUNPACK candidates could have been implemented in portable programs, if based on the Gauss arithmetic-geometric mean process. Unless new algorithms were developed, the other functions had to be evaluated from polynomial or rational approximations having coefficients that changed with machine word length. And because the Gauss scheme converges slowly near the extremes of the argument range and is sometimes inaccurate there, polynomial approximations were chosen for the elliptic integrals as well.

Truncated analytic expansions such as Taylor series, Padé approximations and Chebyshev series [Gautschi, 1975] could have been used for FUNPACK instead of the non-analytic, nearly minimax approximations mentioned above. These have an advantage that coefficients are the same for all machines; only the number of coefficients and the number of digits in each must change with machine word length. On the other hand, analytic expansions are inefficient for arguments remote from the point of expansion, and usually have poor accuracy near zeros that are not expansion points. This latter characteristic sometimes makes it difficult to guarantee full significance in the function value for remote arguments without resorting to higher precision arithmetic.

Properly chosen minimax approximations are nearly always superior to analytic expansions. Because minimax approximations provide the smallest maximum error of approximation among all approximating functions of the selected form, they usually achieve a desired accuracy with fewer coefficients than the corresponding analytic form. In addition, they can be selected to approximate a function to either a given number of digits or a given number of *significant* digits over the entire interval of approximation, thereby providing good accuracy near zeros. Their main disadvantage is that the coefficients change when the interval of approximation or required accuracy changes. A new set of coefficients is probably needed when a function program is moved to a different computer architecture. Everything considered, however, minimax approximations were the optimal choice for FUNPACK I.

We expected that computation of a particular function could follow the same basic algorithm on different machines; only the details of the implementation would have to change to reflect machine dependencies. Even so, this approach required an independent set of programs for each combination of machine and arithmetic precision supported. Coverage was therefore limited to those contemporary machines and precisions normally used for large-scale scientific computation, namely, CDC 6000/7000 series computers in single precision and IBM 360-370 series and Univac 1108 computers in double precision. These precisions range from about 14S (significant decimal digits) on the CDC equipment to about 18S on the Univac 1108.

The problem of how best to handle error conditions was a sticky one. Each strategy considered had advantages and disadvantages. Printed error messages, error parameters in the calling sequence, forced termination of execution and continued execution with default function values all seemed reasonable strategies under some conditions, but not under others. The desirable action depended on user sophistication and problem context.

The final design provided error facilities corresponding to varying degrees of user sophistication. For the unsophisticated user, default procedures printed an error message identifying the program and argument the first time an error occurred, and suppressed later error messages for that program. A count of the number of errors was maintained, and execution continued with reasonable default function values. The more sophisticated user could set internal flags to control the printing of error messages, control termination of execution, and reset or interrogate error counters. In addition, standard systems routines provided traceback information as part of forced termination on Univac and most IBM systems; traceback was not possible under standard CDC systems, however.

Originally, FUNPACK elements were self-contained FUNCTION type subprograms with secondary entries for error monitoring, exponential scaling or alternative arguments where appropriate. Each program thus had two or more entries and included I/O statements for error reporting. The IBM and Univac programs also contained system dependencies for traceback.

Early field tests showed that this organization was poor. Standard Fortran did not include alternative entries, and the syntax varied from compiler to compiler. Inconsistencies were found in I/O formatting conventions and in the designation of standard I/O units. Some IBM installations unexpectedly reported trouble because they did not support the optional system traceback facilities. Local modifications sometimes were necessary for almost every program in the FUNPACK package.

Many of these problems were eliminated and the others localized by completely reorganizing the package into *function packets*. Individual subprograms with multiple entry points were replaced by packets of related subprograms with only one entry each. The remaining system dependencies mentioned above were localized by removing error monitoring facilities from the individual packets and placing them in a separate packet. This cost some flexibility in the wording of error messages, but reorganization enabled necessary system-dictated changes to I/O or traceback facilities to be made only once in the error packet rather than in every subroutine in the package.

A typical function packet was now built around a computational module, a SUBROUTINE that performs all numerical computations for the packet.

Computational entries under the previous organization were implemented as FUNCTION subprograms that invoke the computational module. They provide the programming interface necessary for the user to reference the function in longer algebraic expressions.

Programming conventions followed in EISPACK and carried over to FUN-PACK were designed to help humans read and understand our programs, and to make it easier for us to maintain and modify them. Some were designed to simplify conversion of programs from one precision to another. For example, variable names were chosen to have some mnemonic meaning and to default to the appropriate type in Fortran, but still explicitly declared in TYPE statements. Intrinsic and library functions were also declared, but in separate TYPE statements that are easily commented out when compilers object to such declarations, as some CDC compilers do.

Other conventions specifically addressed program comprehension. For example, every packet was prefaced with comments documenting its usage and liberally sprinkled with other comments explaining major logical and computational points. Statement labels were chosen strictly increasing in value, and no unreferenced labels were used. Indentations were used to set off major control structures such as loops and "if ... then ... else" constructions, and blanks were used within statements to improve readability.

Finally, documentation was designed to be machine-readable, a single document covering all machine versions of a particular packet. The documents were complete and precise. In addition to the usual information on purpose, calling sequences, length of source decks, etc., they detailed error conditions and responses and discussed test results. In particular, they listed the test sites and the machines, operating systems and compilers used in successful testing. Perhaps the most important part of the document was the NATS certification statement [Boyle et al., 1972], a carefully worded pledge to support the software through its useful lifetime.

Everything related to FUNPACK was machine-readable. We used text editing programs in a time-sharing system to prepare and maintain master files for all source code, documentation and test material. It was never necessary to use card decks, although we sometimes did use them for convenience.

The basic design features just outlined carried over to FUNPACK II, except that many of the new packets were based on new, specially developed algorithms instead of previously existing programs.

IMPLEMENTATION

The implementation work on FUNPACK began a little later than the work on EISPACK. As stated above, we produced software for a particular function by first perfecting the program for our local IBM equipment. We originally planned simply to impose our packaging standards on programs already available in the Argonne library. We quickly discovered, however, that more extensive changes were needed. At times it seemed as if we could salvage little from the original programs except the coefficients.

Major changes were required to clean up the logical design of the programs to meet our new understanding of software packaging. For example, we now required all computational paths to proceed smoothly through a program without doubling back. We also preferred logical IF statements to arithmetic IF and computed GO TO statements, although some of the latter persist in early elements of the package. Stylistic changes such as these occurred throughout the FUNPACK effort. We felt even more strongly about these constructions as work progressed, but stylistic improvements in later elements of the package were not easily retrofitted to earlier elements. Each version of a program was handcrafted, and so we could not conveniently use the automated programming aids developed for EISPACK to recover different versions of a particular program from one master copy. The package thus lacks uniformity of style.

We avoided algebraic constructions leading to known numerical problems. For example, when floating-point arithmetic is truncated or rounding occurs before renormalization, the expression $2.0 \times x$ often contains error. We used the expression $x+x$ instead, because it is always error-free on binary machines and never worse than $2.0 \times x$ on other machines. Similarly, the expression $1.0-x$ was replaced, on machines that lacked guard digits, by the expression $(0.5-x)+0.5$ to protect precision for $0 < x < 1.0$.

To simplify planned conversion to other equipment, previously *ad hoc* thresholds of various kinds were precisely defined with machine-dependent parameters. For example, asymptotically

$$E_i(x) \approx e^x/x < e^x, \quad x \gg 1.$$

The Argonne ancestor of DEI prematurely terminated because of overflow in e^x for large x. The new programs were algorithmically restructured to compute good function values until $E_i(x)$ itself overflowed, and then to give an error return when x became too large. The overflow threshold then became a parameter initialized in a DATA statement. Unfortunately, the methods for determining such parameters were never documented for the user. Prologue comments in programs new to FUNPACK II define such constants without telling how to determine them, but even this meager information is missing from FUNPACK I programs. The oversight complicates possible program conversion now for additional machines.

Additional restructuring was necessary to avoid noisy underflow. At one time our local IBM system quietly replaced underflow with zero while other systems did not. Where we previously may have computed small numbers not really caring if results vanished, we now had to specifically test magnitudes and bypass computations where results would be too small. Otherwise, our programs would sometimes trigger spurious system-generated underflow messages.

Most of the remaining changes for IBM equipment were in error detecting and reporting. The error facilities described earlier were completely new and had to be inserted intact into each of the first few function programs. The later design change to function packets simplified the error facilities needed in each packet, but by then only completely new programs were being written anyway.

Draft documents and test material, discussed in detail in the next section, were prepared at the same time as the IBM prototype programs. There were many interactions between the three tasks. Documentation is easiest, of course, when programming details are fresh, but the interactions went deeper than that. The machine-readable documents and the prologue comments in each source program duplicated information that was easily shared. Specifying program actions for documentation often uncovered subtle loopholes in program coverage that led to programming changes and corresponding modifications of test material. Similarly, the test material often uncovered problems leading to program and document changes. Preparation of the IBM prototype programs was thus a complicated and iterative procedure simultaneously refining the documentation, source programs and test material into a coherent whole.

Once the IBM versions were working satisfactorily, prototype source and test materials were prepared for CDC and Univac machines. Problems encountered in preparing the IBM versions helped us to anticipate, isolate and solve many of the design problems for the other versions before sending them out for tuning tests, even though the versions differed substantially from one another.

The first step in conversion was a simple translation of the IBM source, a task no more difficult than the initial preparation of the IBM prototypes from their Argonne ancestors. Many of the necessary changes were routinely made using text editors. Double precision declarations, constants and intrinsic functions had to be changed to single precision for the CDC versions, and all the minimax approximations had to be replaced. Other changes, particularly those reflecting hardware and compiler characteristics, were more significant. Dawson's integral, for example, has the asymptotic behavior

$$F(x) \approx 1/2x, \quad |x| \gg 1.$$

Error returns for this function are unnecessary on IBM and Univac machines because the function is computable and representable for all machine-representable arguments. In particular, the quantity $1/2x$ is representable as a floating-point number on these machines whenever x is. On CDC equipment, however, the distribution of floating-point numbers is biased, and $1/2x$ underflows for sufficiently large machine-representable x. Therefore the CDC version of this function must contain threshold tests for large $|x|$ and provide error facilities not needed in the other versions.

The most difficult conversion tasks were the representation of approximation coefficients and the precise determination of various thresholds. Most of the original Argonne codes specified coefficients and important constants in hexadecimal. This permitted "tweaking" the low-order bits in the coefficients to compensate for rounding and otherwise improve the numerical quality of the programs. As an experiment, we decided to represent these quantities in octal in the CDC and Univac programs and see whether we could tune the programs at Argonne without direct access to CDC and Univac machines. Only the CDC and Univac programs for the exponential integral were written with every constant specified in decimal to permit comparison of the implementation efforts and results for the two methods.

The first problem, of course, was to determine the correct octal representation of quantities known to us only in decimal or hexadecimal. The IBM equipment had shorter exponent range than either the CDC or Univac equipment, and at that time could not provide adequate precision for determining or representing Univac constants. Fortunately we still had a CDC 3600 available that did have adequate exponent range and precision. The CDC Fortran system also provided ENCODE and DECODE commands for formatted core-to-core I/O operations. We therefore wrote a suite of programs for this machine that would convert floating-point numbers back and forth between the internal CDC 3600 representation and the internal representation for each of the other machines. Anticipating later needs, we also provided programs to write binary magnetic tapes on the 3600 that were readable by the other machines.

Halfway through the project the CDC 3600 was replaced by an upgraded IBM system with extended precision arithmetic. The new system had adequate precision but still lacked exponent range and the ENCODE and DECODE commands. Nevertheless, we rewrote the suite of conversion programs for this machine, adding subroutines for formatted core-to-core I/O operations. The few constants with exponents too large to be handled by these programs were scaled by powers of 2 before conversion, and the exponents later modified by hand. Fortunately, we never had to write tapes containing such large numbers.

With these tools we were able to determine threshold parameters for CDC and Univac machines to within a unit or two in the last bit position. Typically, a threshold was determined by solving a nonlinear equation on our local machines, and then verifying the solution by perturbing it in the last bit position for the target machine. These thresholds were later refined during fine tuning at The University of Texas for the CDC versions, and the University of Wisconsin and Jet Propulsion Laboratory for the Univac versions. That work is discussed in the next section.

Most of the algorithms and coefficients used in FUNPACK are now available in the open literature (see Table II). The coefficients used for $J_0(x), J_1(x)$ and $Y_\nu(x)$ are unpublished, as are certain coefficients for Univac programs where previously published approximations were not accurate enough. These unpublished approximations were generated locally with existing software [Cody and Stoer, 1966; Cody, Fraser and Hart, 1968]. Where appropriate, as in $J_0(x), J_1(x)$, and $E_i(x)$, zeros of the functions are built into the approximations. The programs retain accuracy near these zeros by simulating higher precision arithmetic in a few critical operations as described by Cody and Thacher [1968]. In a few cases, the numerator and denominator polynomials of an approximation were transformed into minimal Newton form [Mesztenyi and Witzgall, 1967] or a continued fraction representation to improve numerical conditioning. Otherwise, the approximations were implemented as given.

Table II

Source of Algorithms

Packet	Reference	Packet	Reference
NATSJ0	Unpublished	DELIPK	[Cody, 1965]
NATSJ1	Unpublished	DELIPE	[Cody, 1965]
NATSI0	[Blair, 1974]	DEI	[Cody and Thacher,
NATSI1	[Blair, 1974]		1968; Cody and
			Thacher, 1969]
NATSK0	[Blair and	DDAW	[Cody, Paciorek
	Russon, 1969]		and Thacher, 1970]
NATSK1	[Blair and	NATSPSI	[Cody, Strecok
	Russon, 1969]		and Thacher, 1973]
NATSBESY	[Cody, Motley and		
	Fullerton, 1977]		

TUNING AND TESTING

Originally the NATS group intended to supply field test sites with only source decks and preliminary documentation. Test sites were to avoid bias from the software developers by preparing tests for this material by themselves. We hoped that after preliminary testing the programs would be offered to local users on a trial basis and that we would receive valuable insight on the intangibles of program usage.

Early experience with EISPACK, however, dashed these hopes. The volume of material to be tested overwhelmed most sites. Personnel assigned to testing often lacked the numerical training necessary to produce independent tests or were busy with other responsibilities. In a few cases, field testing was carried out as we had hoped, but most site responses were simply reports that the programs had compiled.

The only way to guarantee useful testing then was to prepare and distribute test material with the programs. For FUNPACK this material fell into two categories. Programs for special tuning and detailed accuracy testing were distributed only to sites cooperating in the pre-field test work. Demonstration programs, designed to accompany FUNPACK in public release, were distributed to all test sites. The supporting test material was prepared as carefully as the individual function packets.

Programs specifically designed for tuning primarily verified and refined threshold parameters. We were confident that the parameters determined on our local machines were close to their correct values, but machine and compiler idiosyncrasies might put our values slightly off. Thus confirmation of the values was essential.

Accuracy testing, using techniques developed years before [Cody, 1969], was the heart of pre-field testing. Briefly, the test sites were provided with binary magnetic tapes containing argument and function value pairs represented in the internal floating-point format for the target machine. These tapes were prepared using the auxiliary high-precision programs described in the last section. Arguments were chosen randomly and ordered algebraically within each test interval. These arguments were considered exact, and corresponding function values were generated and correctly rounded to the precision of the target machine.

Accompanying driver programs compared results from the programs under test with the standard tape values. For each test interval the programs reported the frequency of error measured in ULPs (binary Units in the Last Place of the floating-point significand) and the maximum and root mean square relative errors. The machine representation of arguments and function values were printed whenever ULP errors exceeded an input threshold value. This feature, combined with the algebraic ordering of arguments, traced tendencies toward large computational error and provided clues for "tweaking" coefficients to reduce such error.

Demonstration programs in the form of stand-alone test drivers briefly checked the accuracy of the subprograms, explored various thresholds to verify their value, triggered each of the error returns and exercised each of the error monitoring options. Accuracy tests compared a few computed function values with standard values and reported the ULP errors. These in turn were compared with the corresponding errors noted during pre-field testing.

When all prototype material for a particular function was ready, CDC versions were sent to The University of Texas and Univac versions to the University of Wisconsin or Jet Propulsion Laboratory for pre-field testing. Results from the preliminary tests were used to debug and tune the programs before full field testing. We anticipated numerous iterations in this step but were pleasantly surprised. No functions required more than two iterations; most were ready after the first iteration. Our thresholds usually proved to be exact; only a few minor errors were uncovered, and not much tuning was required.

Perhaps the most elegant example of error pinpointing occurred in initial tests at Wisconsin for the complete elliptic integral $K(m)$. For m close to 1.0, function accuracy depends critically upon accurate evaluation of the expression $(1.0-m)$. As described earlier, this expression was replaced by the more stable expression $(0.5-m)+0.5$. Despite this tactic, the tests detected large errors for m close to 1.0. Using data from the error trace, we were able to show analytically that the results were consistent with an evaluation of the original expression rather than the more stable one (see Cody [1975] for details). We correctly surmised that Univac's optimizing compiler had helpfully restored the earlier expression (a previously unsuspected compiler characteristic), and designed a simple countermeasure. The revised program passed the second round of tests, thus demonstrating the value of carefully designed, sensitive test programs.

The final step before public release was field testing. The only ground rule here was that test sites were not to modify the programs; we wanted all sites testing the same programs. Twenty-one different computing centers with a variety of operating systems and compilers participated at various times. All ran the demonstration programs, and many performed additional testing and made the programs available to users.

We have already discussed the major reorganization into packets because of problems uncovered during field testing. Test sites suggested some minor changes as well, some of which were made and some of which were not. For example, we changed the names of some CDC programs because CDC sites felt that program names beginning with a "D" implied that the programs were double precision. Most CDC compilers truncated decimal constants to 14S before compiling them, but one site had a locally produced compiler that aborted compilation when constants were specified to more than 14S. The latter site wanted the coefficients in the exponential integral packet converted from decimal to machine representation. We rejected the request because the compiler was one-of-a-kind. These experiences ended our experiment with decimal representation, however. They also vindicated our earlier decision to use minimax approximations rather than analytic expansions whose only advantage, increased portability, was predicated on decimal representation. Although a few documentation errors were uncovered, most of the other problems reported were either unimportant or beyond our control. Computations in a few programs lacked monotonicity in low-order bit positions across boundaries separating approximations, for example; also, some libraries contained bad elementary function programs.

RETROSPECT

In our opinion, NATS achieved all its goals in software research. EISPACK and FUNPACK showed that high-quality software could be produced in large systematized packages for diverse computing systems. In the process they set high standards for others to follow.

Five years after the completion of FUNPACK we are still pleased with it. As of March 31, 1981, 227 copies of FUNPACK II have been distributed by NESC. Although we have no information about how many of these copies have been used, only one error of any consequence has been reported. W. Mammel discovered that a boundary between regions in the $Y_\nu(x)$ packet was assigned to the wrong region, and that computations on that boundary were bad. A logical IF statement in the packet was corrected, and all package recipients were notified. We have also compiled a short list of three or four minor inconsistencies between documentation and prologue comments for packets. Most of these involve definitions of argument thresholds or specification of default responses for invalid arguments.

FUNPACK I is estimated to have cost about $20 per source card [Cody, 1975]. This cost represents only the software effort and is independent of the significant underlying research to develop algorithms. We have no reason to suspect that FUNPACK II costs were much different. Initially this figure may seem astronomical, but it is really not much greater than figures quoted for contemporary numerical software projects with less concern for quality.

The figure is also misleading. NATS was not funded to produce software; it was funded to study *how* to produce software. Some of the expense must be attributed to our efforts to understand production processes and is therefore not directly attributable to production itself. Finally, we must weigh cost against benefits. We can never know all the benefits, but undoubtedly the availability of our software saved some people the direct cost of writing programs for themselves, and saved other people the indirect cost of using whatever else might have been available.

If we were to start over, knowing what we know now, would we do things differently? Probably. Although we sought advice from our NATS colleagues, final design decisions were ours. That may have been a mistake. We do not think FUNPACK turned out badly, but it might have been even better if we had been forced to convince others of the wisdom of various decisions. As a general principle, we believe that control of a large numerical software project should not be vested in those with a personal interest in the algorithms being implemented, because that interest might adversely affect packaging decisions.

It was a useful experiment in NATS II to develop a new algorithm for $Y_\nu(x)$ specifically for implementation in FUNPACK, but we would not like to do that again. Research on numerical algorithms cannot follow the same time schedule as software implementation. We believe that algorithm research should be independent of software implementation when the latter faces rigid deadlines.

We also feel that special function programs can now be written more portably than FUNPACK without sacrificing quality. We have had numerous complaints because the package is not available on particular machines and is difficult to transport. Moving high-performance function programs may never be so easy as moving linear algebra programs, but the task can be simpler than it is now. FUNPACK, Fullerton's FNLIB [Fullerton, 1977] and Schonfelder's contributions to the NAG library [Schonfelder, 1977] represent three extreme views on function software: FUNPACK emphasizes accuracy and robustness, FNLIB emphasizes portability and NAG emphasizes speed and transportability. There is a useful middle ground that we hope to explore in the near future.

ACKNOWLEDGMENTS

The programs in FUNPACK were the result of the combined efforts of many people. Key people in the NATS projects were J. Boyle, W. Cowell, B. Garbow, Y. Ikebe, C. Moler, E. Ng, B. Smith and H. C. Thacher, Jr. Source decks were prepared by temporary and permanent staff at Argonne and Jet Propulsion Laboratory with computational advice from staff at The University of Texas at Austin and the University of Wisconsin. In addition to the author, the following individuals contributed directly to the programming and tuning effort for FUNPACK: T. Bennett, K. Gillig, K. Paciorek, R. Hopkins, Y. Ikebe, D. Kincaid, K. LaPlante, R. Motley, E. Ng, H. C. Thacher, Jr. and W. Wallace. Field testing was carried out by individuals at the following computing centers: Ames Laboratory (Iowa State University), Argonne National Laboratory, ICASE/NASA Langley Research Center, Illinois Institute of Technology, Jet Propulsion Laboratory, Kirtland Air Force Base/AFWL, Lawrence Berkeley Laboratory, Lawrence Livermore Laboratory, National Center for Atmospheric Research, Northwestern University, Oak Ridge National Laboratory, Purdue University, Stanford University, Stockholm Data Center, The University of Chicago, The University of Texas at Austin, University of Illinois at Urbana-Champaign, University of Michigan, University of New Mexico, University of Toronto and University of Wisconsin.

We are grateful to W. Cowell, B. Garbow and G. Pieper for their constructive criticism of this manuscript.

REFERENCES

Abramowitz, M., and I. A. Stegun [1964]. *Handbook of Mathematical Functions with Formulas, Graphs, and Mathematical Tables.* Nat. Bur. Standards Appl. Math. Series, 55, Washington, D. C.

Blair, J. M. [1974]. "Rational Chebyshev approximations for the modified Bessel functions $I_0(x)$ and $I_1(x)$." *Math. Comp.* 28:581-583.

Blair, J. M., and A. E. Russon [1969]. *Rational Function Minimax Approximations for the Bessel Functions $K_0(x)$ and $K_1(x)$.* Atomic Energy of Canada Ltd. Report AECL-3461, Chalk River, Ontario.

Boyle, J. M., W. J. Cody, W. R. Cowell, B. S. Garbow, Y. Ikebe, C. B. Moler and B. T. Smith [1972]. "NATS, A Collaborative Effort to Certify and Disseminate Mathematical Software." *Proceedings 1972 National ACM Conference, Vol. II.* Association for Computing Machinery, N.Y., pp. 630-635.

Cody, W. J. [1965]. "Chebyshev approximations for the complete elliptic integrals K and E." *Math. Comp.* 19:105-112.

Cody, W. J. [1969]. "Performance testing of function subroutines." *AFIPS Conf. Proc., Vol. 34, 1969 SJCC.* AFIPS Press, Montvale, N.J., pp. 759-763.

Cody, W. J. [1975]. "The FUNPACK package of special function subroutines." *ACM Trans. on Math. Soft.* 1:13-25.

Cody, W. J., W. Fraser and J. F. Hart [1968]. "Rational Chebyshev approximations using linear equations." *Num. Math.* 12:242-251.

Cody, W. J., R. M. Motley and L. W. Fullerton [1977]. "The computation of real fractional order Bessel functions of the second kind." *ACM Trans. on Math. Soft.* 3:233-239.

Cody, W. J., K. A. Paciorek and H. C. Thacher, Jr. [1970]. "Chebyshev approximations for Dawson's integral." *Math. Comp.* 24:171-178.

Cody, W. J., and J. Stoer [1966]. "Rational Chebyshev approximation using interpolation." *Num. Math.* 9:177-188.

Cody, W. J., A. J. Strecok and H. C. Thacher, Jr. [1973]. "Chebyshev approximations for the psi function." *Math. Comp.* 27:123-127.

Cody, W. J., and H. C. Thacher, Jr. [1968]. "Rational Chebyshev approximations for the exponential integral $E_1(x)$." *Math. Comp.* 22:641-649.

Cody, W. J., and H. C. Thacher, Jr. [1969]. "Chebyshev approximations for the exponential integral Ei(x)." *Math. Comp.* 23:289-303.

Fullerton, L. W. [1977]. "Portable special function routines." *Lecture Notes in Computer Science, Vol. 57; Portability of Mathematical Software.* Ed. W. Cowell. Springer-Verlag, New York, pp. 452-483.

Garbow, B. S., J. M. Boyle, J. J. Dongarra and C. B. Moler [1977]. *Matrix Eigensystem Routines — EISPACK Guide Extension, Lecture Notes in Computer Science, Vol. 51.* Springer-Verlag, New York.

Gautschi, W. [1975]. "Computational methods in special functions — a survey." *Theory and Application of Special Functions.* Ed. R. A. Askey. Academic Press, New York, pp. 1-98.

Mesztenyi, C., and C. Witzgall [1967]. "Stable evaluation of polynomials." *NBS Jour. of Res. B,* 71B:11-17.

Schonfelder, J. L. [1977]. "The production and testing of special function software in the NAG library." *Lecture Notes in Computer Science, Vol. 57; Portability of Mathematical Software.* Ed. W. Cowell. Springer-Verlag, New York, pp. 425-451.

Smith, B. T., J. M. Boyle and W. J. Cody [1974]. "The NATS approach to quality software." *Software for Numerical Mathematics*. Ed. D. J. Evans. Academic Press, New York, pp. 393-405.

Smith, B. T., J. M. Boyle, J. J. Dongarra, B. S. Garbow, Y. Ikebe, V. C. Klema and C. B. Moler [1976]. *Matrix Eigensystem Routines — EISPACK Guide, Lecture Notes in Computer Science, Vol. 6, 2nd ed.* Springer-Verlag, New York.

Wilkinson, J. H., and C. Reinsch [1971]. *Handbook for Automatic Computation, Volume II, Linear Algebra, Part 2.* Springer-Verlag, New York.

Chapter 4

EISPACK
A PACKAGE FOR SOLVING MATRIX EIGENVALUE PROBLEMS

J. J. Dongarra *
Mathematics and Computer Science Division
Argonne National Laboratory
Argonne, Illinois 60439

C. B. Moler †
Department of Computer Science
University of New Mexico
Albuquerque, New Mexico 87131

INTRODUCTION

EISPACK is a collection of Fortran subroutines that compute the eigenvalues and eigenvectors of matrices and matrix systems. The package can determine the eigensystem of complex general, complex Hermitian, real general, real symmetric, real symmetric band, real symmetric tridiagonal, and special real tridiagonal matrices, and generalized real and generalized real symmetric matrix systems. In addition, there are two routines that compute the singular value decomposition, useful in solving certain least squares problems.

The subroutines are based mainly on the Algol procedures published in the *Handbook* series of Springer-Verlag by Wilkinson and Reinsch [1971] and the *QZ* algorithm of Moler and Stewart [1973]. The algorithms have been adapted to Fortran and thoroughly tested on a wide range of different computers. The software has been certified and is supported by the developers.

The software for EISPACK can be obtained from either

IMSL
Sixth Floor, NBC Building
7500 Bellaire Boulevard
Houston, Texas 77036
(Phone: 713-772-1927)
Cost: $75.00 (Tape included)

or

---

* This work was supported in part by the Applied Mathematical Sciences Research Program (KC-04-02) of the Office of Energy Research of the U.S. Department of Energy under Contract W-31-109-Eng-38.
† This work was supported in part by the Computer Science Section of the National Science Foundation under Contract MCS 7603297.

National Energy Software Center (NESC)
Argonne National Laboratory
9700 South Cass Avenue
Argonne, Illinois 60439
(Phone: 312-972-7250)
Cost: Determined by NESC policy.

Requesters in European Organization for Economic Cooperation and Development (OECD) countries may obtain the software from

NEA Data Bank
B.P. No. 9 (Bat. 45)
F-91191 Gif-sur-Yvette
France
Cost: Free.

The documentation for the codes is contained in the following books:

B.T. Smith, J.M. Boyle, J.J. Dongarra, B.S. Garbow,
    Y. Ikebe, V.C. Klema, and C.B. Moler.
Matrix Eigensystem Routines — EISPACK Guide,
Lecture Notes in Computer Science, Volume 6, 2nd Edition,
Springer-Verlag (1976)
Price: $21.00

B.S. Garbow, J.M. Boyle, J.J. Dongarra, and C.B. Moler.
Matrix Eigensystem Routines-EISPACK Guide Extension,
Lecture Notes in Computer Science, Volume 51,
Springer-Verlag (1977)
Price: $20.00.

HISTORY

EISPACK is primarily based on Algol procedures developed in the 1960's by nineteen different authors and published in the journal *Numerische Mathematik*. J. H. Wilkinson and C. Reinsch edited a collection of these procedures, together with some background material, into a volume entitled *Linear Algebra* in the *Handbook for Automatic Computation* series. This volume was not designed to cover every possible method of solution. Algorithms were chosen on the basis of their generality, elegance, accuracy, speed, or economy of storage.

Prior to the actual publication of the *Handbook,* as the Wilkinson-Reinsch collection has come to be known, Virginia Klema and others at Argonne National Laboratory had begun translating many of the Algol procedures into Fortran.

In early 1970 a group of researchers from Argonne, the University of Texas, and Stanford University proposed to the National Science Foundation (NSF) a project to "explore the methodology, costs, and resources required to produce, test, and disseminate high-quality mathematical software and to test, certify, disseminate, and support packages of mathematical software in certain problem areas." The project, funded by NSF in 1971, came to be known as the NATS project. The initials originally stood for NSF, Argonne, Texas, and Stanford, but later for the National Activity to Test Software.

The participants decided very early in the project not to produce a comprehensive library of mathematical software but to concentrate on certain fundamental areas. The two areas that were selected — matrix eigensystems and functional approximation — were natural to the participating individuals and organizations (see Chapters 1 and 3).

During 1971 at a University of Michigan Summer Conference, Wilkinson presented material on the methods used to carry out the calculations involved in the eigenvalue problem, and Cleve Moler discussed the actual implementation of the algorithms into software. By this time the NATS project had developed Fortran versions of selected Algol procedures from the *Handbook.* These Fortran routines were sent to people who had an interest in this kind of software and had agreed to test the programs.

There were about 20 test sites representing universities, industrial organizations, and government laboratories and covering a wide range of computers and operating systems. The test sites were responsible for compiling and testing the EISPACK routines and making them available to site users. Reports on performance were sent to Argonne, which served as the NATS project center.

By May 1972 the collection of Fortran routines was ready for public use. Arrangements were made with the Argonne Code Center (now called the National Energy Software Center) to distribute the collection. The package was available in five versions: IBM 370-360, CDC 6600-7600, Univac 1108, Honeywell 635, and PDP-10. The package was said to be *certified* in the sense that information on testing was available and reports of poor or incorrect performance on the machines and operating systems it was tested on "would gain immediate attention from the developers."

In early 1974 the documentation for the first release of the software was completed. This documentation describes the usage of the various routines and gives detailed timing information on a wide variety of computers. In addition, the user guide describes in detail the alternative paths a user can choose to gain the most efficient solution for various problems.

Work had already started on extensions to the package in 1972. By 1976 the extensions had been tested and were ready for public distribution. The new package, which consisted of 70 routines, offered the capability of handling the generalized eigenvalue problem directly. An additional user guide was subsequently prepared and published to cover the new material in the second release of EISPACK, and the previous guide was updated.

To simplify the solution of the standard and generalized eigenproblems, the control program EISPAC was created. EISPAC is a program written in Fortran and IBM OS 360/370 assembly language and developed to aid users on IBM machines in solving their eigenvalue problems. The control program proved to be valuable for users on IBM machines during the first release of the package. But, because of the highly machine-dependent nature of the program and the addition of driver subroutines to EISPACK, interest in the program has diminished.

Since EISPACK was first made available in 1972, over 1000 copies of the collection have been distributed worldwide.

Recently (1983) some minor changes have been made to eliminate machine-dependent constants, reduce the possibilities of overflow and underflow, and incorporate the modifications of the Algol procedures recommended by Hammarling *et al.* [1981]. This new version, called Edition 3, is described at the end of this chapter.

The cost of developing the package is hard to measure. There are many facets of the project that cannot be assigned precise dollar values. The following figures give very rough estimates of costs [Cowell and Fosdick, 1977]:

EISPACK Edition 1
  Size: 34 routines with 6,000 source cards
  Duration: 34 months
  Total effort: 112 man-months
  Cost: $528,000

EISPACK Edition 2
  Size: 70 routines with 11,130 source cards
  Duration: 41 months (16 months simultaneously with Edition 1)

Total effort: 92 man-months
Cost: $371,000

EISPACK Edition 3
Size: 78 routines with 11,769 source cards
Duration: 12 months
Total effort: 10 man-months
Cost: $100,000.

By these estimates, the total cost of EISPACK is about one million dollars.

ORGANIZATION

The organization of EISPACK can be described from several different points of view. The simplest is that of a "casual" user — someone encountering EISPACK for the first time or someone wanting a quick and easy solution to a matrix eigenvalue problem. Such a user can concentrate on the driver subroutines.

There are 13 drivers, intended for matrices of different forms. Twelve of the drivers provide two options: compute all eigenvalues, or compute all eigenvalues and eigenvectors. One of the drivers provides for all the eigenvalues and some of the eigenvectors of a symmetric matrix. Seven of the drivers are for the "standard" eigenvalue problem involving a single real matrix, A, of various forms:

Driver	Problem	Matrix A
RG	$Ax = \lambda x$	general
RS	$Ax = \lambda x$	symmetric
RSM	$Ax = \lambda x$	symmetric; all values, some vectors
RSB	$Ax = \lambda x$	symmetric band
RSP	$Ax = \lambda x$	symmetric, packed*
RST	$Ax = \lambda x$	symmetric tridiagonal
RT	$Ax = \lambda x$	sign-symmetric tridiagonal [†]

Two of the drivers solve the standard eigenvalue problem for complex matrices:

---

* A packed array stores the lower triangle of an $n \times n$ real symmetric or complex Hermitian matrix in a one-dimensional array with only $n(n+1)/2$ elements.

[†] A sign-symmetric matrix is a real tridiagonal matrix whose offdiagonal elements have matching signs; that is, for all i, $sign(a_{i,i+1}) = sign(a_{i+1,i})$.

Driver	Problem	Matrix A
CG	$Ax = \lambda x$	general
CH	$Ax = \lambda x$	Hermitian

Four of the drivers solve the "generalized" eigenvalue problem involving two real matrices, A and B, without directly inverting either matrix:

Driver	Problem	Matrix A	Matrix B
RGG	$Ax = \lambda Bx$	general	general
RSG	$Ax = \lambda Bx$	symmetric	positive definite
RSGAB	$ABx = \lambda x$	symmetric	positive definite
RSGBA	$BAx = \lambda x$	symmetric	positive definite

The driver subroutines provide easy access to many of EISPACK's capabilities. The user who is satisfied with these capabilities, and whose problems do not make heavy demands on computer time or storage, need not be concerned with any further details of EISPACK organization.

The drivers are actually just "shell" subroutines which call from one to five other EISPACK subroutines to do the computations. Several of these other subroutines are used by more than one driver. For example, TQLRAT — a subroutine that computes all the eigenvalues of a real, symmetric, tridiagonal matrix — is used by several of the drivers. On the other hand, there are some subroutines that are not used by any of the drivers. These routines provide alternative methods for doing some of the computations, as well as specialized capabilities not covered by the drivers.

In addition to the drivers, there are 58 subroutines in EISPACK. This modular organization greatly reduces the amount of both source and object code that must be handled. It also provides opportunities for using EISPACK facilities in computations not envisioned during the original development. But it means that the user who desires to access these facilities is faced with a formidable list of subroutines. Since most of EISPACK consists of Fortran translations of the Wilkinson-Rensch collection, the names of these subroutines are the ones chosen by the original authors of the Algol procedures.

The efficient and accurate solution of a matrix eigenvalue problem usually involves at least two of the following steps:

Initial scaling: An operation known as "balancing" that is applied to nonsymmetric matrices to reduce the roundoff errors in subsequent calculations.

Reduction: Similarity transformation to a matrix with zeros below the subdiagonal. The reduced matrix is known as a Hessenberg matrix. A Hessenberg matrix that is also symmetric has zeros above the first superdiagonal and thus is tridiagonal.

Eigenvalue iteration: Any of several iterative processes to compute the eigenvalues of the reduced matrix (which are the eigenvalues of the original matrix).

Eigenvector calculation: Any of several different methods to find eigenvectors of the reduced matrix.

Back transformation: Application of the inverse of the original reduction transformation to the matrix of eigenvectors.

Back scaling: Application of the inverse of the balancing transformation to the matrix of eigenvectors.

Table I summarizes the capabilities of the 58 EISPACK subroutines that are not drivers. The table is not intended to be a complete description, just a brief overview.

Table I

EISPACK Subroutines

Subroutine	Matrix form	Capability
BAKVEC	sign-symmetric	back transformation
BALANC	real	scaling
BALBAK	real	back scaling
BANDR	symmetric band	reduction
BANDV	symmetric band	some eigenvectors
BISECT	symmetric tridiagonal	some eigenvalues
BQR	symmetric band	some eigenvalues
CBABK2	complex	back scaling
CBAL	complex	scaling
CINVIT	complex Hessenberg	some eigenvectors
COMBAK	complex	back transformation
COMHES	complex	reduction

COMLR	complex Hessenberg	all eigenvalues
COMLR2	complex Hessenberg	all eigenvalues and vectors
COMQR	complex Hessenberg	all eigenvalues
COMQR2	complex Hessenberg	all eigenvalues and vectors
CORTB	complex	back transformation
CORTH	complex	reduction
ELMBAK	real	back transformation
ELMHES	real	reduction
ELTRAN	real	reduction
FIGI	sign-symmetric	reduction
FIGI2	sign-symmetric	reduction
HQR	real Hessenberg	all eigenvalues
HQR2	real Hessenberg	all eigenvalues and vectors
HTRIB3	Hermitian packed	back transformation
HTRIBK	Hermitian	back transformation
HTRID3	Hermitian packed	reduction
HTRIDI	Hermitian	reduction
IMTQL1	symmetric tridiagonal	all eigenvalues
IMTQL2	symmetric tridiagonal	all eigenvalues and vectors
IMTQLV	symmetric tridiagonal	all eigenvalues
INVIT	real Hessenberg	some eigenvectors
MINFIT	real	singular value decomposition
ORTBAK	real	back transformation
ORTHES	real	reduction
ORTRAN	real	reduction
QZHES	generalized	reduction
QZIT	generalized	reduction
QZVAL	generalized	all eigenvalues
QZVEC	generalized	all eigenvectors
RATQR	symmetric tridiagonal	some eigenvalues
REBAK	symmetric generalized	back transformation
REBAKB	symmetric generalized	back transformation
REDUC	symmetric generalized	reduction
REDUC2	symmetric generalized	reduction
SVD	real	singular value decomposition
TINVIT	symmetric tridiagonal	some eigenvectors
TQL1	symmetric tridiagonal	all eigenvalues
TQL2	symmetric tridiagonal	all eigenvalues and vectors
TQLRAT	symmetric tridiagonal	all eigenvalues
TRBAK1	symmetric	back transformation
TRBAK3	packed symmetric	back transformation
TRED1	symmetric	reduction
TRED2	symmetric	reduction
TRED3	packed symmetric	reduction

| TRIDIB | symmetric tridiagonal | some eigenvalues |
| TSTURM | symmetric tridiagonal | some eigenvalues and vectors |

It should be noted that two of the subroutines given in the table above, MINFIT and SVD, do not solve eigenvalue problems. Instead they use algorithms closely related to eigenvalue algorithms to analyze and facilitate solution of the least squares problem for overdetermined linear systems of equations.

ALGORITHMS

This is not the place to describe all the algorithms employed in EISPACK, or to assess their relative performance. We will confine ourselves to presenting a little mathematical background and then showing two examples that illustrate some of the more frequently used algorithms.

Almost all the algorithms used in EISPACK are based on similarity transformations. Two matrices A and B are *similar* if there is a nonsingular matrix S for which

$B = S^{-1}AS.$

When two matrices are similar, they have the same eigenvalues, and the eigenvectors of one are easily obtained from the other. If B were diagonal, then its diagonal elements would be the eigenvalues of A. Moreover, the columns of S would be the eigenvectors of A.

It is not necessary, and numerically it may not even be desirable or possible, to diagonalize a matrix completely. If B is merely triangular, then the eigenvalues are still on the diagonal, and the eigenvectors can be computed by a fairly straightforward substitution process.

A real matrix S is called *orthogonal* if

$S^T S = I$

or, equivalently,

$S^T = S^{-1}.$

(Here, S^T is the transpose of S.) A complex matrix S is called *unitary* if these same conditions hold with S^T replaced by S^H, the complex conjugate transpose.

Similarity transformations based on orthogonal and unitary matrices are particularly attractive from a numerical point of view because they do not magnify any errors that may be present in the input data or that may be introduced during the computation. Thus, numerical linear algebra is fortunate to have available the following theorem of Schur [1909]:

Any matrix can be triangularized by a unitary similarity transformation.

Most of the techniques employed in EISPACK are constructive realizations of variants of Schur's theorem. It is usually not possible to compute Schur's transformation with a finite number of rational arithmetic operations. Instead, the algorithms employ a potentially infinite sequence of similarity transformations

$$A_{k+1} = S_k^{-1} A_k S_k$$

for which A_{k+1} approaches an upper triangular matrix. The sequence is terminated when all of the subdiagonal elements of a particular A_{k+1} are less than the roundoff errors involved in the computation. These elements can then be set to zero without introducing any more perturbations in the eigenvalues than have already been caused by the previous transformations. The diagonal elements of the resulting A_{k+1} are then the desired approximations to the eigenvalues of the original matrix. The corresponding eigenvectors can be readily computed if they have been requested.

It is important for several reasons to have special algorithms that deal with real symmetric matrices:

- Many of the eigenvalue problems encountered in practice involve symmetric matrices.

- The eigenvalues of a symmetric matrix are real.

- The eigenvectors of a symmetric matrix can be chosen to be orthogonal.

- The eigenvalues of symmetric matrices are usually less sensitive to perturbation.

- Algorithms designed specifically for symmetric matrices have better convergence properties.

- Algorithms for symmetric matrices usually require less computer time and storage.

The only similarity transformations that also preserve symmetry are those based on orthogonal matrices. The following example outlines the basic steps used in the algorithms employed by EISPACK's driver subroutine RS, which computes all the eigenvalues and optionally all the eigenvectors of a real symmetric matrix. The input matrix is

$$
\begin{bmatrix}
5 & 4 & 3 & 2 & 1 \\
4 & 5 & 4 & 3 & 2 \\
3 & 4 & 5 & 4 & 3 \\
2 & 3 & 4 & 5 & 4 \\
1 & 2 & 3 & 4 & 5
\end{bmatrix}.
$$

The initial orthogonal similarity transformations are carried out by subroutine TRED1 or TRED2. An $n \times n$ matrix requires $n-2$ transformations, each of which introduces zeros into a particular row and column of the matrix, while preserving symmetry and preserving the zeros introduced by previous transformations. In the case of our 5×5 example, the result of the first transformation is a matrix that has three zeros in the last row and column. The next transformation introduces two more zeros in the fourth row and column. The final transformation places one more zero in the third row and column:

$$
\begin{bmatrix}
0.6594 & -0.1438 & 0 & 0 & 0 \\
-0.1438 & 0.9687 & 0.5678 & 0 & 0 \\
0 & 0.5678 & 5.3052 & 4.4192 & 0 \\
0 & 0 & 4.4192 & 13.0667 & -5.4772 \\
0 & 0 & 0 & -5.4772 & 5.0000
\end{bmatrix}.
$$

Since the result of the initial reduction is a symmetric tridiagonal matrix, it can be stored in just two vectors: one with n components for the diagonal and one with $n-1$ components for the offdiagonal. Subroutine TRED1 returns these two vectors only. Subroutine TRED2 also returns the orthogonal matrix that transforms the original matrix to this tridiagonal matrix.

EISPACK includes several routines for computing the eigenvalues of a real, symmetric tridiagonal matrix. Many, but not all, of these routines are variants of the QR algorithm, originally published by J. G. F. Francis in 1961 and perfected by Wilkinson, Reinsch, Dubrulle, and other authors of chapters in the *Handbook*. The driver RS uses two variants of the QR algorithm: subroutine TQLRAT if the eigenvectors are not requested and subroutine TQL2 if they are. (For technical reasons associated with the scaling of so-called "graded" matrices, the indexing within the algorithm is carried out in the opposite order from the original Francis algorithm; and the resulting version is the QL, rather than the QR, algorithm. The subroutine names thus reflect details that are of interest to the numerical analysts who were the intended audience of the papers in *Numerische Mathematik*, but that are not important to most users of EISPACK.)

The symmetric, tridiagonal QR algorithm produces a sequence of similar matrices whose offdiagonal elements are decreasing in magnitude and whose diagonal elements are approaching the desired eigenvalues. With the QL variant, the shift calculation is intended to reduce the first offdiagonal element most rapidly. The first three iterates are

$$
\begin{bmatrix}
0.5667 & 0.0609 & 0 & 0 & 0 \\
0.0609 & 0.7645 & 0.0989 & 0 & 0 \\
0 & 0.0989 & 1.3832 & 0.7501 & 0 \\
0 & 0 & 0.7501 & 7.0326 & -4.4112 \\
0 & 0 & 0 & -4.4112 & 15.2530
\end{bmatrix}
$$

$$
\begin{bmatrix}
0.5484 & -0.0003 & 0 & 0 & 0 \\
-0.0003 & 0.7655 & 0.0285 & 0 & 0 \\
0 & 0.0285 & 1.2751 & 0.1085 & 0 \\
0 & 0 & 0.1085 & 5.4084 & -1.4364 \\
0 & 0 & 0 & -1.4364 & 17.0025
\end{bmatrix}
$$

$$
\begin{bmatrix}
0.5484 & 0.0000 & 0 & 0 & 0 \\
0.0000 & 0.7641 & 0.0085 & 0 & 0 \\
0 & 0.0085 & 1.2737 & 0.0167 & 0 \\
0 & 0 & 0.0167 & 5.2501 & -0.4103 \\
0 & 0 & 0 & -0.4103 & 17.1637
\end{bmatrix}.
$$

Notice that all the offdiagonal elements have generally decreased with each iteration and that the first offdiagonal element has decreased a great deal. Indeed, the first offdiagonal element has now reached a size that is comparable to the roundoff errors made during the calculation. Thus, setting it to zero can be regarded as simply another roundoff error. The first diagonal element is now an accurate approximation to one of the eigenvalues of the original matrix.

The next two iterations affect only the lower 4 × 4 submatrix:

$$
\begin{bmatrix}
0.5484 & 0 & 0 & 0 & 0 \\
0 & 0.7639 & -0.0000 & 0 & 0 \\
0 & -0.0000 & 1.2738 & 0.0003 & 0 \\
0 & 0 & 0.0003 & 5.2362 & -0.0315 \\
0 & 0 & 0 & -0.0315 & 17.1777
\end{bmatrix}.
$$

The second offdiagonal element is now negligible, and the second diagonal element approximates another eigenvalue. Three more iterations, which we do not show, are needed to reduce the remaining offdiagonal elements to roundoff level. The final diagonal matrix is

$$
\begin{bmatrix}
0.5484 & 0 & 0 & 0 & 0 \\
0 & 0.7639 & 0 & 0 & 0 \\
0 & 0 & 1.2738 & 0 & 0 \\
0 & 0 & 0 & 5.2361 & 0 \\
0 & 0 & 0 & 0 & 17.1778
\end{bmatrix}.
$$

In this example, a total of 11 similarity transformations were required, 3 for the initial reduction and 8 for the QR iterations. The later transformations involved considerably less arithmetic than the earlier ones because they were done on submatrices of decreasing order. Let S denote the product of all the orthogonal similarity transformations that are required, and let D denote the final diagonal matrix. Then

$$
S^{-1}AS = D,
$$

and hence

$$
AS = SD.
$$

This relation shows that the columns of S are the eigenvectors of A. Moreover, since S is the product of orthogonal matrices, it must also be orthogonal. Subroutines TRED2 and TQL2 accumulate eigenvectors S as a by-product of the original reduction and eigenvalue iteration. Since S is a full matrix, most of the execution time is used in its calculation. If only the eigenvalues are desired, then S need not be computed, and TRED1 and TQLRAT could be used instead.

Several of the papers and Algol procedures in the *Handbook* resulted from research that refined and perfected the *QR* algorithm. A paper by Reinsch, which was published after the *Handbook* was completed, describes the fastest routine, TQLRAT. Many of these procedures were taken over to EISPACK, so that there are 10 different subroutines like TQLRAT and TQL2 which compute the eigenvalues of a symmetric tridiagonal matrix. Eight of the ten are variants of the *QR* algorithm; the other two employ bisection algorithms based on Sturm sequences. Ten or fifteen years ago, it was not clear which of these alternative approaches was preferable. Today, experience gained with EISPACK allows us to make such decisions.

Nonsymmetric matrices involve somewhat different techniques, although the general approach of an initial reduction followed by some *QR*-type iteration is still followed. Since there is no symmetry to be preserved, similarity transformations can be based on nonorthogonal matrices. Algorithms that employ elimination methods require less arithmetic and, hence, are potentially faster than those that use orthogonal transformations. However, such algorithms may produce somewhat less accurate results and, in extreme examples, may be completely unstable because of element growth. Deciding between the two classes of algorithms involves comparing execution speed with numerical reliability. When designing a general-purpose library, such decisions are very difficult to make.

The following example illustrates the usual behavior of the "real, general" driver RG. The input matrix is

$$\begin{bmatrix} 5 & 4 & 3 & 2 & 1 \\ 9 & 5 & 4 & 3 & 2 \\ 8 & 9 & 5 & 4 & 3 \\ 7 & 8 & 9 & 5 & 4 \\ 6 & 7 & 8 & 9 & 5 \end{bmatrix}.$$

Subroutine ELMHES uses nonorthogonal elimination methods to produce the following Hessenberg matrix:

$$\begin{bmatrix} 5.0000 & 8.8889 & 4.0174 & 4.6055 & 2.0000 \\ 9.0000 & 12.2222 & 6.2902 & 6.4082 & 3.0000 \\ 0 & 16.2963 & 11.7891 & 10.9525 & 7.0000 \\ 0 & 0 & -2.5925 & -2.6545 & -1.9705 \\ 0 & 0 & 0 & 1.3901 & -1.3569 \end{bmatrix}.$$

In this case, there has been very little growth in the size of the elements during the initial reduction, so there has been very little roundoff error magnification.

RG now uses subroutine HQR2 to carry out an "implicit, double-step, Hessenberg" *QR* iteration. The shift calculations are designed to reduce the size of the last subdiagonal element. After two iterations we have:

$$
\begin{bmatrix}
24.8549 & 10.9746 & 3.7663 & -13.3527 & 6.1188 \\
-0.2000 & 3.5305 & -1.7923 & 5.3528 & -5.6901 \\
0 & 0.0538 & -1.2472 & -1.7953 & 0.0899 \\
0 & 0 & 0.9872 & -0.5516 & -2.6015 \\
0 & 0 & 0 & 0.2848 & -1.5867
\end{bmatrix}.
$$

Notice that the first two subdiagonal elements are actually decreasing more rapidly than the last one, which is the target of the shift calculation. This is an important property of the *QR* algorithm: It effectively "works on" all eigenvalues simultaneously. Five more iterations lead to

$$
\begin{bmatrix}
24.7514 & -11.1821 & -3.6155 & 12.9960 & 7.0641 \\
0 & 3.6275 & -1.3330 & 4.9235 & 6.0036 \\
0 & 0 & -1.2203 & -1.8985 & -0.4090 \\
0 & 0 & 1.2977 & -0.6818 & 2.4406 \\
0 & 0 & 0 & 0 & -1.4768
\end{bmatrix}.
$$

Three of the eigenvalues are revealed in diagonal positions 1, 2, and 5. The 2×2 submatrix that includes diagonal positions 3 and 4 is the source of a pair of complex conjugate eigenvalues

$$-0.9511 \pm 1.5463 i.$$

The entire computation has proceeded using real arithmetic, even though the final results are complex.

NUMERICAL PROPERTIES

The algorithms used by EISPACK that are based on orthogonal transformations are always numerically stable in the sense that they produce the exact answer to an eigenvalue problem involving a matrix $A + E$, which is a small perturbation of the given matrix A. The norm of the perturbation E is roughly the size of roundoff error when compared to the norm of A. The same can be said for the algorithms based on nonorthogonal transformations if no exceptional growth in the size of the matrix elements occurs during the computation.

One immediate consequence of this numerical stability is that the computed results will produce small residuals. If λ and x are a computed eigenvalue and eigenvector of a given matrix A, then Ax will always be close to λx. More precisely, the size of the *relative residual*

$$\frac{||Ax - \lambda x||}{||A|| \, ||x||}$$

can always be expected to be roughly equal to the relative accuracy of floating point arithmetic on the computer being used.

But what about accuracy? How close are the computed eigenvalues to the exact eigenvalues? The answers to these questions depend more on the matrix involved than on the particular EISPACK subroutine used to do the computation. To see why, assume that A has a complete, linearly independent set of eigenvectors X, and let D denote the diagonal matrix of eigenvalues. Then

$$X^{-1}AX = D.$$

Suppose that A is perturbed somehow, either by errors in its initial formation or by roundoff errors generated by EISPACK. Then

$$X^{-1}(A + E)X = D + X^{-1}EX.$$

The resulting perturbation to D is not diagonal, but this equation makes it plausible that the damage done by E to the eigenvalues in D could be as large as $||X^{-1}|| \, ||E|| \, ||X||$, rather than merely $||E||$. The quantity

$$\kappa(X) = ||X|| \, ||X^{-1}||,$$

which is the condition number of X, occurs in the perturbation analysis for systems of simultaneous equations. Note that here with the eigenvalue problem, it is the condition of X — the matrix of eigenvectors — and not the condition of A itself that is relevant. If the eigenvector matrix is nearly singular, then the eigenvalues are potentially sensitive to perturbation.

If A is real and symmetric — or more generally, if A commutes with its complex conjugate transpose — then X can be taken to be unitary and $\kappa(X)$ (with respect to the 2-norm) is 1. In this case, a small change in the matrix causes a correspondingly small change in the eigenvalues. In other words, the eigenvalues of such matrices are always well-conditioned.

In the extreme case where A does not have a full set of eigenvectors, so that its Jordan Canonical Form is not diagonal, then $\kappa(X)$ should be regarded as infinite. The eigenvalues are infinitely sensitive to perturbation in the sense that they are no longer analytic functions of that perturbation.

With EISPACK, the consequences of this perturbation theory are the following:

For real symmetric matrices and for complex Hermitian matrices, the eigenvalues are always computed with an accuracy that corresponds to a few units of roundoff error in the largest eigenvalue of the matrix. The small eigenvalues of such matrices will not necessarily be computed to high accuracy relative to themselves, but they will be computed to full accuracy relative to the norm of the matrix. For such matrices, the presence of multiple eigenvalues has little effect on the accuracy of the computed results.

For general, nonsymmetric matrices, the effect of roundoff errors on the computed eigenvalues will increase as the condition number of the eigenvector matrix increases. If a nonsymmetric matrix has multiple eigenvalues or is close to a matrix with multiple eigenvalues, and if its eigenvector matrix has a large condition number, then the computed eigenvalues may be accurate to less than full precision.

As an example, consider the following matrix:

$$A = \begin{bmatrix} -64 & 82 & 21 \\ 144 & -178 & -46 \\ -771 & 962 & 248 \end{bmatrix}.$$

This matrix was constructed in such a way that its exact eigenvalues happen to be 1, 2, and 3 . When the eigenvalues are computed using EISPACK subroutines on a computer with a 24-bit floating point fraction, the results are

$$1.00195$$
$$2.00113$$
$$2.99736 \ .$$

Such a computer has a relative floating point accuracy of better than 10^{-6}, so the computed eigenvalues have lost half the available figures.

The difficulties lie with the matrix itself, not with EISPACK. The matrix of computed eigenvectors, renormalized so that the last component of each vector is one, is

$$X = \begin{bmatrix} 0.090922 & 0.075114 & 0.111016 \\ -0.181810 & -0.196554 & -0.166741 \\ 1.000000 & 1.000000 & 1.000000 \end{bmatrix}.$$

The condition number of X is greater than 10^3. Thus a single roundoff error in

A, which on this computer affects the sixth significant figure, could cause changes in the third significant figure of the eigenvalues. EISPACK has computed the eigenvalues as accurately as is possible using floating point arithmetic of this precision.

As another measure of accuracy, let D be the matrix whose diagonal elements are the computed eigenvalues. Then

$$\frac{||AX - XD||}{||A|| \, ||X||} = 0.13 \times 10^{-6}.$$

In other words, the relative residuals are on the order of roundoff error, even though the eigenvalues are "accurate" to only three figures. For further information on the perturbation theory for the eigenvalue problem, see Stewart [1973] and Wilkinson [1965].

EISPACK 3

In this section we describe the latest version of the package, EISPACK 3, a limited set of modifications to the second edition of EISPACK. The modifications eliminate the machine-dependent constants and reduce the probability of underflow/overflow difficulties. They also may improve the execution time of a few subroutines. However, they do not introduce any new capabilities, nor do they change any calling sequences. The resulting collection is thus readily portable from machine to machine and somewhat more robust and efficient, but the two EISPACK guides continue to serve as the basic documentation.

Writing or revising code is the easy part of any software project. Proper testing, certification, and distribution were key features of the original EISPACK activity. We have used the same procedures in testing this version as in the past.

Until the current release, there has been no "official" double precision version of EISPACK. Several of the different machine versions are single precision. The IBM version uses the nonstandard declaration REAL*8 because we regard this as the appropriate working precision for such machines and did not want to call it "double." We now give in to generally accepted usage and provide two machine-independent versions, one for single precision and one for double precision.

Three routines which use inverse iteration to find selected eigenvectors of symmetric matrices — BANDV, TINVIT and TSTURM — have been modified in a way that may reduce the size of the groups involved in the reorthogonalization process. This should make the routines significantly faster for matrices where the original grouping criterion produced larger groups than necessary. The size of these groups also affects the orthogonality of the computed eigenvectors, but we expect that the new versions should still produce acceptable orthogonality. These inverse iteration routines were primarily intended to find just a few selected eigenvectors. But, particularly with this new grouping calculation, in many cases they also provide the *fastest* way to find *all* the eigenvectors.

A new driver for real symmetric matrices, RSM, has been added to the collection. It provides easy access to the recommended routines in the inverse iteration path. It has an additional integer parameter M which must be less than or equal to the order N . It computes all the eigenvalues, and the M eigenvectors associated with the M smallest eigenvalues, of a real symmetric matrix. It should be considered for use even when M = N so that all the eigenvectors are computed. It is difficult to summarize the tradeoffs between RSM and RS, the other real symmetric driver which computes all the eigenvectors. RSM is usually faster, and may be up to 40% faster on some matrices. The computed eigenvalues are usually of comparable accuracy. But the computed eigenvectors provided by RS may be somewhat more accurate in two senses: the residuals may be smaller, and the departure from orthogonality may be smaller.

Fortran standards require storage of two-dimensional arrays by column. Many modern computer systems employ cache or paged memories where access to columns of a matrix is much more efficient than access to rows. Most Algol implementations store two-dimensional arrays by rows. Accordingly, the inner loops of TRED1 and TRED2 access rows of the matrix; and an inner loop in TRED3 contains an IF statement resulting from its one-dimensional subscripting of a symmetric two-dimensional array.

We have rewritten TRED1, TRED2, and TRED3 so that their inner loops involve sequential access to memory. The improvement in efficiency of the new versions of the TREDs will be very dependent on the order of the matrix and the nature of the computer and operating system being used. For small matrices with conventional architecture and memory management, there will be little change. But for large matrices that require the use of virtual memory, the improvement can be significant.

EISPACK is now over 10 years old. It was never designed as a uniform "package" as we now use that term today. The overall organization, the choice of algorithms, the subroutine names, and the structure of the code itself are all inherited from the Algol collection in the *Handbook*.

We believe that it is time for a major new edition of EISPACK. A complete revision should have expanded capabilities, new user interfaces, and uniform naming conventions. Some algorithms should be altered to take account of vector machine architectures and paging operating systems. The programs should be written in Fortran 77 with an eye to future versions of Fortran. Such a project would require careful planning and extensive resources.

REFERENCES

Cowell, W. R., and L. D. Fosdick [1977]. "Mathematical software production." *Mathematical Software III*. Ed. J. R. Rice. Academic Press, New York, pp. 195-224.

Garbow, B. S., J. M. Boyle, J. J. Dongarra, and C. B. Moler [1977]. *Lecture Notes in Computer Science, Vol. 51, Matrix Eigensystem Routines — EISPACK Guide Extension*. Springer-Verlag, New York.

Hammarling, S. J., P. D. Kenward, and H. J. Symm [1981]. *Amendments to Handbook for Automatic Computation, Volume II*. NPL Report DNACS 41/81.

Moler, C. B., and G. W. Stewart [1973]. "An algorithm for generalized matrix eigenvalues problems." *SIAM J. of Numer. Anal.* 10:241-256.

Schur, I. [1909]. "Uber die characteristischen Wurzeln einer linearen Substitution mit einer Anwendung auf die Theorie der Integral Gleichungen." *Math. Ann.* 66:488-510.

Smith, B. T., J. M. Boyle, J. J. Dongarra, B. S. Garbow, Y. Ikebe, V. C. Klema, and C. B. Moler [1976]. *Lecture Notes in Computer Science, Vol. 6, 2nd Edition, Matrix Eigensystem Routines — EISPACK Guide*. Springer-Verlag, New York.

Stewart, G. W. [1973]. *Introduction to Matrix Computations*. Academic Press, New York.

Wilkinson, J. H. [1965]. *The Algebraic Eigenvalue Problem*. Oxford University Press (Clarendon), Oxford.

Wilkinson, J. H., and C. Reinsch, eds. [1971]. *Handbook for Automatic Computation, Vol. II, Linear Algebra*. Springer-Verlag, New York.

Chapter 5

THE MINPACK PROJECT

Jorge J. Moré, Danny C. Sorensen, Burton S. Garbow, and Kenneth E. Hillstrom *
Mathematics and Computer Science Division
Argonne National Laboratory
Argonne, Illinois 60439

INTRODUCTION

The MINPACK project is a research effort whose goal is the development of a systematized collection of quality optimization software. The first step towards this goal has been realized in MINPACK-1, a package of Fortran programs for the numerical solution of nonlinear equations and nonlinear least squares problems. Interest in the MINPACK project currently centers on unconstrained and linearly constrained minimization problems and large-scale optimization problems. Thus, the next version of the package will probably include algorithms for these problems.

Our main goals during the development of MINPACK software have been reliability, ease of use, and transportability. These goals are not easy to achieve, and the lessons learned are sometimes forgotten. Therefore, one of the purposes of this paper is to describe some of the techniques used to achieve these goals.

The reader should be aware that optimization is a rapidly changing subject. This is in contrast to the areas of linear algebra covered by EISPACK and LINPACK. In those areas there has been relative stability, and there is general agreement on the algorithms that should be implemented. In contrast, there is little agreement on the optimization algorithms that should be implemented. Our attitude has been to implement state-of-the-art algorithms in the best possible way. This means that there is a heavy research component to the project, as well as a keen interest in the practicalities of software implementation. We hope that the advantages of this dual interest are reflected in this paper.

We begin our paper with an overview of the history behind the MINPACK project. We describe the circumstances that led to the project and the various changes that occurred during the next four years.

The MINPACK project has been influenced by the development of trust region methods as powerful algorithms for the solution of nonlinear optimization problems. The MINPACK-1 algorithms are trust region methods, and it seems likely that much of the MINPACK software will be based on the trust region

* This work was supported by the Applied Mathematical Sciences Research Program (KC-04-02) of the Office of Energy Research of the U.S. Department of Energy under Contract W-31-109-Eng-38.

concept. We have thus provided an introduction to the basic ideas behind trust region methods.

The discussion of trust region methods is followed by a description of the techniques we have used to produce reliable, easy-to-use, transportable software. We discuss only those topics that are relevant to the development of software optimization libraries: scale invariance, robustness, and convergence testing. An understanding of these topics is a prerequisite to the development of reliable optimization software; yet these topics are rarely discussed in the optimization literature. We also discuss user interface and documentation standards because of their importance in the development of easy-to-use software.

HISTORY

The MINPACK project was started by James C. T. Pool, then head of Applied Mathematical Sciences at Argonne. The project effectively started on June 8, 1976. This is the date of an internal planning document which described the plans that eventually led to MINPACK-1. Previous to this date there had been work at Argonne, mainly by Ken Hillstrom, on the design and testing of optimization software.

It is interesting and important to recall the state of mathematical software in June of 1976: EISPACK had been released in 1972, with the user guide available early in 1974; the NATS ideas [Smith, Boyle, and Cody, 1974] on robustness, reliability, and transportability that were used in EISPACK and FUNPACK were not then being applied to the development of optimization software; LINPACK was not available, and would only become available two years later.

In June 1978, a letter requesting assistance with the testing of MINPACK-1 was sent to 16 sites; 14 agreed to become test sites. About a month later the first version of MINPACK-1 was sent to the test sites. This version contained four algorithmic paths (*brent, hybrd, lmder, lmdif*) and a testing package. The *brent* and *hybrd* paths for systems of nonlinear equations were based, respectively, on the work of Brent [1973] and Powell [1970]. The *lmder* and *lmdif* paths for nonlinear least squares problems were based on the work of Moré [1978]. An algorithmic path consisted of an easy-to-use driver, a core subroutine, and a number of auxiliary programs. Each program had extensive prologue comments on its purpose and the roles of its parameters; in-line comments introduced major blocks in the body of the program. There was no user guide.

The initial testing did not uncover any deficiencies in the codes. There were, however, a number of suggestions and strong encouragement for the production of a user guide.

The suggestions by users and test sites led to a number of modifications in the original version of MINPACK-1. In the 1979 version of MINPACK-1, we decided to delete the *brent* path from the package, to add two other paths (*hybrj*, *lmstr*), to include subroutine *chkder* for checking the consistency of the Jacobian matrix with the function values, and to add machine-readable documentation.

The *brent* path was dropped because the *hybrd* path had proved to be more reliable. The *hybrj* path was added to allow users to specify the Jacobian matrix. The main advantage of providing the Jacobian matrix is increased reliability; for example, performance of the algorithm is then much less sensitive to function values corrupted by noise. The *lmstr* path was added in response to requests from several minicomputer users. The difference between the *lmstr* and *lmder* paths is that in the *lmder* path the m by n Jacobian matrix is computed all at once, while in the *lmstr* path the user is required to supply the Jacobian matrix one row at a time; with the latter arrangement it is then possible to reduce the required storage from mn to n^2. In least squares problems m can be considerably larger than n, and then this reduction becomes significant.

Machine-readable documentation was sent to 20 test sites in July 1979 with a request for comments. In November, the final version of MINPACK-1 was sent to the test sites. At that time we started work on the user and implementation guides. The user guide contained, in particular, introductory documentation with information on the solution and analysis of nonlinear problems, for example, information on selecting algorithms and scaling problems.

The MINPACK-1 package was completed in July 1980; the announcement of availability was made in October. The package consists of the programs, their documentation, and the testing material. The package comprises approximately 28,000 card images and is transmitted on magnetic tape. The tape is available from the following sources:

National Energy Software Center (NESC)
Argonne National Laboratory
9700 South Cass Avenue
Argonne, Illinois 60439

or

IMSL
Sixth Floor, NBC Building
7500 Bellaire Boulevard
Houston, Texas 77036

European requesters in countries belonging to the Organization for Economic Cooperation and Development (OECD) may obtain the tape from

NEA Data Bank
B.P. 9 (Bat 45)
F-91191 Gif-sur-Yvette
France

The package includes both single precision and double precision versions of the programs; and for those programs normally called by the user, machine-readable documentation is provided in both single and double precision. The user guide [Moré, Garbow, and Hillstrom, 1980] and implementation guide [Garbow, Hillstrom, and Moré, 1980] are included with the tape.

ALGORITHMS

The MINPACK-1 algorithms for the solution of systems of nonlinear equations and nonlinear least squares problems are based on the idea of a trust region. Algorithms that use the trust region concept have also been proposed for unconstrained and linearly constrained minimization problems, and for large-scale optimization problems. Interest in these algorithms derives, in part, from the availability of strong convergence results and from the development of software for these methods which is reliable and amazingly free of *ad hoc* decisions. In view of these advantages, it seems likely that much of the MINPACK software will incorporate the trust region concept.

The basic ideas behind trust region methods can be illustrated by discussing algorithms for the solution of systems of nonlinear equations and nonlinear least squares problems. Consider a mapping $F: R^n \rightarrow R^m$, where $m = n$ for systems of nonlinear equations and $m \geqslant n$ for nonlinear least squares problems. Given a starting vector $x_0 \in R^n$, an algorithm for these problem areas generates a sequence of iterates $\{x_k\}$ which, under suitable conditions, converges to a solution x^* of the problem

$$min\{\| F(x)\| : x \in R^n\}, \tag{1}$$

where $\| \cdot \|$ is the Euclidean norm. In general, all that can be expected from an algorithm for problem (1) is that x^* satisfy the first-order conditions for a local minimizer, that is,

$$F'(x^*)^T F(x^*) = 0, \tag{2}$$

where $F'(x)$ is the Jacobian matrix of F at x. In particular, if $m = n$ and $F'(x^*)$ is nonsingular then (2) implies that $F(x^*) = 0$.

A trust region method for problem (1) generates a sequence $\{x_k\}$ of iterates, where the step s_k between iterates is a solution of the subproblem

$$min\{\| F(x_k) + F'(x_k) w \| : \| D_k w \| \leqslant \Delta_k \} \tag{3}$$

for some bound Δ_k and scaling matrix D_k. The motivation for this choice of step s_k is that in a neighborhood of x_k we expect the linear model

$$L_k(w) \equiv f_k + J_k w , \tag{4}$$

where

$$f_k = F(x_k) , \quad J_k = F'(x_k) ,$$

to provide a reasonable prediction of the behavior of F. In (3) we are assuming that the region of trust for the linear model — the trust region — can be taken to be of the form

$$\{ w \in R^n : \| D_k w \| \leqslant \Delta_k \} .$$

This is an n-dimensional hyperellipsoid. If D_k is a diagonal matrix, then the axes of the hyperellipsoid are parallel to the standard coordinate axes.

Given an iterate x_k, a bound Δ_k, and a scaling matrix D_k, a trust region method computes a tentative step s_k. The reduction produced by the step s_k is measured by the ratio

$$\rho_k \equiv \frac{\| F(x_k) \|^2 - \| F(x_k + s_k) \|^2}{\| L_k(0) \|^2 - \| L_k(s_k) \|^2} . \tag{5}$$

Note that ρ_k is the ratio of the actual reduction in F to the reduction predicted by the step s_k, and that $\rho_k \geqslant 1$ if and only if

$$\| F(x_k + s_k) \| \leqslant \| L_k(s_k) \| .$$

Thus values of ρ_k near unity are desirable. Also note that it is possible to show that $|\rho_k - 1|$ can be made arbitrarily small by choosing Δ_k sufficiently small. The trust region method, however, attempts to keep ρ_k close to unity while keeping Δ_k relatively large. If the step is satisfactory in the sense that s_k produces a sufficient reduction, then Δ_k can be increased; presumably a bigger step will yield an even bigger reduction. If the step is unsatisfactory, then Δ_k should be decreased. Algorithm (6) expresses these ideas in more detail.

Algorithm (6). Updating of Δ_k.

Let $0 < \mu < \eta < 1$ and $0 < \gamma_1 < \gamma_2 < 1 < \gamma_3$ be specified constants.

    a) If $\rho_k \leqslant \mu$ then $\Delta_{k+1} \in [\gamma_1 \Delta_k , \gamma_2 \Delta_k]$.

    b) If $\rho_k \in (\mu, \eta)$ then $\Delta_{k+1} \in [\gamma_2 \Delta_k , \Delta_k]$.

    c) If $\rho_k \geqslant \eta$ then $\Delta_{k+1} \in [\Delta_k , \gamma_3 \Delta_k]$.

Algorithm (6) has the proper qualitative behavior. If the step is satisfactory in the sense that $\rho_k \geqslant \eta$, then Δ_k can be increased, while if $\rho_k \leqslant \mu$, then Δ_k is decreased. Typical values for these constants are $\mu = 0.25$ and $\eta = 0.75$. Values for the other constants are usually $\gamma_1 = 0.1$, $\gamma_2 = 0.5$, and $\gamma_3 = 2$.

We have now discussed the main ingredients of a trust region method for problem (1). Algorithm (7) summarizes these ingredients.

Algorithm (7). Trust Region Method.

Let $x_0 \in R^n$ and $\Delta_0 > 0$ be given. Choose $\sigma > 0$.

For $k = 0, 1, \ldots,$

  a) Compute $F(x_k)$ and the model L_k.

  b) Determine an approximate solution s_k to subproblem (3).

  c) Compute ρ_k from (5).

  d) If $\rho_k > \sigma$, then $x_{k+1} = x_k + s_k$. Otherwise $x_{k+1} = x_k$.

  e) Update Δ_k with Algorithm (6).

  f) Update the model L_k and the scaling matrix D_k.

We have assumed that J_k is the Jacobian matrix $F'(x_k)$, but Algorithm (7) is unchanged if we choose J_k as a difference approximation or a quasi-Newton approximation to the Jacobian matrix. The choice of J_k, however, does affect the choice of the step.

A reasonable step s_k can be obtained if we determine an approximate solution of subproblem (3). The requirements that we impose on the step s_k can be expressed in terms of the quadratic model

$$\psi_k(w) \equiv \tfrac{1}{2} \left(\| L_k(w) \|^2 - \| L_k(0) \|^2 \right). \tag{8}$$

If we desire a nearly optimal solution of subproblem (3), then we require that

$$\psi_k(s_k) \leqslant \sigma_1 \min\{\psi_k(w) : \| D_k w \| \leqslant \Delta_k\}, \quad \| D_k s_k \| \leqslant \sigma_2 \Delta_k,$$

for some positive constants σ_1 and σ_2. This is a reasonable requirement if J_k is the Jacobian matrix $F'(x_k)$ or a difference approximation to $F'(x_k)$. In these situations, L_k provides a reasonable prediction of the behavior of F in a neighborhood of x_k and thus it usually pays to obtain a nearly optimal solution. This argument does not apply if J_k is a quasi-Newton approximation to the Jacobian matrix. An alternative is to obtain the step as a convex combination of the Newton direction and a scaled gradient direction. Specifically, we define

$$p_k(\alpha) \equiv \alpha p_k^C + (1 - \alpha) p_k^N,$$

where p_k^C is the Cauchy step and p_k^N is the Newton step, and require that

$$\psi_k(s_k) = \min\{\psi_k(p_k(\alpha)) : \| D_k p_k(\alpha) \| \leqslant \Delta_k\}.$$

We have mentioned two ways of choosing the step; references and further information on other possible choices of step in a trust region method are provided by Moré [1983].

Note that we update x_k only if $\rho_k > \sigma$. We certainly don't want to update if $\rho_k \leqslant 0$, and thus it is also reasonable not to update if $\rho_k \leqslant \sigma$ for some small $\sigma > 0$. A sensible value for σ is 0.0001.

Also note that if J_k is $F'(x_k)$ or a difference approximation to $F'(x_k)$, then the model L_k is updated only when $\rho_k > \sigma$. If J_k is a quasi-Newton approximation to the Jacobian matrix, then L_k is updated at each iteration because quasi-Newton methods use the information at $x_k + s_k$ to improve the approximation J_k. It is always possible to combine the two approaches and use a quasi-Newton approximation until the algorithm fails to make reasonable progress, and then compute J_k as a difference approximation.

The material here should provide an introduction to the concept of a trust region method. As mentioned in the introduction to this section, these methods have been extended to unconstrained and linearly constrained optimization problems and to large-scale optimization problems. A trust region method for these areas is similar to Algorithm (7). One of the main differences is that instead of viewing L_k as a model of F, we now view the quadratic model (8) as a model of the reduction

$$\tfrac{1}{2} \left(\| F(x_k + w) \|^2 - \| F(x_k) \|^2 \right)$$

in a neighborhood of x_k. Consider, for example, the unconstrained minimization problem

$$\min \{ f(x) : x \in R^n \}.$$

For this problem we assume that the reduction $f(x_k + w) - f(x_k)$ is modeled by a quadratic of the form

$$\psi_k(w) \equiv g_k^T w + \tfrac{1}{2} w^T B_k w, \tag{9}$$

where g_k is an approximation to the gradient $\nabla f(x_k)$ and B_k is an approximation to the Hessian matrix $\nabla^2 f(x_k)$. The step s_k is computed as an approximate solution of the subproblem

$$min\{\psi_k(w) : \| D_k w \| \leqslant \Delta_k \}, \tag{10}$$

and the reduction produced by the step s_k is measured by the ratio

$$\rho_k \equiv \frac{f(x_k) - f(x_k + s_k)}{- \psi_k(s_k)}. \tag{11}$$

Note that the quadratic model (9) reduces to the model (8) if we let

$$g_k = J_k^T f_k, \quad B_k = J_k^T J_k.$$

Here g_k is the gradient of the nonlinear least squares function f defined by

$$f(x) = \tfrac{1}{2} \| F(x) \|^2,$$

but B_k is not the Hessian matrix of f. It is certainly possible to use other models in the nonlinear least squares problem. Dennis, Gay, and Welsch [1981], for example, present a nonlinear least squares algorithm in which model (8) is replaced by (9) with B_k a clever quasi-Newton approximation to the Hessian of the nonlinear least squares function. If

$$g_k = \nabla f(x_k), \quad B_k = \nabla^2 f(x_k),$$

then model (9) leads to a trust region version of Newton's method. The step calculation requires considerable numerical attention to deal with the possibility of an indefinite Hessian, but recent research has resulted in satisfactory algorithms for this problem. In many respects, Newton's method for unconstrained minimization provides an excellent introduction to some of the more recent work in trust region methods; for an introduction along these lines, see the paper of Moré and Sorensen [1982]. Trust region methods are explored further in the books of Fletcher [1980 and 1981] where they are called restricted step methods, and in the survey papers of Sorensen [1982] and Moré [1983].

SCALE INVARIANT ALGORITHMS

The reliability of an optimization algorithm is greatly enhanced if the algorithm is scale invariant. This is a natural requirement which can have a significant impact on the implementation and performance of the algorithm.

We first consider algorithms for the solution of systems of nonlinear equations and nonlinear least squares problems. Given a function $F: R^n \rightarrow R^m$ and a starting vector $x_0 \in R^n$, consider problems related by the transformation

$$\hat{F}(x) = \alpha F(D^{-1}x), \quad \hat{x}_0 = D x_0, \tag{12}$$

where D is a nonsingular matrix and α is a positive scalar. If the iterates $\{x_k\}$ and $\{\hat{x}_k\}$ generated by the algorithm satisfy

$$\hat{x}_k = D x_k, \quad k > 0, \tag{13}$$

then the algorithm is *scale invariant*.

The concept of scale invariance extends easily to the general constrained minimization problem

$$\min\{f(x) : c_i(x) \leqslant 0, \, i \in I; \, c_i(x) = 0, \, i \in E\}.$$

Given a constrained minimization problem defined by the objective function f, the constraint functions c_i, and the starting point x_0, we can consider a transformed problem where $\hat{f}$ and $\hat{x}_0$ are defined by

$$\hat{f}(x) = \alpha f(D^{-1}x), \quad \hat{x}_0 = D x_0,$$

while $\hat{c}_i$ is defined by setting

$$\hat{c}_i(x) = \beta_i \, c_i(D^{-1}x)\,, \quad i \in I \bigcup E\,,$$

for some constants β_i with $\beta_i > 0$ for $i \in I$ and $\beta_i \neq 0$ for $i \in E$. If the iterates $\{x_k\}$ and $\{\hat{x}_k\}$ generated by the algorithm satisfy (13), then the constrained minimization algorithm is scale invariant.

If the algorithm is scale invariant, then the convergence properties of the algorithm are independent of the scaling of the variables. This is clearly a desirable property. In contrast, the performance of a scale-dependent algorithm usually deteriorates on badly scaled problems. Comparison of a scale-dependent program with a scale-invariant program shows that they behave similarly on well scaled problems but that the scale-invariant program is usually superior on badly scaled problems.

Scale invariance for the trust region method specified by Algorithm (7) can be achieved by a careful choice of scaling matrix D_k and initial bound Δ_0. We can choose, for example,

$$D_k = diag(\delta_{i,k})\,, \quad \delta_{i,k} = max\{\delta_{i,k-1}, \| F'(x_k)e_i \| \}$$

and

$$\delta_{i,0} = \| F'(x_0)e_i \|\,.$$

If $\delta_{i,0} = 0$ for some i, then the i-th column of the Jacobian matrix is zero at the starting point x_0. In this case we can hold the i-th variable fixed until we obtain an iterate for which the i-th column of the Jacobian matrix is not identically zero. For the bound Δ_0 we can choose

$$\Delta_0 = \gamma \| F(x_0) \|$$

for some constant $\gamma > 0$. Other choices of D_k and Δ_0 are possible, but these illustrate a reasonable way to achieve scale invariance.

Scale invariance of the trust region method can be verified by showing that if we apply the algorithms to problems related by the transformation (12), then (13) holds. The proof is by induction. Assume that $\hat{x}_l = Dx_l$ holds for $0 \leqslant l \leqslant k$. Since

$$\hat{F}'(x) = \alpha F'(D^{-1}x)D^{-1}\,,$$

we have that $\hat{D}_k = \alpha D_k D^{-1}$. Moreover, it is clear that $\hat{\Delta}_0 = \alpha\Delta_0$. We now show that $\hat{s}_k = Ds_k$ under the assumption that the step is the solution of subproblem (3) and that J_k is the Jacobian matrix at x_k. In this case note that

$$\| \hat{f}_k + \hat{J}_k Dw \| = \alpha \| f_k + J_k w \|$$

holds for any vector w, and thus $\hat{s}_k = Ds_k$. Hence, $\hat{x}_{k+1} = Dx_{k+1}$ as desired.

In this derivation the assumption that J_k is the Jacobian matrix at x_k was needed to guarantee that

$$\hat{J}_k = \alpha J_k D^{-1}.$$

This observation can be used to show that if J_k is chosen by a modification of the rank-1 update of Broyden [1965], then scale invariance still holds. In this modification we update J_k by setting

$$J_{k+1} = J_k + \frac{(y_k - J_k s_k)(D_k^T D_k s_k)^T}{\| D_k s_k \|^2},$$

where $y_k = F(x_k + s_k) - F(x_k)$. The diagonal scaling matrices D_k could be chosen, for example, by setting

$$D_k \equiv diag\left(\| F'(x_0) e_i \|\right).$$

With these choices it is now straightforward to show that the trust region method of Algorithm (7) is scale invariant.

It is not necessary to assume that s_k is the solution of subproblem (3) in order to obtain scale invariance of Algorithm (7). For example, a nearly optimal solution of subproblem (3) can be obtained if we choose a parameter λ_k such that if s_k is the solution of

$$(J_k^T J_k + \lambda_k D_k^T D_k) s_k = -J_k^T f_k,$$

then either $\lambda_k = 0$ and $\| D_k s_k \| \leqslant \Delta_k$, or $\lambda_k > 0$ and

$$|\| D_k s_k \| - \Delta_k | \leqslant \sigma_0 \Delta_k$$

for some σ_0 in $(0,1)$. These conditions are invariant under transformation (12), and thus the scale invariance of Algorithm (7) can be preserved with this choice of step.

The nonlinear least squares programs in MINPACK-1 obtain the step as outlined above. An advantage of this choice of step is that it can be shown that if ψ_k is the quadratic model (8), then

$$\psi_k(s_k) \leqslant (1 - \sigma_0)^2 \, min\{\psi_k(w) : \| D_k w \| \leqslant \Delta_k\}, \quad \| D_k s_k \| \leqslant (1 + \sigma_0) \Delta_k.$$

Thus s_k is a nearly optimal solution of (3).

For the unconstrained minimization problem it is possible to choose s_k as a nearly optimal solution of (10). In this case, however, the quadratic ψ_k is not necessarily convex, and this makes the model problem more difficult to solve. Moré and Sorensen [1981] present an algorithm that is guaranteed to find a nearly optimal solution of (10) in a finite number of steps.

In this section we have shown that the trust region method specified by

Algorithm (7) is scale invariant. In the next two sections we shall show that scale invariance also plays a role in the implementation of the algorithm and in the design of the convergence tests.

ROBUST SOFTWARE

A robust implementation of an optimization algorithm extends the domain of the algorithm so that it copes with as many problems as possible without a serious loss of efficiency. A robust implementation, therefore, is a prerequisite for reliability.

An important requirement of robustness — usually ignored by optimization software — is the avoidance of unnecessary overflows and underflows. In general it is not possible to avoid all overflows and underflows without a serious loss of efficiency; MINPACK algorithms try to avoid *destructive* underflows and overflows, that is, overflows or underflows that damage the accuracy of the computation. We illustrate this requirement of robustness by discussing the computation of the Euclidean norm of an n-vector.

In our discussion we assume that the bounds on the allowable (working precision) floating point numbers can be specified by the smallest magnitude *dwarf* and the largest magnitude *giant*. An attempt to calculate a floating point number f with $0 < |f| < dwarf$ causes an underflow, while an attempt to calculate a floating point number f with $|f| > giant$ causes an overflow.

The Euclidean norm $\| x \|$ of an n-vector x can be computed by setting

$$\| x \| = sqrt\left(\sum_{j=1}^{n} x_j^2\right). \tag{14}$$

If the components of x are suitably restricted, then (14) computes the Euclidean norm of x accurately and efficiently. However, if

$$abs(x_j) > sqrt(giant)$$

for some index j, then the computation overflows even though the Euclidean norm of x may be within the range of the machine. Similarly, if

$$dwarf < abs(x_j) < sqrt(dwarf), \quad j = 1,2,...,n,$$

then (14) incorrectly sets $\| x \|$ to zero. In essence, the problem is that the algorithm specified by (14) is not scale invariant, that is, $\| \alpha x \|$ is not necessarily $|\alpha| \| x \|$. This problem can be avoided by computing a suitable scaling of the vector. For example, we can set

$$\sigma = max\{abs(x_j) : 1 \leqslant j \leqslant n\},$$

and, if $\sigma \neq 0$, then compute $\|x\|$ using the formula

$$\|x\| = \sigma \; sqrt\left(\sum_{j=1}^{n}\left(\frac{x_j}{\sigma}\right)^2\right). \tag{15}$$

Destructive overflows and underflows do not occur in (15). An underflow may occur in the computation of $(x_j/\sigma)^2$, but this underflow is not destructive. Similarly, (15) may overflow, but this only happens if $\|x\|$ exceeds *giant*.

Some researchers claim that the scaling required by (15) is unnecessary. This claim rests on the belief that the occurrence of overflows or underflows signals a deficiency in the formulation of the problem, and that in these situations it is the user's responsibility to reformulate the problem. We disagree with this view and believe that if the optimization software is to be useful, then it should attempt to solve the problem as provided by the user. Avoiding destructive overflows and underflows extends the class of problems that can be solved by the optimization program; it is particularly important on those machines with a small range of numbers, for example, on machines such as the VAX where *dwarf* and *giant* are on the order of 10^{-38} and 10^{38}, respectively.

A valid objection to the scaling required by (15) is that it requires two passes over the vector x, and thus slows down the computation of the Euclidean norm. For small- to medium-scale problems this is not usually important because norm computations do not tend to dominate the computing cost. It is possible to have a robust computation of the Euclidean norm without two passes over the vector x at the cost of a somewhat more elaborate code. The interested reader should consult, for example, the Euclidean norm program in MINPACK-1.

Avoiding destructive overflows and underflows is not the only important requirement of robustness, although from the user's viewpoint it is quite important. Another important requirement of robustness is the formulation of the algorithm to avoid an unnecessary loss of accuracy. Consider, for example, the computation of the reduction

$$\|f_k\|^2 - \|f_k + J_k s_k\|^2$$

predicted by the step s_k. In exact arithmetic this predicted reduction is positive for a reasonable choice of s_k, but roundoff problems can cause this reduction to be negative. The trust region method requires that this reduction be positive so that $\rho_k > 0$ if and only if

$$\|F(x_k + s_k)\| < \|F(x_k)\|.$$

It is therefore important to formulate this computation so that the result is guaranteed to be positive. This can be done if, as mentioned at the end of the previous section, the step s_k satisfies

$$(J_k^T J_k + \lambda_k D_k^T D_k) s_k = -J_k^T f_k$$

for some $\lambda_k \geqslant 0$. In this case it is not difficult to verify that

$$\| f_k \|^2 - \| f_k + J_k s_k \|^2 = \| J_k s_k \|^2 + 2\lambda_k \| D_k s_k \|^2,$$

and that this reformulation guarantees that the reduction is nonnegative, even in the presence of roundoff. However, this reformulation may still suffer from destructive overflows and underflows. This drawback can be eliminated by noting that the predicted reduction is only needed to determine the ratio ρ_k in (5), and therefore it suffices to compute the scaled reduction

$$\left(\frac{\| J_k s_k \|}{\| f_k \|} \right)^2 + 2 \left(\frac{\lambda_k^{\frac{1}{2}} \| D_k s_k \|}{\| f_k \|} \right)^2.$$

Given this scaling, we can formulate the tests involving ρ_k so that destructive overflows and underflows are avoided. The details can be found in the MINPACK-1 algorithms for nonlinear least squares. Similar difficulties must be overcome in order to successfully calculate ρ_k in (11) for unconstrained minimization. Details of these calculations may be found in Moré and Sorensen [1981].

A robust implementation should also have checks on the input parameters, especially those that lead to a disastrous failure if improperly set. This type of checking is essential in providing robust software.

It may not be reasonable to check all of the input data. For example, programming errors may lead the user to provide a Jacobian matrix that is not consistent with the function. Tests for detecting this situation should be provided, and the user should be encouraged to use them. Unfortunately, these tests are not completely reliable; they almost always detect an inconsistent Jacobian, but sometimes mistakenly classify a consistent Jacobian as inconsistent. For this reason, tests of this type should not be forced on the user.

An exhaustive list of the requirements for a robust implementation is not possible because the requirements depend on the implementor's vision of the algorithm's domain. We hope that the discussion here conveys the sense of what is required from a robust implementation in the MINPACK package.

CONVERGENCE TESTING

A reliable optimization program must be able to converge from a wide range of starting values, and if convergence takes place, the program must be able to decide whether the current approximation to the solution of the optimization program has the desired accuracy. The purpose of a convergence test is to make this decision.

If the optimization problem is not within the domain of the algorithm, then it is also important for the program to terminate with the appropriate information. For example, the optimization problem may be a system of nonlinear equations without a solution, or a minimization problem without a finite minimizer. Tests for detecting these non-convergent situations are just as important as convergence tests. Here, however, we shall discuss only convergence tests.

The construction of a reliable convergence test requires attention to several issues. Consider, for example, nonlinear least squares problems. A common test is to require that

$$\| J^T f \| \leqslant gtol, \tag{16}$$

where J is the current approximation to the Jacobian matrix, f is the function value at the current approximation to the solution, and $gtol$ is the tolerance. One objection to this test is that it is not scale invariant; as a result, it is difficult to choose the tolerance $gtol$. A scale-invariant version of (16) is

$$max\{ \frac{|a_i^T f|}{\| a_i \|} : 1 \leqslant i \leqslant n \} \leqslant gtol \| f \|, \tag{17}$$

where a_i is the i-th column of the Jacobian matrix. Note, however, that (16) and (17) are satisfied in a neighborhood of any critical point of the nonlinear least squares problem; ideally, a convergence test should be satisfied only near a local minimizer. Also note that although this test is scale invariant, it is still difficult to choose the tolerance $gtol$ because there is no clear relationship between the size of $gtol$ and optimality of the residual norm $\| f \|$.

Another objection to (17) is that it usually fails whenever the Jacobian matrix has a zero column at the solution. This objection may seem to be based on an unlikely situation, but the situation is actually not uncommon. For example, consider the problem of Kowalik and Osborne [1968]. The model that underlies this problem describes the chemical reactions produced by an enzyme. The problem has $n = 4$ parameters and $m = 11$ component functions with

$$f_i(x) = y_i - \frac{\xi_1 (\mu_i^2 + \xi_2 \mu_i)}{(\mu_i^2 + \xi_3 \mu_i + \xi_4)}; \tag{18}$$

the data μ_i and y_i are given in their paper. Numerical results show that algorithms started near

$$z = (0.25, 0.39, 0.415, 0.39)^T$$

tend to converge towards a minimizer with a residual norm of about 0.017, but that algorithms started near $10z$ tend to converge towards a minimizer with a residual norm of about 0.032. This second minimizer arises because

$$\lim_{\alpha \to \infty} f_i(\alpha, \xi, \alpha, \alpha) = y_i - \frac{\mu_i}{\mu_i + 1} (\xi + \mu_i);$$

as a consequence, it can be shown that if ξ^* is the solution of the linear least squares problem

$$min\sum_{i=1}^{m}\left[y_i - \frac{\mu_i}{\mu_i + 1} \, (\xi + \mu_i) \right]^2$$

and α is large enough, then (17) fails in a neighborhood of $(\alpha,\xi^*,\alpha,\alpha)$.

Given a positive tolerance $ftol$, a reasonable convergence test should attempt to guarantee that the current approximation x is near a local minimizer x^* with

$$\| F(x) \| \leqslant (1 + ftol) \| F(x^*) \| . \tag{19}$$

We attempt to achieve this condition by requiring that

$$\left| \| F(x_k) \| - \| F(x_k + s_k) \| \right| \leqslant ftol \| F(x_k + s_k) \| , \tag{20}$$

where s_k is the solution of the linear least squares problem (3) for some $\Delta_k > 0$. An advantage of this test is that if F is linear and x^* is in the current trust region, then (20) implies (19). If ρ_k is the ratio (5) and if, say,

$$|\rho_k - 1| \leqslant \tfrac{3}{4} , \tag{21}$$

then we can assume that the behavior of F is nearly linear. If, in addition,

$$\| D_k F'(x_k)^t F(x_k) \| \leqslant \Delta_k , \tag{22}$$

where A^t denotes the Moore-Penrose generalized inverse of the matrix A, then we can also assume that x^* is in the current trust region. If F is linear and $F'(x_k)$ has full rank, then this last condition is equivalent to requiring that x^* be in the current trust region.

The tests (20), (21), and (22) are all invariant under transformation (12). If the three conditions are satisfied, then it can be safely assumed that x_k satisfies (19). In practice, not all three conditions are satisfied, and then it is reasonable to terminate the algorithm and inform the user that x_k may not satisfy (19).

In the Kowalik and Osborne problem (18), all three tests are satisfied when the algorithm converges towards the minimizer with residual norm of about 0.017, but (22) fails near the other minimizer. Note that (22) is likely to fail whenever $F'(x^*)$ is rank deficient because then $F'(x_k)^t$ can be difficult to compute. The other tests can also fail. For example, if $F(x^*) = 0$ then it may not be possible to satisfy (20), while (21) fails whenever the reduction predicted by the model is not reasonable. An alternative test is to require that

$$\| F(x_k) \| \leqslant (1 + ftol) \| F(x_k) + F'(x_k) s_k \| , \tag{23}$$

where s_k is the solution of the linear least squares problem (3). It is not unusual for a step s_k to satisfy (20) and (23), but not (21).

The ideas presented in this section provide the basis for a reliable convergence test. The convergence tests used in MINPACK-1 follow these ideas. Details can be found in the user guide. These ideas can be extended to other areas; Gay [1982], for example, has a nice discussion of convergence testing in trust region methods for unconstrained minimization.

USER INTERFACE

The development of easy-to-use software requires a carefully designed user interface. The user of an optimization program prefers a short calling sequence because then the program is easier to use. On the other hand, an optimization program with a short calling sequence may not provide the necessary flexibility.

This conflict between flexibility and ease-of-use is resolved in the MINPACK-1 software by providing driver and core programs. The driver has a short calling sequence; its purpose is to call the core program with an appropriate set of default parameters. The core program can be directly called by the user if additional flexibility is needed.

The construction of the driver programs requires some care. We illustrate this process with the *lmder* program for nonlinear least squares problems. This program has 24 parameters, while the *lmder*1 driver program has 12.

One of the techniques used to reduce the number of parameters is to assign default values to a parameter. For example, in *lmder* the parameter *mode* specifies either automatic scaling or fixed scaling by the array *diag*. Automatic scaling is used in *lmder*1 because, in our experience, it is usually superior to fixed scaling.

The choice of default values, however, can sometimes be difficult. In *lmder* the parameter *factor* is used to determine a bound on the initial step. The default setting in *lmder*1 is *factor* = 100, but on several occasions this value has allowed a step that is too large. A more appropriate setting may be *factor* = 1, although there is certainly no guarantee that this setting will avoid all cases of poor performance.

The number of parameters can also be reduced by coalescing two or more parameters. For example, the calling sequence of *lmder* has an array *diag* for the diagonal scaling matrix, an array *qtf* for information associated with the orthogonal factorization of the Jacobian matrix, and four work arrays. In *lmder*1 these arrays are replaced by a single array *wa* and a length parameter *lwa*.

The decision as to which parameters should be retained in the driver is usually easy. In *lmder*1, the parameters *fcn, m, n, x, fvec, fjac, ldfjac,* and *info* are retained. The parameter *fcn* is the name of the user-supplied subroutine; *m* and *n* are the number of equations and variables, respectively; *x* is the vector of variables; *fvec* is the vector of function values; *fjac* is the $(ldfjac,n)$ array for the Jacobian matrix; and *info* is used to report the completion state of the algorithm. It is clear that all of these parameters are essential.

Developments in the Fortran language may have a profound effect on the user interface; Du Croz [1982] has an interesting discussion of the impact of programming languages on numerical subroutine libraries.

The use of reverse communication also affects the user interface. An optimization program with a reverse communication interface returns to the calling program whenever there is a need to evaluate the problem functions. The calling program must then evaluate the problem functions and return control to the optimization program. This process can be repeated at the discretion of the calling program. A reverse communication interface eliminates the need to transfer information to the program that evaluates the problem functions but makes the software harder to use. The MINPACK-1 derivative checker uses reverse communication because then the same program can be used to check gradients, Jacobians, and Hessians. For similar reasons, Gill, Murray, Picken, and Wright [1979] use reverse communication in their line search. Gay [1980] satisfies demands for both types of interface by using reverse communication in the core program and a standard interface in the driver.

DOCUMENTATION

The development of easy-to-use software also requires carefully designed documentation. Introductory documentation with information on the solution and analysis of nonlinear problems should be provided, as well as documentation for the programs normally called by the user and for all the auxiliary programs.

In MINPACK-1 the documentation for the programs normally called by the user (single precision and double precision versions) is supplied in machine-readable form with the package. This avoids the need of providing each user with a user guide and is particularly useful if the user likes to obtain his information from the terminal. The documentation for each program is divided into nine sections:

Purpose.

Subroutine and Type Statements.

Parameters.

Successful Completion.

Unsuccessful Completion.

Characteristics of the Algorithm.

Subprograms Required.

References.

Example.

This subdivision is fairly common, although documentation standards do not usually provide for sections on the successful and unsuccessful completions of the algorithm. Instead, this information tends to be hidden in other sections.

The description of the purpose of the program is brief; it tells the user if this is the program needed. The subroutine and type statements define the calling sequence and the parameters required by the program. By convention, the parameters that describe the problem appear first in the calling sequence, followed by the parameters that specify the algorithm, and finally the parameters that return information to the user.

The parameters are then described in the order they appear in the calling sequence. The description of each parameter gives its type (integer variable, logical variable, variable, integer array, logical array, array) and states whether it is an input or an output parameter. Input parameters must always be specified by the user. Output parameters are always determined by the algorithm. For the other parameters we specify both their input and output values. We decided not to split the descriptions of the parameters into input and output sections because we found this separation awkward.

The section on successful completion describes the convergence tests of the algorithm. The section on unsuccessful completion details those situations that lead to unsuccessful termination of the algorithm and provides advice on how to remedy them.

The section on characteristics of the algorithm contains a brief description of the algorithm. In addition, this section mentions the storage required by the algorithm and provides timing information in terms of the number of arithmetic operations needed by the algorithm per call to the user-supplied function fcn. The timing information shows that unless fcn can be evaluated quickly, the timing of the algorithm is dominated by the time spent in fcn.

The documentation of the programs ends with an example of the use of the program. This is probably the most important section of the documentation because users tend to follow the example in setting up their programs.

Each MINPACK-1 program has detailed prologue comments on its purpose and the roles of its parameters, and in-line comments for each major block in the body of the program. In addition, the logical structure of the program has been delineated with the TAMPR system of Boyle and Dritz [1974].

Documentation standards that follow the guidelines suggested in this section should make the software easier to use. Additional documentation for the user who needs to know the details of the algorithm would also be useful. We plan to produce documentation of this type; at present, this information is contained in the references and in the user guide provided with the programs.

TRANSPORTABILITY

Transfer of software packages to a machine should require only a small number of well-defined alterations; the modified software should then perform satisfactorily. Software that satisfies these two requirements is *transportable*.

Transportable mathematical software can be produced by writing it in terms of a few machine-dependent parameters which can be obtained with a call to a function subprogram. It is necessary to prepare single precision and double precision versions of this subprogram. Two versions of the other programs are also required, but these can be generated automatically. The TAMPR system of Boyle and Dritz [1974] was used to generate the single precision version of MINPACK-1 from the master (double precision) version.

The MINPACK-1 software requires three machine-dependent parameters. If the machine has τ base β digits, and its smallest and largest exponents are *emin* and *emax*, respectively, then these parameters are

$$epsmch = \beta^{1-\tau}, \textit{ the machine precision},$$

$$dwarf = \beta^{emin-1}, \textit{ the smallest positive magnitude},$$

$$giant = \beta^{emax-1}(1 - \beta^{-\tau}), \textit{ the largest magnitude}.$$

We have already mentioned *dwarf* and *giant* in the discussion of robust software; here they are defined in terms of the parameters τ and β.

In the MINPACK-1 package the correct values of these constants are encoded into *data* statements in functions *spmpar* (single precision) and *dpmpar* (double precision). The constants for many of the systems were obtained from the corresponding PORT library subprogram [Fox, Hall, Schryer, 1978]. We have added others, and clearly the list can be expanded further. The current list includes the constants for the systems noted in the table below:

IBM 360/370	CDC 6000-7000
Univac 1100	Cray-1
Burroughs 6700	VAX 11
Dec PDP-10	Honeywell 6000
Itel AS/6	Prime 400
ICL 2980	

It has been our custom to have the constants for the IBM 360/370 system activated in the master versions of *spmpar* and *dpmpar*; all other constants are deactivated by the presence of *C* in column 1. If the IBM constants are not appropriate for the machine, then these constants must be deactivated and the appropriate set of constants activated.

The MINPACK-1 testing package uses the single and double precision versions of the environmental inquiry program of W. J. Cody to test the constants supplied with *spmpar* and *dpmpar*. The test program prints

$$rerr(1) = (epsmch - eps)/epsmch$$

$$rerr(2) = (dwarf - xmin)/dwarf$$

$$rerr(3) = (giant - xmax)/giant,$$

where *eps*, *xmin*, and *xmax* are the values returned from Cody's program.

The constants specified in the MINPACK-1 package are conservative and do not take into account special features of the system. Therefore, components of *rerr* are not necessarily zero. For example, rounded arithmetic is reflected by *eps* being half as large as *epsmch*, and this produces a value of 0.5 for *rerr*(1). Values with magnitude as large as unity are suspicious, but we have encountered the following exceptions:

a)   In Honeywell and Prime systems, the use of extra length registers results in an overly small value of *eps*, and so the value of *rerr*(1) is close to unity; this occurs only in single precision on the Prime but in both precisions on the Honeywell.

b)   In CDC systems, the treatment of small double precision numbers requires a value of *dwarf* that is large relative to *xmin*, and thus the value of *rerr*(2) is close to unity.

c)   Honeywell and Prime systems do not allow a proper determination of *xmax* in double precision and produce a value of *rerr*(3) close to minus unity.

These exceptions are not troublesome to the MINPACK-1 software because the required machine-dependent parameters are set rather than computed during installation. If it were possible to compute the required parameters reliably, then the software could be made fully portable. The above exceptions show that this is not possible at present. In particular, exception a) shows that if we define *eps* as the smallest floating point number such that $1 + eps > 1$ in working precision, then *eps* may be much smaller than *epsmch*. If the software depends on *eps* rather than on *epsmch*, then this may lead to a disastrous failure.

The second requirement of transportability — satisfactory performance of the modified software — is more difficult to satisfy. We recommend that the guidelines of Smith [1977] be followed during the preparation of the software. We also recommend the use of BRNANL [Fosdick, 1974] and the PFORT verifier [Ryder, 1974]. BRNANL provides execution counts for each block of a program, while the PFORT verifier checks that the program adheres to a portable subset of Fortran 66. If the software is written in Fortran 77, then these tools cannot be used; but the TOOLPACK project [Osterweil, Hague, and Miller, 1982] should have suitable replacements available soon. Another recommendation is the use of a software tool that checks run-time errors such as out-of-range subscripts and uninitialized variables.

We also suggest that the software be developed on a machine with a relatively small set of floating point numbers. The MINPACK-1 software was developed on the IBM 360/370 system, and the limited range of this system made us aware, early on, of several potential problems. We are currently developing software on a VAX 11/780 which has an even more restricted range of floating point numbers.

The measurement of satisfactory performance requires a suitable set of test problems. Moré, Garbow, and Hillstrom [1981] have produced an easy-to-use collection of test problems for systems of nonlinear equations, nonlinear least squares, and unconstrained minimization. For constrained minimization, Hoch and Schittkowski [1981] have produced an extensive and easy-to-use collection of test problems. There is still a need for collections of test problems in other areas, for example, linearly constrained and large-scale optimization.

Given a suitable set of test problems, it is then possible to test the software on any given machine. See, for example, the work of Hiebert [1981 and 1982]. The measurement of satisfactory performance on a representative set of machines, say those listed earlier, can be done by test sites. The assistance of the MINPACK-1 test sites was invaluable to the project. They tested the software and the documentation, and suggested improvements. In addition, some of the sites made the software available to their users on a limited basis and thus provided feedback from the users, while other sites provided the values for the machine constants needed by *spmpar* and *dpmpar*.

As a final recommendation, we suggest that the software be accompanied by an implementation guide. The MINPACK-1 guide [Garbow, Hillstrom, and Moré, 1980] provides, for example, a description of the tape contents and advice on how to interpret the test results.

CONCLUDING REMARKS

We have discussed some of the techniques that we have used to produce reliable, easy-to-use, transportable optimization software. There has been no attempt at completeness in these discussions; rather, we have tried to focus on those techniques that we have found particularly useful. Other papers that discuss the development of optimization software include Gill, Murray, Picken, and Wright [1979], Moré [1982], and Gill, Murray, Saunders, and Wright [1982].

We are currently applying these techniques to the development of software for unconstrained, linearly constrained, and large-scale optimization. We are also working on a portable symbolic differentiation program. We have found these techniques useful in these new areas; we hope that they prove useful to other researchers interested in the development of optimization software.

REFERENCES

Boyle, J. M., and K. W. Dritz [1974]. "An automated programming system to facilitate the development of quality mathematical software." *Proceedings IFIP Congress.* North-Holland, Amsterdam, pp. 543-546.

Brent, R. P. [1973]. "Some efficient algorithms for solving systems of nonlinear equations." *SIAM J. Numer. Anal.* 10:327-344.

Broyden, C. G. [1965]. "A class of methods for solving nonlinear simultaneous equations." *Math. Comp.* 19:577-593.

Dennis, J. E., D. M. Gay, and R. E. Welsch [1981]. "An adaptive nonlinear least-squares algorithm." *ACM Trans. on Math. Soft.* 7:348-368.

Du Croz, J. J. [1982]. "Programming languages for numerical subroutine libraries." *The Relationship Between Numerical Computation and Programming Languages.* Ed. J. K. Reid, North-Holland, Amsterdam, pp. 17-32.

Fletcher, R. [1980]. *Practical Methods of Optimization, Volume 1: Unconstrained*

Optimization. John Wiley & Sons, New York.

Fletcher, R. [1981]. *Practical Methods of Optimization, Volume 2: Constrained Optimization.* John Wiley & Sons, New York.

Fosdick, L. D. [1974]. *BRNANL, A Fortran Program to Identify Basic Blocks in Fortran Programs.* University of Colorado Computer Science Report 40.

Fox, P. A., A. D. Hall, and N. L. Schryer [1978]. "The PORT mathematical subroutine library." *ACM Trans. on Math. Soft.* 4:104-126.

Garbow, B. S., K. E. Hillstrom, and J. J. Moré [1980]. *Implementation Guide for MINPACK-1.* Argonne National Laboratory Report ANL-80-68, Argonne, Illinois.

Gay, D. M. [1980]. *Subroutines for Unconstrained Minimization Using a Model/Trust Region Approach.* Massachusetts Institute of Technology Center for Computational Research in Economics and Management, Report 18, Cambridge, Massachusetts.

Gay, D. M. [1982]. *On Convergence Testing in Model/Trust Region Algorithms for Unconstrained Optimization.* Bell Laboratories Computing Science Report 104, Murray Hill, New Jersey.

Gill, P. E., W. Murray, S. M. Picken, and M. H. Wright [1979]. "The design and structure of a Fortran program library for optimization." *ACM Trans. on Math. Soft.* 5:259-283.

Gill, P. E., W. Murray, M. A. Saunders, and M. H. Wright [1982]. "Software for constrained optimization." *Nonlinear Optimization 1981.* Ed. M. J. D. Powell. Academic Press, New York, pp. 381-393.

Hiebert, K. L. [1981]. "An evaluation of mathematical software that solves nonlinear least squares problems." *ACM Trans. on Math. Soft.* 7:1-16.

Hiebert, K. L. [1982]. "An evaluation of mathematical software that solves systems of nonlinear equations." *ACM Trans. on Math. Soft.* 8:5-20.

Hoch, W., and K. Schittkowski [1981]. *Test Examples for Nonlinear Programming Codes.* Lecture Notes in Economics and Mathematical Systems 187, Springer-Verlag, Berlin.

Kowalik, J. S., and M. S. Osborne [1968]. *Methods for Unconstrained Optimization Problems.* Elsevier, New York.

Moré, J. J. [1978]. "The Levenberg-Marquardt algorithm: Implementation and theory." *Proceedings of the Dundee Conference on Numerical Analysis*. Ed. G. A. Watson. Lecture Notes in Mathematics 630, Springer-Verlag, Berlin, pp. 105-116.

Moré, J. J. [1982]. "Notes on optimization software." *Nonlinear Optimization 1981*. Ed. M. J. D. Powell. Academic Press, New York, pp. 339-352.

Moré, J. J. [1983]. "Recent developments in algorithms and software for trust region methods." *Mathematical Programming Bonn 1982 - The State of the Art*. Springer-Verlag, Berlin.

Moré, J. J., B. S. Garbow, and K. E. Hillstrom [1980]. *User Guide for MINPACK-1*. Argonne National Laboratory Report ANL-80-74, Argonne, Illinois.

Moré, J. J., B. S. Garbow, and K. E. Hillstrom [1981]. "Testing unconstrained optimization software." *ACM Trans. on Math. Soft.* 7:17-41.

Moré, J. J., and D. C. Sorensen [1981]. *Computing a Trust Region Step*. Argonne National Laboratory Report ANL-81-83, Argonne, Illinois.

Osterweil, L. J., S. Hague, and W. Miller [1982]. *TOOLPACK Architectural Design: The Users' Perspective*. Mathematics and Computer Science Division, Argonne National Laboratory, Toolpack Project Report, Argonne, Illinois.

Powell, M. J. D. [1970]. "A hybrid method for nonlinear equations." *Numerical Methods for Nonlinear Algebraic Equations*. Ed. P. Rabinowitz. Gordon and Breach, New York, pp. 87-114.

Ryder, B. G. [1974]. "The PFORT verifier." *Software Practice and Experience* 4:359-377.

Smith, B. T. [1977]. "Fortran poisoning and antidotes." *Portability of Numerical Software*. Ed. W. R. Cowell. Lecture Notes in Computer Science 57, Springer-Verlag, Berlin, pp. 178-254.

Smith, B. T., J. M. Boyle, and W. J. Cody [1974]. "The NATS approach to quality software." *Software for Numerical Mathematics*. Ed. D. J. Evans. Academic Press, New York, pp. 393-405.

Sorensen, D. C. [1982]. "Trust region methods for unconstrained optimization." *Nonlinear Optimization 1981*. Ed. M. J. D. Powell. Academic Press, New York, pp. 29-38.

Chapter 6

SOFTWARE FOR ORDINARY DIFFERENTIAL EQUATIONS

L. F. Shampine and H. A. Watts *
Sandia National Laboratories
Albuquerque, New Mexico 87185

INTRODUCTION

In this chapter we take up mathematical software for the initial value problem for a system of ordinary differential equations (ODEs). This is an interesting area in mathematical software because the algorithms blend the fruits of a long, intensive numerical analysis effort with a great deal of art. Although the better codes are quite effective, significant portions of their algorithms have been only partially analyzed. This situation poses a real challenge to numerical analysis, but it will be touched upon only in passing in this chapter. There are a great many codes available, a few of which certainly qualify as mathematical software, but we do not intend to survey them. Our principal object is to discuss some aspects of software as related in particular to ODEs.

Naturally, software for ODEs shares many concerns and characteristics with software in general. We shall mention only a few illustrative examples and then restrict our attention to ODEs. Evolution of high-quality software is a slow process. The early stages usually are referred to as *research* codes and contain numerical methods that work satisfactorily most of the time. These stages are followed by further analysis of the underlying problems as well as further understanding of the methods being used. Coupled with development of the software interface and attention to questions of use, this leads to the development of preliminary *production* codes. Next comes detailed examination and comparison of available software. And finally, after further understanding of theoretical aspects, additional algorithmic improvements, and software refinements, a robust general-purpose software item emerges.

Many important codes were written before ideas about programming methodology crystallized. The important code DIFSY1 [Hussels, 1973], in its most widely disseminated form, has no comments in its body at all. In view of the fact that it is machine dependent in several important ways, a lack of explanation severely hinders transportability. The important code DIFSUB [Gear, 1971] is notoriously difficult to understand because of the undisciplined use of GO TOs in its FORTRAN. This has led to a great many rewrites of the code, which in itself represents a severe software problem of maintenance. A related

* This work was supported by the U.S. Department of Energy.

example is the code STEP [Shampine and Gordon, 1975], which in its published version has an error. The official outlet for the code, the National Energy Software Center, distributes the correction, but a great many people have gotten the code from other sources. The code RKF45 [Shampine and Watts, 1977 and 1979] was intended to be an alternative to the code ODE [Shampine and Gordon, 1975] in appropriate circumstances, but because the codes were written at different times and by (some) different authors, they are not entirely consistent. This lack of consistency has led to errors of use and is partially responsible for development of the *package* of codes DEPAC [Shampine and Watts, 1980].

MATHEMATICAL PROBLEM

The first question we address is, what is the mathematical problem? Initial value problems for ODEs arise in the most diverse forms. Fortunately, they can usually be cast into a standard form as a system of first order equations

$$y'=f(x,y),\ a\leqslant x\leqslant b,\ y(a)\ \text{given},\tag{1}$$

where we need not explicitly indicate the vectors. This simple form is obtained by the introduction of new variables in a way usually taken up in classes on ordinary differential equations. It is this form of the problem that the vast majority of ODE codes treat. However, there are technical advantages of treating some higher order equations directly, e.g.,

$$y''=f(x,y).\tag{2}$$

DVDQ [Krogh, 1971 and 1975] and its descendants allow systems of equations in which each individual equation can involve derivatives of order as high as four. It is the only widely known general-purpose code to do this. Just the specification of the problem is an annoying software issue. In the contexts where more specialized forms like (2) arise frequently, special codes are common.

Solution of *boundary value problems* (BVPs) for ODEs has much in common with *initial value problems* (IVPs), but is much more difficult in a technical sense because a richer set of phenomena is possible. It is also more difficult in a software sense because there is no simple standard form like (1) that is comparably useful. Ascher and Russell [1981] recently described how to prepare many classes of BVPs in order to apply some typical codes. We shall say no more about BVPs on grounds of length and coherence of this chapter.

The mathematical problems are conventionally categorized as stiff or nonstiff. Unfortunately, stiffness is a complex phenomenon and certainly depends on the numerical method as well as the mathematical problem. Fuller discussions of the question from a user's point of view with many references to codes can be found in Shampine and Gear [1979], Seifert [1980], and Watts [1980].

There are two essential qualifications for the mathematical problem to be stiff. The problem must possess a solution or subspace of solutions that are very easy to approximate. The problem must be extremely stable so that all solution curves starting near the "easy" ones must converge very rapidly to an "easy" solution curve. There are effective codes for both stiff and non-stiff problems, but at the present time it is important that the user recognize stiffness and select a suitable code. In an attempt to aid the user in this dilemma, some non-stiff codes, e.g., those in DEPAC, have algorithms that "test for stiffness" and inform the user when the problem is judged to be stiff. If he should desire, the user then can switch to an appropriate stiff code.

NUMERICAL METHODS AND CODES

Before continuing with the description of the mathematical problem and how it differs from the computational problem, we need to comment briefly about the most popular numerical methods for the solution of (1). They all generate a sequence of solution values by stepping through the interval $[a, b]$. They begin with the given value $y(a)$ at $x_0 = a$. When they have obtained an approximation y_m to $y(x_m)$, they step a distance h to obtain an approximation y_{m+1} at $x_{m+1} = x_m + h$. They all vary their step size to get through the whole interval as quickly (cheaply) as possible while still meeting an accuracy requirement. The Runge-Kutta methods evaluate the function f at several arguments in the interval $[x_m, x_{m+1}]$ and form y_{m+1} as a linear combination of these values. An important example is RKF45. The Adams methods make use of function evaluations at previous points $x_{m-1}, x_{m-2}, \ldots$ to obtain an accurate value for y_{m+1} with a small number of new evaluations in the step. Important examples are the ODE and DVDQ codes already mentioned. These codes vary the number of previous values used — vary their order — to match the behavior of the solution as well as possible. The extrapolation methods repeatedly integrate to a desired point x_m with a simple method of the same general nature as an Adams method, and then combine these results to yield a very accurate result at x_m. Because the number of integrations to the point x_m is varied, this method is also of variable order. The best code of this kind at present is probably DIFEX1 [Deuflhard, 1980].

By far the most popular codes for stiff problems are based on the backward differentiation formulas (BDF). These formulas, like the Adams formulas, use previously computed solution values, varying the number used (the order of the formula) for the sake of efficiency. A crucial difference is that they are implicit, so that a set of nonlinear algebraic equations must be solved for y_{m+1}. This procedure is very expensive compared to what formulas for non-stiff problems do and is cost-effective for stiff problems only because they can use step sizes very much larger than codes based on Runge-Kutta, Adams, and extrapolation methods. As remarked, the DIFSUB code spurred a host of rewrites and improved versions; an effective one is LSODE [Hindmarsh, 1980]. Another important code based on the BDF is FACSIMILE [Curtis, 1978].

SOFTWARE MATTERS FOR SOLUTION OUTPUT

One aspect of the computational problem is what is meant by a solution. Sometimes one is interested in $y(b)$ only. At other times one wants solution approximations for a set of points in $[a,b]$. This set of output points may be specified explicitly by the user, or perhaps any set defined by the algorithm and spread out in a "reasonable" way is acceptable. For example, the ODE solvers tend to bunch the $\{x_m\}$ where the solution changes rapidly and spread them out where the solution changes slowly. This behavior is especially true of the fixed order Runge-Kutta codes. Finally, one might want an approximate solution everywhere. This matter affects the choice of method and the software design profoundly. The Adams and BDF codes are based on representation of the solution by a polynomial. If one wants output values everywhere, or at a great many specific points, these codes are the most efficient. An important software issue is involved here. The code must be allowed to select its step size to be as efficient as possible and obtain answers at specific points by evaluating the underlying polynomial. Thus if a user wants to start at $x=0$ and obtain an answer at $x=1$, the code must be allowed to proceed past $x=1$ internally. But how far past? Output points convey scale information to the code, so it should not completely ignore where the user specifies them. More important, what if the equation changes for $x>1$, or what if it is not even possible to integrate past $x=1$? The software interface must give the user the capability of preventing the code from integrating past a given point. This capability can be provided in various ways, some of which have defects. For details see Shampine and Watts [1980].

There are two basic types of software design for ODE solvers. One advances a single step towards $x=b$ and returns control to the user. The other returns control only when it produces an answer at a specified point or determines that it is unable to do so for reasons that it reports. The *step−oriented* codes provide a degree of control essential to some applications, but present the typical user with an unacceptable complexity. It is a difficult software question how to provide flexibility to the user who needs it and still provide a code easy to understand and apply. The most popular approach is to have different levels of code. An example is the ODE/STEP,INTRP suite [Shampine and Gordon, 1975]. STEP is a step-oriented Adams code providing great flexibility. ODE is an *interval−oriented* code which is easy to use. ODE is actually a driver for STEP and INTRP, but as far as the user is concerned it is a different code entirely. At our computing laboratory, the use of these codes has been monitored for many years. Practically all usage from the mathematical subroutine library is of ODE. STEP is mostly used as a dedicated subroutine, perhaps slightly modified, in packages. In the following we shall concentrate on interval-oriented codes.

We describe the codes in DEPAC as interval oriented although we have made provision for the user to inspect intermediate results following each step on the way to b. We refer to this as the *intermediate−output* mode of operation, which differs from the usual step-oriented code in that it is not intended that the user assume control of the integration. With this design it is always necessary for the user to provide a future output point b, and even in the intermediate-output mode the codes return with a solution at b. This decision has an important algorithmic consequence. When the user specifies an interval of interest to the code, he provides it with a very useful piece of information about the scale of the problem. There is a related software matter in how b is obtained. Some older codes preferred to have the user enter an increment Δx and formed b internally from $b=a+\Delta x$. However, it has now become standard for the user to supply the value of b directly. This approach fits in better with our way of proceeding in the intermediate-output mode as well as with our general philosophy of easily continuing the integration task when it has been interrupted before getting to the output point.

When using an interval-oriented code, the user may cause the code to turn around. For example, after the integration is started at $x=0$, the user asks for an answer at $x=2$, and then at $x=1$. There is a serious conceptual difficulty here. As we hinted in our comments about stiffness, ODE codes have strong directional effects whereby integrating from $x=0$ to $x=1$ may be surprisingly different from integrating from $x=0$ to $x=2$ and then to $x=1$. We have not taken up the specification of accuracy so far, but we shall see that the computational problem may depend on the history of the integration. As a technical matter, all the methods have to restart if the integration changes direction (with a minor qualification for Adams and BDF codes). Depending on the method, this restart can be relatively expensive. The codes RKF45 and ODE automatically restart as necessary. When we modified them to form part of the DEPAC package, we attached so much importance to the implications, both conceptual and practical, of turning around that we forced the user to do the restart himself. The point is that a restart is too important to be done without the user's consent.

The step-oriented Adams code DIFSUB does not give explicit instructions about how output should be obtained. We have seen every possibility we could conceive of used, and then saw one we did not even imagine. Some have obtained output by manipulating the step size to land on the desired output point, but this is inefficient and has bad effects on the order and step size selection and possibly causes difficulties with precision. Some check to see if the next step will take them past the output point and then evaluate the interpolating polynomial at the output point. This approach should be avoided because one does not know that the code will in fact successfully step past the output point. The best procedure is to wait until the code steps past and then evaluate the interpolating polynomial. To our amazement, we read a report of substantial tests in which the author had the code integrate past the output point, turned around with a restart and then adjusted the step size to land on the output point, and finally turned around again with another restart to continue the integration. These observations make it clear that one must carefully instruct users of step-oriented codes or, in our opinion, preferably not leave this matter up to users.

Whether the particular choice of output points reduces the efficiency of Runge-Kutta or extrapolation codes is very dependent on the problem. Two devices, which have been employed in Runge-Kutta codes to prevent unduly small step sizes from occurring when landing exactly on an output point, are "looking ahead" and "stretching" [Shampine and Watts, 1979]. These schemes have proven quite satisfactory at reducing the impact of moderate output in such codes. Even so, we believe that this is a matter that software should monitor further. In RKF45 (and its DEPAC counterpart DERKF), for example, we inform the user if his output is severely reducing the efficiency of the code and advise him what he should do. It is suggested that using the intermediate-output mode is a better way to proceed if the behavior of the solution curve is desired, but if the user must have the solution at this many *specific* points, it would be best to use the companion Adams code.

There are other output considerations too. We believe that a code should take reasonable action for *any* meaningful interval $[a,b]$, no matter how close together a and b are. We have advanced the solution using Euler's method in certain special circumstances, for example, in RKF45 when $b-a$ is smaller than the machine dependent minimum allowable step size. The case of $a=b$ is debatable. It is convenient to permit $a=b$ so that initial values can be printed out easily without having to distinguish them as a special case. In DEPAC we do allow this at the initial point of integration and return $y'(a)$ too, but do not allow it on subsequent calls.

ERROR CONTROL

Having said where the approximate solutions are wanted, we must now consider how accurate they should be. This is a knotty problem as regards both technical and software issues. Virtually all codes control the *local error*, which is a very natural concept technically but which practically no user fully understands. When the integration has reached x_m, the local solution $u(x)$ is defined as the solution of $u'=f(x,u)$, $u(x_m)=y_m$. The codes try to approximate this solution, *not* $y(x)$. Thus the local error is $y_{m+1}-u(x_{m+1})$. How closely $y(x)$ is approximated depends on the behavior of the solution curves of the differential equation, in particular the stability of the problem. Practically all users think that when they ask a code to produce, say, five digits correct, they are asking that the approximation to $y(x)$ have five digits correct. The fact that the code may yield either more or less accuracy they attribute to vagaries of the code. This is a fundamental misconception, fostered by the fact that the codes are "tuned" to produce *true or global errors* (i.e., the errors in approximating $y(x)$) comparable to the local error tolerance for "typical" problems.

In recent years, for example in DEPAC, we have emphasized a different viewpoint in the code documentation. The group led by T. E. Hull at the University of Toronto has also vigorously supported this view; see DVERK [Hull et al., 1976]. For non-stiff problems, the local error control is essentially equivalent to producing a continuous, piecewise differentiable, approximate solution $v(x)$ such that

$$v(x_m)=y_{m'}\ v(x_{m+1})=y_{m+1}\quad \text{and}$$
$$v'(x)=f(x,v(x))+r(x),$$

where the size of the residual function $r(x)$ has been controlled to satisfy essentially the same bound as the local error. We feel that this is much easier for the user to understand, and the relationship of $v(x)$ and $y(x)$ due to the perturbing term $r(x)$ is easier to appreciate. This approach to the problem properly communicates to the user what the code attempts.

A popular scheme for assessing global errors, when it is done at all, is the technique of reintegration. The problem is solved again with the tolerances reduced, say by a factor of ten, and the answers compared. However, this idea should be approached with caution, because even with the best codes wrong conclusions can be drawn. The basic pitfall lies with the algorithm and software design. In order to reliably assess global errors by repeated integration, the code must be constructed with the design goal of achieving uniform proportionality of the true errors of the computed solutions as tolerances are changed (reduced). In the present generation of codes this was, at best, a secondary goal.

Another approach to error control is to derive algorithms that actually give the user what he expects, namely, global error control. There has been a lot of attention given this matter lately; see, for example, the survey of Prothero [1980]. We, ourselves, have written a companion GERK [Shampine and Watts, 1976] to the RKF45 code and have years of experience with it. There are some serious difficulties. In the first place, one cannot control global error, even in principle, in a single integration with the kinds of methods we are discussing. The best one can do is monitor it during integration controlled by, say, the local error and report if the global error gets too large. Because the task is difficult, the procedures increase the cost significantly with respect to codes using local error control alone. To hold down the costs, the procedures make much stronger assumptions about the problem; and the codes, or rather the estimates of true error, are much less robust and reliable. There are points of contact with the solution of systems of linear equations. In this simpler problem it is now generally appreciated that algorithms like Gaussian elimination produce results that have small residuals, but which do not necessarily have small true errors. Techniques like residual correction are available for estimation of the true errors, but they are often not considered worth the cost. This last observation coincides with our experience with GERK. We believe that because the users do not fully understand that they are not *already* controlling true errors, they are unwilling to spend more to solve their problems and obtain solution error estimates.

One sees a remarkable variety of error controls permitted users. The matter is quite important. If the user does not request enough accuracy, he may obtain a solution qualitatively different from that desired, because only local error control is done. This is a very dangerous area for ODE software. Reliability of error estimation and step size adjustment procedures is severely strained. One important difficulty is that of getting on scale so that the formula actually "sees" the behavior of the solution. The fundamental algorithms are all based on asymptotic arguments which depend on the step size h being "small enough." We shall look more closely at this issue shortly.

Having discussed error control in general, we proceed to more specific matters. The user may specify a tolerance that is to hold on all solution components simultaneously or may specify a vector of tolerances that are to apply to the corresponding solution components. The former leads to a more convenient user interface and is ordinarily quite adequate for non-stiff problems. Stiff problems very often involve solution components differing in size by many orders of magnitude so that the vector error control is almost necessary. The variety of error controls arises from how the error is to be measured — specifically, in the case of a relative error control, what it is relative *to*. We are convinced that the variety seen is due to users not understanding the indirect relation of local error control to true error. The most popular controls are a mixed relative-absolute control,

$$|\text{local error}| \leqslant RE^*|\text{solution}| + AE, \tag{3}$$

and a relative-threshold control,

$$|\text{local error}| \leqslant RE^* max(|\text{solution}|, TH). \tag{4}$$

The former is a pure relative error, relative to the solution, when $AE=0$ and a pure absolute error when $RE=0$. The latter is a pure relative error when the solution is greater in magnitude than the threshold TH and a pure absolute error otherwise. Obviously, these two controls are very closely related. The first control is used in DEPAC where, for simplicity, the user supplies both RE and AE as scalars or both as vectors.

Another popular control (see, e.g., DIFSUB, DIFEX1) is to take the error relative to the largest magnitude attained by the solution in the integration from a to the current point x. In some respects this is a safe relative control (except for starting the integration) but may not be stringent enough when solutions decay. It is a clear example of how some error controls depend on the history of the integration and so show directional effects.

A serious difficulty occurs when the user requests a pure relative error control and the solution vanishes because of a zero crossing or, what is worse, underflow. One should not simply introduce an AE internally (some do, of course) because a suitable value depends on the scale of the solution; e.g., if the solution component is smaller in magnitude than 10^{-50} on $[a,b]$, it is meaningless to permit an absolute error of 10^{-20} at a zero crossing even if 10^{-20} seems like a "small" number. Also a *very* small AE may force the code to work unduly hard at a zero crossing. A better approach is to adopt a reasonable definition of the quantity |solution| in (3) and (4). For example, one could use the average magnitude of the solution on $[x_m,x_{m+1}]$, estimated perhaps from the value at x_m and the tentative value at x_{m+1}. This is quite convenient in Runge-Kutta codes and is done, for example, in RKF45. The extrapolation codes form approximate solutions of low accuracy at (possibly) many points in $[x_m,x_{m+1}]$, and they typically form an average or maximum magnitude using these values. Adams and BDF codes take, as a rule, much smaller steps than Runge-Kutta or extrapolation codes, so normally use $|y_m|$ in (3) or (4). Ordinarily, this causes problems only when a component of the initial vector $y(a)$ vanishes and pure relative error is specified. Any of these remedies and others such as that of Curtis [1978] could fail, so mathematical software must monitor the weights used in the error control to prevent a divide check.

We have already mentioned some problems associated with absolute error control. Any absolute error control involves an *a priori* judgment about the scale of the solution. A bad judgment can lead to meaningless results because the control is too lax or can result in unnecessary expense because the control is too stringent. In DEPAC we caution the user about pure absolute error control and tell him that to acquire the appropriate scale information he may have to solve the problem more than once.

The form of the error control in (3) and (4) is referred to as error per step. Another popular formulation is error per unit step, in which the right-hand sides contain a factor of the step size h. Good arguments support each criterion [Shampine and Gordon, 1975, and Shampine and Watts, 1979]. From a theoretical viewpoint, error per unit step might be preferred. However, on practical grounds, error per step appears to be the better choice. At least for us it has been more efficient, and certain difficulties are handled more effectively.

As our observations should make clear, there are technical difficulties in deciding what kind of control is appropriate, software difficulties in providing a safe and reliable control, and communication difficulties in allowing user specification of a control and then reporting back problems with his choice.

CHOICE OF STEP SIZE

Practically all of the first variable step size codes were based on one-step methods and, traditionally, the user has had to supply the initial step size. This is, in fact, the crucial step because many algorithms in the codes depend on slow variation of quantities, including the solution, and well-written codes enforce heuristics to ensure this. Thus, if a good code starts on scale, it is unlikely to lose the scale. Guessing the initial step size is at best inconvenient and at worst disastrous, because a suitable step size depends on the method and implementation details of the code as well as the problem. In our opinion, the matter should not even arise in an interval-oriented code. We have always regarded automatic selection of initial step size as a question of numerical analysis important for both software and numerical reasons. The STEP code was probably the first piece of software to select the step size automatically. It uses the initial values of the first derivative, which is especially relevant in this variable order code which starts at order 1. It can be fooled, as Shampine and Gordon [1975] show by example, so the code allows the user to *limit* the first step. This device also permits the user to supply his experience with similar problems. In interval-oriented codes we have argued [Shampine and Watts, 1979] that where a user asks for output conveys valuable information about the scale of the problem, so in ODE and RKF45 (and in the newer codes of DEPAC) the first step is limited by the first output point. Hull and his associates have asked that the user supply some global information, specifically an estimate for a Lipschitz constant, in DVERK [Hull *et al.*, 1976]. The difficulty is that users generally do not understand this concept and do not have the information necessary for a decent estimate even if they do understand it. Recently, a number of authors [Lindberg, 1973; Shampine, 1980a; and Watts, 1981] have devised schemes for the automatic estimation of a Lipschitz constant with the aim of using it to select the initial step size. As a dividend, a "large" Lipschitz constant is a strong signal that the problem should be solved with a code suitable for stiff problems (at least locally).

Subsequent step size control is generally based on use of the "locally optimal" step size formula

$$h_{m+1} = h_m \left| \frac{\epsilon}{est} \right|^{1/p+1},$$

where p is the order of the method, ϵ is a tolerance parameter, and *est* represents an estimate of the local error. This same formula is to be used following both successful and unsuccessful steps. (A step is accepted or rejected depending on whether or not $|est| \leqslant \epsilon$.) To write an effective code, one must exercise great caution in using the locally optimal step size. Its derivation might not be valid because it requires certain smoothness assumptions and a principal error function that does not change rapidly. Nearly all codes employ a safety factor by taking a fixed fraction, say, 0.9 of the locally optimal step size, which amounts to aiming at a certain fraction of the tolerance. The closer this fraction is to one, the greater will be the number of rejected steps; the smaller the fraction chosen, the more conservative the code will be. There is a balance to be struck which depends on the method and error estimation procedures being used. Also, the global error is likely to have more regular behavior for values of the fraction closer to one.

Further limitations on the step size are still required. An effective code will restrict the rate of increase and decrease of the step size, $h_{m+1} = \mu \, h_m$, where, for example, $1/10 \leqslant \mu \leqslant 5$ as in RKF45. The purpose of this control is protection against a breakdown of underlying assumptions, a breakdown that is nearly always signaled by the prediction that a large change is needed. The "chattering" problem discussed by Shampine and Watts [1979] clearly points out the benefits of such controls. In RKF45 we also keep $\mu \leqslant 1$ following a step that involved a rejection. It is also necessary to keep h bigger than a machine-dependent, minimum allowable step size, as discussed in the next paragraph. We have not been in favor of a user-supplied, maximum allowable step size, but prefer to argue that the user's selection of output points rather naturally conveys important scale information. Beyond this, we think it might be useful for the code to estimate a Lipschitz constant and base a maximum step size on it.

ACCURACY CONSIDERATIONS

Returning now to the matter of how stringent the error control is, we observe that the opposite situation of requesting too much accuracy also poses software problems. Years ago it was pointed out that certain elementary requirements should be built into the codes. One is that x_m and $x_m + h$ should be different machine numbers. Actually, the codes may evaluate f for arguments between x_m and x_{m+1}. Clearly, if the code believes it needs an h so small that these arguments cannot be distinguished, one is at limiting precision, and many algorithms lose all justification. The older codes require users to specify a minimum step size. We have argued [Shampine, 1974, and Shampine and Watts, 1979] that this does not accomplish the goals commonly offered in its support and involves users in a matter of no interest or use to them. Rather, a minimum allowable step size quantity can be dynamically computed by the code and should be used for purposes just indicated.

The other requirement can be simply described as follows: It is senseless to ask for a result more accurate than the correctly rounded true local solution. It might be incredible to the reader that a user would do this; but it is, in fact, easy to do accidentally. If a user specifies a pure absolute error control and the solution grows sufficiently, this situation will occur. A remedy seen in many codes, e.g., RKF45, is simply to insist that the user's error control permit a relative error sufficiently large to stay out of trouble. The code STEP is intended to yield great accuracy and so proceeds more delicately. It monitors the condition at every step, informs the user if too much accuracy was requested, and reports the maximum possible accuracy. When these codes were modified for DEPAC, RKF45 was altered to provide the more delicate control. Error tolerances are increased as deemed necessary, and the situation is reported to the user. If the user wants to continue with the relaxed tolerances, he need only call the code again. This way of proceeding was followed consistently with a number of informative or warning returns. We found that too many users did not even examine the returns; they simply continued calling the code until it reported success. In DEPAC we force the user to reset a variable after such a return in order to state that he has chosen to accept, say, relaxed tolerances and wishes to continue.

Near the limits of the machine precision, results can be seriously affected by roundoff. It is often thought that requests for too much accuracy will cause some kind of error return, but this need not be so, especially in the older codes. More commonly one finds that the cost goes up and the true error becomes worse as the tolerances approach limits imposed by the machine. Only a few codes attempt to detect this problem or try to do something about it. F. T. Krogh, in several codes, has tried to detect the problem by monitoring differences. Several codes — for example, STEP — in effect augment the precision near the limit. This is a task for numerical analysis arising from the software problem that the user needs protection but is unable to spot limiting precision without aid. We might remark that, in our opinion, the best way to achieve great accuracy is by use of high order methods. There are some other simple precautions that one can take to enhance the numerical behavior of the algorithm when working near limiting precision. For example, where possible, quantities can be grouped to take advantage of cancellation, sums can be performed in order of increasing magnitudes, careful scaling may preserve significance, and difficulties from premature underflows can sometimes be avoided by careful programming.

Use of local extrapolation to achieve a higher order of accuracy is also a rather popular device. In effect, the estimate of the local error is subtracted out of the basic solution approximation, yielding a more accurate approximation, at least asymptotically. This is not merely a software decision, however, as the reliability and general effectiveness of the error estimating scheme are clearly important. In addition, stability questions and tolerance range of applicability must be examined. We believe that these numerical analysis issues are important considerations in choosing whether to use local extrapolation. Although we advocate its use when appropriate, we must remember that the estimated local error may bear little relationship to the error in the extrapolated solution.

SPECIFYING THE DIFFERENTIAL EQUATIONS

A fundamental question is how the user tells the solver what the mathematical problem is. Written as (1) it is perhaps obvious that f should be a subroutine. However, there are many high-level packages that allow the user to specify his problem in a way more natural to him. Many packages exist, for example, for chemical kinetics in which the user specifies a reaction mechanism and the package forms the differential equations (1). We mention as examples those of Deuflhard *et al.* [1980], Edsberg [1973], and Carver and Boyd [1979]. Some packages are fairly general purpose; see, for example, Chang *et al.* [1979], Halin [1979], and the CSSL Ref. Manual [1973]. More specialized examples are in Hyman [1976] and Schiesser [1976]. Some possibilities are exotic: SPEAK-EASY does not even have functions in the usual sense, working instead with arrays [Shampine, 1978]. For mathematical subroutine libraries there are two major choices. Most common by far is a subroutine for f. The other approach, exemplified by DVDQ [Krogh, 1971], is called reverse communication. In this approach the solver returns control to the calling program with the arguments t, v and a request that $f(t, v)$ be supplied. The matter is complicated by the fact that these evaluations may be necessary from several places in the solver. Krogh handled the returns with multiple entries in the Fortran routine DVDQ. Unfortunately, this feature of Fortran is not at all portable. A more portable and popular implementation uses a suitable indicator variable and a COMPUTED GO TO.

The reverse communication technique is undoubtedly the more complicated software interface and involves the user in matters he would ordinarily prefer to avoid. For this reason it is uncommon in ODE solvers. On the other hand, it provides flexibility in applications coupling several pieces of software, which is difficult or impossible to obtain with the more conventional approach. At the present time we favor a software design using levels of codes, with the lowest level based on reverse communication and a higher level based on a subroutine call.

There are a variety of software questions about the implementation and documentation of the f subroutine which we take up elsewhere [Shampine and Watts, 1980]. Let us mention only a couple here. Some codes, e.g., DIFSUB, use a fixed name for the subroutine to evaluate f. Because it is common to solve more than one problem in a single main program, consensus has it that passing the name of the subroutine through the call list of the ODE solver is more convenient. It is true that users are prone to forget to declare the name in an EXTERNAL statement in Fortran. Unless one documents carefully, users are likely to think that the arguments t, v presented to the f subroutine are accurate approximate solutions, which they are not. They must also be told sternly not to alter these arguments in the f subroutine because Fortran then causes them to be altered in the ODE solver. It would be easy enough to protect the solver from this for the scalar variable x, but the cost in time and storage to protect the vector variable y (or worse, a matrix in the solution of stiff problems) is not considered acceptable.

There is some controversy about what ought to be in the call list of the subroutine for f. For instance, there has been a long-standing disagreement about the necessity of having the number of differential equations appear in the argument list. There are applications in which the main program must convey even more information to the f routine but for which COMMON is inconvenient or cannot be used. Although these situations are not typical, in DEPAC we decided to include REAL and INTEGER arrays in the call list of f for this purpose. The user can dimension them to his convenience, and the solver ignores them.

Codes for stiff problems also require the Jacobian matrix $\partial f_i/\partial y_j$. This poses a number of software problems. It is quite a lot of work for the user to program the evaluation of this set of functions if there are many equations, which is a reason packages like those for chemical kinetics are so popular. An alternative is to generate the Jacobian by numerical differentiation. This may or may not be more expensive than analytical evaluation, but it is extremely popular. Reliable numerical formation of Jacobians is a difficult numerical analysis question. Large problems pose special difficulties. If one has N equations, the Jacobian has N^2 entries. The storage demands become prohibitive for values of N regarded as routine for non-stiff problems. Fortunately, the Jacobians for large problems are, it seems, almost always sparse. The software difficulty arises in how the user is to specify the structure and provide the Jacobian. A useful, and simple, possibility is that of banded Jacobians, whose structure is easy to specify. There is a special algorithm for numerical differentiation which takes full advantage of the structure. Specifying a general sparse matrix is not trivial. Discovering the sparsity pattern numerically has been unsuccessful.

Following the convention of putting the name of f in the call list of the ODE solver, we also suggest including the name of the subroutine for evaluating the Jacobian. Again, it must be declared in an EXTERNAL statement. Here a possible software difficulty might arise. The user may choose to have the code generate the Jacobian by numerical differentiation, in which case he would just ignore the parameter in the call list by treating it as a dummy argument. Unfortunately, for some compilers it may be necessary to write a dummy subroutine with the same name in order to avoid problems associated with missing EXTER-NAL routine names. In DEPAC the Jacobian subroutine closely models the f subroutine and, in particular, REAL and INTEGER parameter arrays are included in the call list for communication with the user's calling program.

STORAGE MANAGEMENT AND LEVELS OF CODES

Storage management is handled in various ways. The simplest and most portable is to restrict the number of equations allowed and dimension all arrays needed with a fixed, maximum size. Establishing such a maximum, however, is wasteful for small problems, and, in a large computing installation, is hardly practical because the largest requirements seen commonly are likely to be quite large indeed if stiff problems are to be solved. Moreover, it seems inevitable that one would want to solve a system a little bigger than the maximum allowed. The PORT library has a dynamic storage allocation scheme, and we refer the reader to the chapter on it. This does not completely resolve the issue because for large problems the system storage must be augmented. The most popular approach is to write the ODE solvers so that the arrays are of variable length. The annoying thing here is that for the code to be understandable, the various arrays must all have their own names. This leads to a very long call list with a set of arrays, mysterious to the user, which must be dimensioned. The standard approach is to introduce another level of code to which the user supplies REAL and INTEGER work arrays. The driver code partitions these arrays and calls the lower level code with the full call list.

It has been stated that a serious problem confronting mathematical software in general is getting people to use it. Our response in developing ODE software is to make the codes as easy as possible to use and still powerful enough to solve most problems. The matters of what should or should not belong in the call list and the design interface structure have been hotly disputed for a number of years. Attempts to arrive at a consensus for an ODE "standard interface" have failed. This lack of success resulted because the design of an ODE package is greatly influenced by the potential users and their computing environment. We refer the reader to Shampine and Watts [1980] for further discussion and one possible approach. In any case, it is absolutely essential to orient the software design towards user convenience; and it is clear that long argument lists are incompatible with this goal.

The approach of creating various levels of code in which storage arrays are carved up is virtually necessary in order that the user be able to concern himself simply with matters of problem definition and use of the code. Not only must these work arrays provide vector storage space used in intermediate computations, but values of certain parameters must also be saved between calls to the code. This is necessary to avoid problems of local retention of variables in an overlay situation. More complicated program structures also allow for optional input and output items of lesser interest to be buried in the work arrays. This is more appealing to the casual user in not requiring extraneous information but still providing flexibility to more sophisticated users needing the features. However, the code usage description must be carefully worded and structured, or else the prospective user will likely despair.

The necessary amount of storage can be rather complicated, especially when solving stiff problems where the structure is paramount. We believe that it is a useful software device to have the user input the length of the work array as well as the array itself. The code can then include this along with other routine argument tests (which every piece of mathematical software should have) to verify that enough storage was provided. The situation needs further attention. Solving stiff problems requires that linear systems of algebraic equations be treated repeatedly. Unfortunately, when working with a general sparse matrix, one does not know in advance how much storage will be needed. The computer bill often depends on the storage required. Progress in numerical analysis is resulting in codes that can automatically recognize and respond to stiffness. Such codes might recognize that a problem is non-stiff and never need the much larger storage for stiff problems. Because the character of a problem can change, the code must be able to get the extra storage when the problem is stiff but could relinquish it when the problem is not stiff. Some kind of dynamic storage allocation and appropriate billing algorithms are needed to take full advantage of this technical advance.

PROVIDING USER ASSISTANCE

We feel strongly that mathematical software for ODEs should provide the user with every possible assistance in recognizing that the task presented the code is one for which it is designed. This assistance ranges from the trivial, like consistency checks on input arguments which protect against accidental misuse of the code, to questions that are still challenging numerical analysts. We have already mentioned some examples, such as monitoring the effect of output and the consequences of the error control.

Another software matter that we deem important is to monitor the amount of computational work expended in solving the problem. Naturally, how long the job is permitted to run on the computer limits the total effort or expenditure. However, to protect the user against unexpected difficulties, incorrect problem formulation, and the like, we think it is prudent to measure the work being carried out and tell the user when this becomes excessive. Of course, this type of interruption of the integration task is to be viewed as an informative message and not as a task failure, so that the user may easily continue the integration if he wants. There is also the possibility that the code is working too hard because the accuracy requirement is too stringent for the method or perhaps some stiffness has been encountered. (In DEPAC we tell the user when the code thinks stiffness is the reason for the excessive work.) Measures of work vary in existing codes; popular controls include counting the number of evaluations of the differential equation and the number of successful steps taken. In DEPAC we count the number of *attempted* steps, arguing that this is a more compatible measure of computational effort for the various methods. In particular, the codes tell the user when 500 steps have been attempted, it having been found that this number represents a significant amount of work for these procedures.

CHOOSING A CODE

We have alluded to the issue of recognizing stiffness, which we consider the most pressing *software* issue facing numerical analysis. There are effective codes for both stiff and non-stiff problems. Stiffness is a complex phenomenon [Shampine and Gear, 1979, and Shampine, 1981], and the user simply does not have the information available to decide reliably whether a problem is stiff even if he knew how. Furthermore, the designation "stiff" is made very loosely. What is meant is that over a significant portion of the interval the problem should be treated as stiff. Dynamic recognition of changes in stiffness during the integration offer the possibility of improved efficiency; from a user's viewpoint a serious deficiency would be eliminated if codes were available that automatically selected a stiff or non-stiff mode of operation appropriate to the problem behavior. Quite a bit of work has been done on this issue [Shampine, 1981b, and Petzold, 1980], but much remains to be done both in terms of the technical issue and in terms of putting software into the hands of users. The next generation of ODE software can be expected to detect special behavior and adapt automatically to the difficulties at hand.

At the present time every user must select a specific code to solve his problem. Unfortunately, it is not clear which is the best code. Certain general observations can be made based on a great deal of experience and the general nature of the basic methods; we provide some guidance of this kind in the prologues to DEPAC. A decisive factor can be the cost of evaluating the differential equation on a particular computer. In our library version of RKF45 we do a real-time measurement of this cost; and using that cost, the number of equations, and the requested accuracy, we warn the user when the companion code ODE would be significantly more efficient. The experience used in the decision procedure is detailed in Shampine *et al.* [1976]. We think this kind of assistance could be developed much further.

Where does one find guidelines as to which code is "best"? Evaluation of mathematical software is a demanding task in its own right. Evaluations of ODE software have appeared regularly for many years. As the difficulties are better appreciated, the evaluations have become more relevant. It was thought at first that the dominant factor was the underlying method, and this was stressed in the tests of Hull *et al.* [1972]. We vigorously disputed this and, in tests we made [Shampine *et al.* 1976], clearly showed that the implementation might be as important as the basic method. If any one conclusion should be drawn from our experience, it is that codes are written to satisfy quite different objectives and simply cannot be compared fairly unless they are very similar in structure and goals. Great emphasis has been given to counting the number of function evaluations because it is machine independent. If f is expensive to evaluate, this gives a reasonable measure of performance. Unfortunately, f is often not sufficiently expensive, and the cost may be quite machine dependent. This affects the choice of algorithm, as we took up in considerable detail in Shampine *et al.* [1976]. Many feel too much stress has been put on function evaluations and not enough on the general overhead of the algorithm; Gupta [1980] provides some quantitative arguments to this effect. The situation with respect to stiff problems is far more serious. How Jacobians are formed, how often they are formed, and exactly how the linear algebra is done can have widely differing effects on the cost of solution. We take a very cautious view of testing. It is certainly useful in developing a code. It provides a useful comparison of very similar codes. It illuminates the weaknesses of codes, in particular the limits of their effectiveness, and provides general guidelines. However, to say code A is "better" than code B is a statement we can make only with stringent qualifications. It is important that the reader appreciate the points we made in Shampine *et al.* [1976] to the effect that efficiency may be the least important factor in choosing a code. The quality of the software documentation, ease of use, reliability, robustness, and portability are very often the deciding factors.

DEPAC SOFTWARE

We have repeatedly referred to DEPAC [Shampine and Watts, 1980] when illustrating certain features of software design. It represents a software effort undertaken at Sandia National Laboratories by the authors to develop a package of ODE solvers. The primary intent of the work was to design a user interface that would make codes in the package as easy as possible to use, to facilitate switching between codes in the package, and to take full advantage of the underlying algorithms. At present three methods are implemented: Runge-Kutta scheme in DERKF, Adams method in DEABM, and backward differentiation formulas in DEBDF. These codes are adapted versions of RKF45, ODE, and LSODE, respectively. An important feature of the design is the detailed prologues which provide the user with carefully laid out instructions about what the codes do, how they should be used, and how the results should be interpreted. We believe that this part of the code must be self-contained and capable of serving as a user's only source of documentation. Also, the interface design was written in a way that easily allows for extensions to future needs (codes).

SUMMARY

In this chapter we have concentrated on software issues associated with the numerical solution of ordinary differential equations. Particular methods and codes implementing the various methods have been mentioned only to illustrate a particular point. An in-depth survey of numerical methods and their distinctive features seemed beyond the scope of this book, certainly beyond the space limitations for this chapter. Likewise, it did not seem appropriate merely to catalog available ODE software, providing information about how to obtain it. Information of this kind is available in a recent survey article [Watts, 1980]. We hope that we have provided the reader with enough detail about ODE software characteristics and design criteria so that he may more fully appreciate the problems confronting efforts in developing "standard" ODE packages.

REFERENCES

Ascher, U., and R. D. Russell [1981]. "Reformulation of boundary value problems into standard form." *SIAM Review* 23:238-254.

Carver, M. B., and A. W. Boyd [1979]. "A program package using stiff, sparse integration methods for the automatic solution of mass action kinetics equations." *Int. J. Chem. Kinetics* 11:1097-1108.

Chang, Y. F., *et al.* [1979]. *Automatic Taylor Series Method (ATSMCC) User Manual.* Computer Science Department, University of Nebraska, Lincoln.

CCSL [1973]. Control Data 6000 Computer Systems Continuous System Simulation Language (CSSL3). *Reference Manual 17304400.*

Curtis, A. R. [1978]. *The FACSIMILE Numerical Integrator for Stiff Initial Value Problems.* AERE-R.9352. A.E.R.E. Harwell, Oxfordshire.

Deuflhard, P. [1980]. *Order and Stepsize Control in Extrapolation Methods.* Preprint No. 93. Institut für Angewandte Mathematik, University of Heidelberg, Germany.

Deuflhard, P., *et al.* [1980]. *LARKIN — a software package for the numerical simulation of LARge systems arising in chemical reaction KINetics.* Institut für Angewandte Mathematik Rept. 100, University of Heidelberg, Germany.

Edsberg, L. [1973]. "Integration package for chemical kinetics." *Stiff Differential Systems.* Ed. R. A. Willoughby. Plenum Press, New York.

Gear, C. W. [1971]. *Numerical Initial Value Problems in Ordinary Differential Equations.* Prentice-Hall, Englewood Cliffs, New Jersey.

Gupta, G. D. [1980]. "A note about overhead costs in ODE solvers." *ACM Trans. on Math. Soft.* 6:319-326.

Halin, H. J. [1979]. "Integration across discontinuities in ordinary differential equations using power series." *Simulation* 32:33-45.

Hindmarsh, A. C. [1980]. "LSODE and LSODI, two new initial value ordinary differential equation solvers." *SIGNUM Newsletter* 15:10-11.

Hull, T. E., *et al.* [1972]. "Comparing numerical methods for ordinary differential equations." *SIAM J. Numer. Anal.* 9:603-637.

Hull, T. E. *et al.* [1976]. *User's Guide to DVERK — A Subroutine for Solving Nonstiff ODEs.* Report 100. Department of Computer Science, University of Toronto, Toronto.

Hussels, H. G. [1973]. *Schrittweitensteuerung bei der Integration Gewöhnlicher Differentialgleichungen mit Extrapolationverfahren.* M. Sc. Thesis. Universität Köln, Germany.

Hyman, J. M. [1976]. *Method of Lines Solution of Partial Differential Equations.* Report C00-3077-139. Courant Institute of Mathematical Sciences, New York University, New York.

Krogh, F. T. [1971]. *Suggestions on Conversion (with Listings) of the Variable Order Integrators VODQ, SVDQ, and DVDQ.* Jet Propulsion Laboratory, California Institute of Technology Technical Memo 278.

Krogh, F. T. [1975]. *Preliminary Usage Documentation for the Variable Order Integrators SODE and DODE.* Jet Propulsion Laboratory, California Institute of Technology Computing Memo 399.

Lindberg, B. [1973]. *IMPEX2: A Procedure for Solution of Systems of Stiff Differential Equations.* Department of Information Processing Report TRITA-NA-7303. Royal Institute of Technology, Stockholm, Sweden.

Petzold, L. [1980]. *Automatic Selection of Methods for Solving Stiff and Nonstiff Systems of Ordinary Differential Equations.* Sandia Laboratories Report SAND80-7328.

Prothero, A. [1980]. "Estimating the accuracy of numerical solutions to ordinary differential equations." *Computational Techniques for Ordinary Differential Equations.* Ed. I. Gladwell and D. K. Sayers. Academic Press, New York.

Schiesser, W. E. [1976]. *DSS/2: An Introduction to the Numerical Method of Lines Integration of Partial Differential Equations.* 2 Vols. Lehigh University, Bethlehem, Pennsylvania.

Seifert, P. [1980]. "Software für steife Differentialgleichungen." *Numerische Behandlung gewöhnlicher Differentialgleichungen.* T. H. Karl-Marx-Stadt, Sektion Mathematik Rept. 20. Karl-Marx-Stadt, D.D.R.

Shampine, L. F. [1974]. "Limiting precision in differential equation solvers." *Math. Comp.* 28:141-144.

Shampine, L. F. [1978]. "Solving ODEs with discrete data in SPEAKEASY." *Recent Advances in Numerical Analysis.* Ed. C. de Boor and G. H. Golub. Academic Press, New York.

Shampine, L. F. [1980]. "Lipschitz constants and robust ODE codes." *Computational Methods in Nonlinear Mechanics.* Ed. J. T. Oden. North-Holland, Amsterdam.

Shampine, L. F. [1981a]. *Stiffness and the Automatic Selection of ODE Codes.* Sandia Laboratories Report SAND81-0525.

Shampine, L. F. [1981b]. "Type-insensitive ODE codes based on implicit A-stable formulas." *Math. Comp.* 36:499-510.

Shampine, L. F., and C. W. Gear [1979]. "A user's view of solving stiff ordinary differential equations." *SIAM Review* 21:1-17.

Shampine, L. F., and M. K. Gordon [1975]. *Computer Solution of Ordinary Differential Equations: The Initial Value Problem.* W. H. Freeman, San Francisco.

Shampine, L. F., and H. A. Watts [1976]. "Global error estimation for ordinary differential equations." *ACM Trans. on Math. Soft.* 2:172-186, 200-203.

Shampine, L. F., and H. A. Watts [1977]. "The art of writing a Runge-Kutta code, part I." *Mathematical Software III.* Ed. J. R. Rice. Academic Press, New York.

Shampine, L. F., and H. A. Watts [1979]. "The art of writing a Runge-Kutta code, Part II." *Appl. Math. and Comput.* 5:93-121.

Shampine, L. F., and H. A. Watts [1980]. *DEPAC — Design of a User Oriented Package of ODE Solvers.* Sandia Laboratories Report SAND79-2374.

Shampine, L. F., *et al.* [1976]. "Solving nonstiff ordinary differential equations — the state of the art." *SIAM Review* 18:376-411.

Watts, H. A. [1980]. "Survey of numerical methods for ordinary differential equations." *Electric Power Problems: The Mathematical Challenge.* Eds. A. M. Erisman, K. W. Neves, and M. H. Dwarakanath. SIAM, Philadelphia.

Watts, H. A. [1981]. *Starting Step Size for an ODE Solver.* Sandia Laboratories Report SAND80-1734.

Chapter 7

SOURCES OF INFORMATION ON QUADRATURE SOFTWARE

D. Kahaner
Scientific Computing Division
National Bureau of Standards *
Washington, D.C. 20234

INTRODUCTION

Quadrature concerns itself with the problem of evaluating a definite integral:

$$I(f) = \int_R f(x)\,dx \ .$$ (1)

This paper is about the numerical approximation of $I(f)$. Equation (1) is very general. We restrict our attention to cases where

- R is an interval on the real line or a hyperrectangle in s dimensional space,

- $f(x)$, the integrand, is available for calculation via formula or procedure almost everywhere in R (as opposed to a preselected set of data values), and

- High-quality Fortran programs are readily available for obtaining estimates in one or more of these subcases.

Our intent is to pull together information that is useful to people who compute, i.e., solve problems. Thus we emphasize software-related issues at the expense of analytic techniques, although, where appropriate, we will point to some analytic techniques. Consequently, we concentrate on one-dimensional techniques and results. The development of methods for multidimensional integrals has been going on for many years. There are a large number of very specialized programs for high-dimensional integration on some particular problems. Similarly, many clever and insightful techniques have been incorporated into programs to make them run faster. However, very few of these techniques have been translated into programs that can be called general-purpose software, and few library-quality programs are available for this general problem. We expect this situation to change dramatically within the next two to three years, and we touch upon this later.

ONE-DIMENSIONAL QUADRATURE

We use the term *quadrature* to refer to Eq. (1), as opposed to *integration,* which we use to mean solution of a differential equation. The concepts are related because the initial value problem

$$\frac{dy}{dx} = g(x,y) \quad y(0) = a \tag{2}$$

can be rewritten as

$$y(x) = \int_0^x g(t,y(t))\,dt + a, \tag{3}$$

suggesting a quadrature. Some early techniques for approximating (1) and approximately integrating (2) were similar, but modern methods have developed differently (see also Chapter 6). Software for (2) is not recommended for approximating (1).

Anyone studying the subject of quadrature should be aware that the number of potential references is staggering. In the references at the end of this chapter, we mention a few basic textbooks, important survey papers, works on automatic quadrature, journals in which new results and programs are published, and software collections or libraries containing good quadrature software.

QUADRATURE FORMULAS

We first consider approximations to (1) of the form

$$\int_A^B f(x)\,dx = \sum_{i=1}^n W_i f(x_i) + R_n f$$
$$= Q_n f + R_n f . \tag{4}$$

Here the W_i, x_i, and $R_n f$ are called quadrature weights, quadrature nodes, and remainder, respectively. For formulas of this type to be useful, the nodes and weights cannot depend on f, so they need be tabulated only once. The remainder $R_n f$ does, of course, depend on f. It is never known for any real problem except in terms of derivatives or norms of the integrand function f. Even then, usually only a bound is available. For example, Simpson's point rule [Davis and Rabinowitz, 1975:46] can be written in the form (4) for n odd,

$$\int_A^B f(x)\,dx = \frac{B-A}{3(n-1)} [f(A)+4f(A+h)+2f(A+2h)+4f(A+3h)$$
$$+2f(A+4h)+...+2f(B-2h)+4f(B-h)$$
$$+f(B)]+R_n f, \tag{5}$$

with $h = \dfrac{B-A}{n-1}$.   Here we take $W_1 = \dfrac{B-A}{3(n-1)}$, $W_2 = 4\dfrac{B-A}{3(n-1)}\dots, x_1 = A$,
$x_2 = A+h$, $\cdots$, and $|R_n f| \leqslant M \cdot (B-A)^5/(180(n-1)^4)$, $|f^{(4)}(x)| \leqslant M$.

Not only does Equation (5) illustrate (4) but, by examining the bound for $|R_n f|$, we see that by increasing the number of points, n, $R_n f$ goes to zero rapidly. That is, Equation (5), which in reality represents a family of formulas depending on the parameter n, converges at least for functions smooth enough to have a bounded fourth derivative. We say that Simpson's rule has fourth order convergence and often write this $R_n f \sim n^{-4}$. Actually, convergence of a sequence of quadrature formulas is usually no great feat. We can see from (5) that Simpson's rule is actually a Riemann sum. As such, convergence is ensured for any Riemann-integrable function. This is a large class and includes step functions and functions such as $x^{\frac{1}{2}}$ or $x \log(x)$ which have infinite derivatives.

Convergence is also ensured for some improper integrals, e.g., $\displaystyle\int_0^1 x^{-\frac{1}{2}} dx$, as long as we omit $f(A)$ in (5). Nevertheless, as the rate of convergence (rapidity of $R_n f$'s decrease toward zero) is strongly affected by the smoothness of the integrand, Equation (5) is not recommended for functions that are not smooth.

Another common formula is the classical $n+1$ point trapezoidal rule,

$$\int_0^1 f(x)\,dx = \frac{1}{n}\,[\frac{1}{2}\,f(0)+f(\frac{1}{n})+f(\frac{2}{n})+\dots+f(\frac{n-1}{n})+\frac{1}{2}\,f(1)]$$

$$-\frac{f''(\zeta)}{12n^2}\qquad 0<\zeta<1$$

$$= T_n f + R_n f,$$

which indicates second order, $1/n^2$, convergence for functions with a bounded second derivative. But if $f(x)$ is periodic with period 1 and sufficiently smooth, much more rapid convergence is common. This can be seen by expanding such a periodic f in a Fourier series

$$f(x) = \sum_{-\infty}^{\infty} C_j e^{2\pi\, ijx}$$

and applying T_n termwise. We get

$$\int_0^1 f(x)\,dx - T_n f = R_n f = C_n + C_{-n} + C_{2n} + C_{-2n} \dots.$$

Now a standard result [Lyness, 1970b] is that if periodic f has at least k continuous derivatives, then these coefficients go to zero rapidly. In fact, $C_m \sim m^{-k}$, so $R_n f \sim 1/n^k$ rather than only $1/n^2$.

What if, as is common, $f(x)$ is not periodic on $[0,1]$? Let $\bar{f}(x)$ be the periodic continuation of f, i.e., $\bar{f}(x)=f(x)$, $0<x<1$, and $\bar{f}(x+1)=\bar{f}(x)$ for all x. We see from the above remarks that the trapezoidal rule is expected to work better when $\bar{f}(x)$ has a high degree of continuity. Consider $f_1(x)=f(|2x-1|)$ and $f_2(x)=(6x-6x^2)\cdot f(x^2(3-2x))$. It is easy to check that $I(f)=I(f_1)=I(f_2)$ on $[0,1]$. But $\bar{f}_2$ has higher continuity than $\bar{f}_1$ which has higher continuity than $\bar{f}$. Thus, $T_n f_2$ or $T_n f_1$ ought to be more accurate than $T_n f$.

Convergence of a sequence of quadrature formulas explains what happens in the limit as the number of nodes increases without bound. A convergence rate, used loosely such as $R_n f \sim n^{-4}$ for Simpson's rule, relates errors in successive members of the sequence, *in the limit*. Such rates are very useful guides to the selection of formulas. They can sometimes be helpful in explaining performance. Of course, in practice we use one or at most a few formulas. Often there is no analytic knowledge about the integrand. We can then only be cautiously optimistic that the values of n selected are large enough that successive numerical estimates have errors behaving as suggested by the convergence rate.

By examining the bound on $R_n f$ in (5), we note that $R_n f = 0$ if $f^{(4)}(x)=0$, i.e., if $f(x)$ is a polynomial of degree 3 or less. Formula (4) is said to be of polynomial order d if it is exact whenever $f(x)$ is a polynomial of degree d or less but not exact for some polynomial of degree $d+1$. This notion, which develops naturally from (5), is the basis of many far-reaching generalizations of (4). Here we mention two important ones: Gaussian quadrature and Kronrod quadrature.

Gaussian Quadrature

We give up (at first) consideration of (4) as a Riemann sum and concentrate only on the concept of polynomial order. Thus, consider the n W's and n x's as free parameters and seek to determine them to give the resulting formula the highest possible order, $2n-1$. It can be shown that this problem has a unique solution [Stroud and Secrest, 1966; Engels, 1980]. The nodes and weights also satisfy

$A < x_i < B$

$W_i > 0$.

Both are pleasing but were not part of the original specifications. Furthermore, the resulting formula can, in fact, be thought of as a Riemann sum [Szego, 1959] (this is not obvious). However, it can be shown that neither the W's nor the x's are simple numbers; either a program or a table is necessary to provide them. Thus, some physical insight is lost. Nevertheless, these formulas are highly recommended and are known as Gauss-Legendre quadrature [Gautschi, 1982].

An important generalization is obtained by considering the integrand as a product

$$\int_A^B w(x)f(x)\,dx = \sum_{i=1}^n W_i f(x_i) + R_n f \, , \qquad (6)$$

where $w(x)$ is a given non-negative integrable function, usually referred to as a weight function, which is known *a priori*. As the $w(x)$ carries some essential information about the integrand, that information is incorporated into the W's and x's. Thus, a part of the integration is essentially done explicitly. For example, with $(A,B)=(0,\infty)$, exponentially decaying integrands are common. A potentially useful formula is then of the form

$$\int_0^\infty e^{-x}f(x)\,dx = \sum_{i=1}^n W_i f(x_i) + R_n f \, . \qquad (7)$$

This formula, known as Gauss-Laguerre quadrature [Stroud, 1974:125], has polynomial order $2n-1$; i.e., if $f(x)$ is any polynomial of degree $2n-1$ or less, $R_n f = 0$.

Many useful weight functions $(w(x) = 1, e^{-x}, e^{-x^2}, \sqrt{1-x^2})$ lead to Gaussian quadrature formulas with easily calculated W's and x's. Others can present computational problems because of instability, overflow, etc.

The disadvantages of Gaussian quadrature are

- Possible difficulty obtaining weights and nodes.
- Non-physical interpretation of formulas.
- Distinctness of nodes for each n.

By distinctness of nodes, we mean that if we wish to compute a sequence of estimates by successively increasing the number of points, then each different n corresponds to n new evaluation points $x_i = x_i^{(n)}$.

Kronrod Quadrature

Most of the time in numerical quadrature programs is spent evaluating the integrand function. Computing a sequence of estimates, or at least two estimates, is of practical value in assessing the accuracy of computations. In Gaussian quadrature the cost of using an n and $n+1$ point formula is $2n+1$ integrand evaluations. The higher order formula is not much more accurate than the lower; thus, computing an error estimate is as expensive as computing a quadrature estimate. Two elegant and useful ideas to obtain more accurate estimates with the same amount of computational effort have been provided by Patterson [1968] and Kronrod [1965]. Kronrod's idea was to begin with Gauss quadrature formula (6), using n points and with $w(x)=1$, and then construct another formula using these same points as well as $n+1$ others:

$$\int_A^B w(x)f(x)\,dx = \sum_{i=1}^{n} c_i f(x_i) + \sum_{j=1}^{n+1} d_j f(\xi_j) + R_{2n+1}f. \tag{8}$$

The points ξ_j and weights c_i, d_j are carefully selected to achieve a formula of polynomial order $(3n+1)$. The result is a pair of estimates at the cost of only the higher order one. The difference between these estimates can then be used to estimate the quadrature error. The Gauss quadrature pair $(n,\ n+1$ points) and Gauss-Kronrod pair $(n,\ 2n+1$ points) both have the same function evaluation cost of $(2n+1)$ points, but the $2n+1$ point Kronrod estimate is usually far more accurate than the $n+1$ point Gauss one. Patterson generalized this by beginning with a 3-point Gaussian quadrature formula, adding 4 Kronrod points, and then continuing by adding 8 more points (total of 15), then 16 more points (total of 31), and so on up to a formula of 511 points. This approach provides a mechanism for obtaining a sequence of improved estimates without wasting any previously computed function values.

These formulas, referred to as Gauss-Kronrod or Gauss-Kronrod-Patterson, have not yet been completely analyzed. But in several of the most important practical cases all the necessary points and weights have been computed. In fact, within the SLATEC Library (see Chapter 11) the following subroutines are available:

Single Precision		Double Precision
QK15	,	DQK15
QK21	,	DQK21
QK31	,	DQK31
QK41	,	DQK41
QK51	,	DQK51
QK61	,	DQK61 (9)

These are simple evaluation routines to form an approximation of type (4) using the Gauss-Kronrod formulas with 15, 21, 31, 41, 51 and 61 total points. In addition, the routines

$$\text{QNG} \quad , \quad \text{DQNG} \tag{10}$$

implement the Gauss-Kronrod-Patterson algorithm.

USING QUADRATURE FORMULAS

Anyone using a table of quadrature formulas will at some time be confronted with the situation that the formulas are given on one interval, often $[-1,1]$ or $[0,1]$, and the interval of interest is, say, $[A,B]$, where both A and B are finite. What is required is a simple change of variable, and the programs mentioned above implement that automatically. However, for completeness we give the formula as (13) below. Not doing this transformation correctly is one of the most common errors made by unsophisticated users.

If the integral of interest is

$$\int_A^B g(x)\,dx, \tag{11}$$

and the quadrature formula is given as

$$\int_\alpha^\beta f(x)\,dx \approx \sum_{i=1}^n W_i f(x_i), \quad \alpha,\beta \text{ finite}, \tag{12}$$

then

$$\int_A^B g(x)\,dx \approx \frac{B-A}{(\beta-\alpha)} \sum_{i=1}^n W_i g\left(\frac{[B-A]x_i+\beta A-\alpha B}{\beta-\alpha}\right). \tag{13}$$

Such a linear change of variable leaves the degree and other properties of the formula unchanged. On the other hand, if either A or B is infinite, then (13) must be replaced with some other transformation which of necessity is nonlinear. For example, the transformation $x \to x/(1-x)$ gives

$$\int_0^\infty g(x)\,dx = \int_0^1 \frac{g(x/(1-x))}{(1-x)^2}\,dx \equiv \int_0^1 f(x)\,dx.$$

The character of f depends both on that of g and on the transformation. Say, as is often the case, that $g(x) = e^{-x} G(x)$ and $G(x) = O(x^k)$ as $x \to \infty$ (G goes to infinity like x^k for large x). Then $f(x)=O((1-x)^{k+2} \exp(-1/(1-x)))$ as $x \to 1$, so that f goes to zero rapidly as $x \to 1$. For an example of another transformation, see Haber [1977].

In many programs an increase in accuracy is obtained not by applying a higher order formula to the original interval but by applying the same formula, transformed as (13), to each half (or third) of the interval and adding the contributions. If this is done with a formula such as (5), which has equally spaced nodes, then all the integrand evaluations can be reused. For Gauss or Gauss-Kronrod formulas, however, no evaluations can be reused.

AUTOMATED QUADRATURE PROGRAMS

Automated quadrature programs take as input the interval endpoints A, B and a tolerance ϵ, or sometimes ϵ_A and ϵ_R (ϵ-absolute and ϵ-relative), and attempt to produce an estimate as accurate as required. Currently, with very few exceptions all such programs are of the adaptive type. By that we mean that the points of evaluation of the integrand are not fixed in advance, but depend on the particular function being integrated.

We illustrate below a typical user-written main program and external function which calls a generic packaged automatic quadrature routine QUAD:

Program 1: Direct Communication

```
C      BEGIN MAIN PROGRAM — DIRECT COMMUNICATION
       EXTERNAL F
       A=
       B=
       EPS=
       CALL QUAD (F,A,B,EPS, RESULT, other
         parameters)
       PRINT *, RESULT, etc.
         .
         .
         .
       END

C      USER'S FUNCTION BEGINS HERE
       REAL FUNCTION F(X)
         .
         .
         .
       RETURN
       END
```

Note that the quadrature routine QUAD knows about the integrand only by sampling it. One important conclusion to be drawn is that no such routine can be foolproof, since at non-sampled points the function can have arbitrarily bad behavior. That is, there is no way to prevent a user from calling QUAD with an integrand function F for which it was not designed. This is different from, say, the sine function, where it is relatively straightforward to determine if the user is attempting to evaluate the function with an unreasonable argument.

We call the above organization a Direct Communication quadrature routine. Direct Communication is the easiest to use, although the EXTERNAL statement is a source of error for many inexperienced programmers (don't forget to include it). Almost all current routines use Direct Communication.

An alternative form is Reverse Communication.

Program 2: Reverse Communication

```
      C     BEGIN MAIN PROGRAM — REVERSE COMMUNICATION
            A=
            B=
            EPS=
            IFLAG = 0
      10    CALL QUAD (X,FX,A,B,EPS,IFLAG, RESULT, other
              parameters)
            IF (IFLAG .EQ. 2) GO TO 20
            FX = F(X)
            GO TO 10
      20    PRINT *, RESULT, etc.
```

In Reverse Communication the integrand function is called not by QUAD but by the user's main program. The directive to "evaluate the integrand at x" is provided to the user by QUAD via the parameter IFLAG, along with the point *X*. This organization makes it harder for users to code but eliminates the need for the EXTERNAL statement, keeps the calculation much more tightly under the user's control, and provides additional flexibility. We think Reverse Communication will soon be an accepted approach to automatic quadrature.

It is useful to trace the history of automatic quadrature programs to appreciate the various algorithms in common use. Some of the early algorithms had their motivation in methods for solving initial value problems for ordinary differential equations.

Consider the differential equation problem

$$y' = f(x,y) \quad y(A) = a \quad A \leqslant x \leqslant B \tag{14}$$

with the solution approximated by Euler's method [Lapidus and Seinfeld, 1971]:

$$y_{n+1} = y_n + h_n f(x_n, y_n) \quad n = 0, 1, \dots \tag{15}$$

$$y_0 = a \quad h_n = x_{n+1} - x_n.$$

It is known that, for sufficiently smooth f, if $y_n = y(x_n)$, then

$$y_{n+1} - y(x_{n+1}) = \frac{h_n^2}{2} y''(\xi_n) \quad x_n < \xi_n < x_{n+1}. \tag{16}$$

As $y''(x) = \dfrac{\partial f(x,y) y'(x)}{\partial y} + \dfrac{\partial f(x,y)}{\partial x}$, we can approximate $y''(\xi_n) \approx y''(x_n)$ in principle, from the differential equation above. This method, although primitive, suggests a variable step marching procedure: Step from x_n to a further point x_{n+1} by (15), choosing h_n to keep the magnitude of our error estimate (16) sufficiently small.

Below we describe the earliest (known to us) adaptive quadrature program, one that was installed on an IBM 7090 at the CERN Laboratory in Geneva. (This and subsequent algorithms are presented without error traps and other important details.)

Program 3: An ODE-like Adaptive Quadrature

Input: $[A,B]$, ϵ, "F"

Output: $Q \approx \int F$, $R \approx |Q - \int F|$

$\alpha = A$, $\beta = B$, $Q = 0$, $R = 0$, $L = B - A$

while $\alpha < \beta$:
   $G_6 = 6$ point Gaussian quadrature of F
   on $[\alpha,\beta]$
   $G_5 = 5$ point Gaussian quadrature of F
   on $[\alpha,\beta]$
   $E = |G_6 - G_5|$

   if $E < \dfrac{\beta - \alpha}{L} \cdot \epsilon$:

      $\alpha = \beta$
      $\beta = B$
      $Q = Q + G_6$
      $R = R + E$
   else:
      $\beta = \dfrac{\alpha + \beta}{2}$

end

This algorithm clearly has the same sense of marching left to right as does (16). Although seemingly primitive, such algorithms are still being used effectively. In fact, ODEQ and DODEQ in the PORT library (see Chapter 13) use an initial value solver directly to integrate a sequence of functions.

A somewhat more modern version which is in the spirit of subsequent developments was given by McKeeman [1963], originally in ALGOL 60. The Fortran version, SIMP [Shampine and Allen, 1973], is very close to the standard 1970 one-dimensional adaptive quadrature program currently available in many places [Forsythe, Malcolm, and Moler, 1977; Haskell, Vandevender, and Walton, 1980; Lyness, 1969]. To allow flexibility, we describe it next in more generality.

With each interval I_j (subinterval of $[A,B]$) we associate three parameters Q_j, E_j, and ACTIVE_j of the form

$$\text{real } Q_j \approx \int_{I_j} f$$

$$\text{real } E_j \approx |\int_{I_j} f - Q_j|$$

logical $ACTIVE_j$ (that is, .TRUE. or .FALSE.).

Q_j and E_j are known as the local quadrature estimate and local error estimate, respectively. We are thinking here that thee local estimates are "easy" to compute; that is, they do not involve any complex strategy. (In the notations of the CERN algorithm we have $Q_j = G_6$, $E_j = E$.) We denote a cell to mean an interval I_j defined by its endpoints and these three parameters. Using this terminology, we consider the following pseudo-algorithm:

Program 4: Model 1970 Adaptive Quadrature

Input: $[A,B]$, ϵ_A, ϵ_R "F"

Output: $Q \approx \int F$, $R \approx |Q - \int F|$

      initialize: Make $[A,B]$ into a cell:
                        $I_1 = [A,B]$, $Q_1 = 0.$, $E_1 = 1.$,
                        $\text{ACTIVE}_1 = $ .TRUE.
                    $Q = 0.$, $R = 1.$

     while   ACTIVE intervals exist:
          Subdivide one ACTIVE interval into p parts,
              called parent and children, respectively.
              Make each child ACTIVE and parent not ACTIVE.
          Update Q,R.
          for each new ACTIVE cell I_k, test:
              if E_k "small," then make I_k not ACTIVE.
     end

Comments and Variations

1. **Initialization:** The initialization is quite arbitrary. We take $Q_1 = 0$, $E_1 = 1$, for convenience, although most software will use more realistic local estimates.

2. **Subdivide step:** The recent implementations subdivide into two equal parts; McKeeman's original algorithm trisected each interval. A few codes subdivide into two unequal parts. This step actually involves not only cutting the interval but supplying for each child, I_k, the local quadrature and error estimates Q_k, E_k. In SIMP these two estimates are obtained by using Simpson's 3-point rule [Davis and Rabinowitz, 1975:45] and Simpson's 7-point rule, at a cost of only 7 integrand evaluations. Further, the subsequent subdivision of an interval into thirds means that the local estimates on each third cost only 4 new evaluations. Thus, there is a natural synergism between the subdivision strategy and the evaluation rules used for the local estimates. There are two disadvantages to this approach: 1) extra storage and a data structure are needed to save the function values so that they can be reused later (but also see comment #5 below), and 2) it relies on formulas based upon equal spacing rather than more accurate ones such as the Gauss-Kronrod formulas discussed previously. A minor problem is that the integrand will be evaluated at the interval end points. Thus the user must decide what to do with a function that is infinite there — a common situation. Current thinking is to use formulas such as Gauss-Kronrod requiring integrand evaluations that cannot be reused in any subinterval. This approach avoids all the difficulties mentioned here, and the inherent accuracy of these formulas compensates for the fact that evaluations are used only once and then are thrown away.

3. **Update Q,R:** This is usually easily accomplished by subtracting from Q and R the local estimates of the parent and adding in those of the children.

4. **Error test:** The software designers have implemented a variety of things here. Common ones are

> "is $E_k < \epsilon_A$?" absolute error test,
>
> "is $E_k < \epsilon_A \cdot \dfrac{|I_k|}{B-A}$?" length prorated absolute error test,
>
> "is $E_k < \epsilon_R \cdot |Q|$?" approximate relative error test,
>
> "is $E_k < \max(\epsilon_A, \epsilon_R|Q|)$?" combined absolute-relative error test.

Of these, the last is probably the most useful, although it leaves many users with an uncomfortable feeling of loss of intuition.

Another useful enhancement is to compute Q,

$$Q \approx \int_A^B |F|,$$

which can be done at the same time as Q at little extra cost, and to use it instead of $|Q|$ in the error tests. In the practical implementations of these tests, some extra coding is included to take into account the finite word length of all computers. Such modifications may be substantial [Lyness, 1970a], are important in the code's overall performance and robustness, and occur in other places in the algorithm. Nevertheless, for expediency, we omit discussion of them here.

5. **Selection of which ACTIVE cell to subdivide:** All the error tests so far are termed *local* because each cell either passes or fails independently of any other. It should be clear that, apart from minor changes in the estimate Q (or $\bar{Q}$), whether a cell will pass or fail the error test is predetermined. That is, the order in which cells are examined plays no role in the final result. Thus, the selection of a cell to subdivide is related entirely to ease of implementation and not to algorithmic efficiency. The workhorse adaptive programs of this type use the rule "subdivide the leftmost ACTIVE cell." The resulting program is efficient of storage and is fairly easy to code [Forsythe, Malcolm, and Moler, 1977]. Cells made not ACTIVE are thrown away in most implementations.

6. **While step:** Comment 5 (which states that the ordering of subintervals is irrelevant to the calculation) is counter-intuitive and suggests the alternative in the "while":

> While $R < \epsilon_A$
>
> (or $R < \epsilon_R \cdot |Q|$,
>
> or $R < \max(\epsilon_A, \epsilon_R |Q|)....$).

With this implementation the calculation may terminate with ACTIVE cells, e.g., cells that do not pass the error test, as long as the total error is sufficiently small. The order in which ACTIVE cells are subdivided now matters. Two possibilities are to order the cells either to make the largest reduction in R at each step or to leave the most difficult ones to the end. Implementing either of these orderings produces a modest improvement at some increase in storage costs.

7. **Banking** [Forsythe, Malcolm, and Moler, 1977:108]: An additional and more substantial enhancement than that of Comment 6 can be made by observing that many cells pass their error test handily while some just barely fail. The spirit of Comment 6 is to drive the total error down; thus, we should "save up" any excess error by which a cell passes to use later on cells that fail. This algorithm is very dependent on ordering; clearly we should build up a bank account of excess error early.

Several other kinds of variations are given by Rice [1975].

All the variants described above share one common problem: Unless special coding is included, they work very poorly if the integrand has a point of discontinuity or singularity [Fritsch, Kahaner, and Lyness, 1981]. In fact, they may not terminate. The essential reason for this is that if a cell fails the error test, it is replaced by a pair of cells each half as long. If the parent cell has a discontinuity, one of the children will too. The local error estimate on that child can, in general, be no less than half that of its parent. Unfortunately, it will be called upon to pass an error test roughly twice as strict, causing it to fail too. The solution is to remove the local error test entirely, suggesting a new algorithm:

Program 5: Modern Global Adaptive Quadrature

Input: $[A,B]$, ϵ_A, ϵ_R, "F"
Output: $Q \approx \int F$, $R \approx |Q - \int F|$
    initialize: Make $[A,B]$ into a cell:
                    $I_1 = [A,B]$, $Q_1 = 0.$, $E_1 = 1.$,
                    $\text{ACTIVE}_1 = .\text{TRUE}.$
                    $Q = 0.$, $R = 1.$
    while $R < \max(\epsilon_A, \epsilon_R \cdot |Q|)$:
               Subdivide worst ACTIVE cell into 2 parts.
               Update Q, R.
    end

Such an algorithm generalizes very well to higher dimensions; the sense of "marching" is entirely gone. Furthermore, it is easy to see that the input parameters ϵ_A, ϵ_R play no role in the calculation other than in the stopping criterion. The practical effect is that programs implementing this algorithm can be written to allow continuation to higher accuracy without repeating the original calculation. Such an algorithm has been termed "global" [Malcolm and Simpson, 1975] because there is no sense of an individual cell passing or failing. The non-termination situation mentioned earlier is not a difficulty for this type of algorithm. Global adaptive programs are among the most effective automatic methods for quadrature. However, the cell selected for subdivision is that one with largest E_i, i.e., the "worst." Thus, a "stack" or another data structure such as a "heap" [Kahaner, 1980] must be maintained, and this increases the overhead in the programs. Furthermore, the number of cells is constantly growing, and storage may be a problem for difficult integrals.

The latest evolutionary step in automatic quadrature development has been to add to this type of algorithm extrapolation capabilities [de Doncker, 1978]. These not only enhance the efficiency of the programs but enable them to integrate difficult, singular functions. However, the details are quite technical; we refer the reader to de Doncker [1978] or Kahaner and Stoer [1982].

MULTIDIMENSIONAL QUADRATURE

As mentioned earlier, software for multiple quadrature is in a much less developed state than that for univariate quadrature. In fact, a glance at the programs listed in Table I in the last section of this chapter shows only a total of six readily available. A few Fortran programs have been published in research journals [Robinson and de Doncker, 1982; Friedman and Wright, 1981; Genz and Malik, 1980]. However, these are often difficult to obtain. Sometimes they are not as portable as one would like, or the user interface leaves something to be desired. Furthermore, the support level of these programs by their authors is quite variable. Usually, the best of them or their constituent algorithms are reworked to fit into a library such as IMSL (see Chapter 10), NAG (see Chapter 14), PORT, or SLATEC, resulting in a much more finely polished piece of work. In an archival article such as this, we have focused on these libraries as sources of programs, because we feel that users should consult such libraries first and consider the research programs only as a backup, and much less desirable, strategy. However, as multidimensional quadrature programs have not yet been incorporated into these libraries in any substantial way, we include here a brief description of the basic methods. Much of this material is condensed from the survey article of Haber [1970] or from unpublished notes of Patterson [1981], who is preparing versions of several of these algorithms for the next mark (Mark 9) of the NAG library.

In one-dimensional problems, intervals are the only possible regions, and singularities can occur at a few points at most. Higher dimensional problems involve arbitrary geometry, but common regions are the cube, sphere, simplex, cone, cylinder, orthant, or all space. Singularities occur on lower dimensional hypersurfaces but most often at a point, an edge, or other simple curve. Software, of course, is unlikely to provide this kind of flexibility. Early programs integrated over a cube, and this is still the favorite region, although in two dimensions the triangle (or a union of them) has now become popular.

ITERATED INTEGRALS

If the problem can be written as an iterated integral, then a sequence of one-dimensional programs can be used. For example, if in two dimensions

$$I = \int_R f(x,y)\,dA = \int_\alpha^\beta \int_{\gamma(x)}^{\delta(x)} f(x,y)\,h(x,y)\,dy\,dx, \tag{17}$$

then

$$I = \int_\alpha^\beta g(x)\,dx, \quad g(x) = \int_{\gamma(x)}^{\delta(x)} f(x,y)\,h(x,y)\,dy. \tag{18}$$

A rather complicated region can thus be the domain, although setting up such a problem is very prone to blunder. For example, given two automatic quadrature routines (Direct Communication) QUADA and QUADB, each suitable for a one-dimensional integral, the following program segment illustrates how to calculate a double iterated integral such as (18).

Program 6: Evaluation of 2-D Iterated Integral

```
C   MAIN PROGRAM 2-D ITERATED INTEGRAL
      EXTERNAL G
      ALFA =
      BETA =
      EPSA =
      CALL  QUADA (G, ALFA, BETA, EPSA, RESULT, other parameters)
      PRINT *, RESULT
          .
          .
          .
      END
C
      REAL FUNCTION G(X)
      COMMON X1
      EXTERNAL F
      X1 = X
      GAM = GAMMA(X)
      DEL = DELTA(X)
      EPSB =
      CALL  QUADB (F, GAM, DEL, EPSB, RESULT, other parameters)
      G = RESULT
      RETURN
      END
C
      REAL FUNCTION  F(Y)
      COMMON X
      evaluate  F = f(X,Y) · h(X,Y)  here
      RETURN
      END
```

For this situation, setting the tolerance parameters EPSA and EPSB is far from trivial [Fritsch, Kahaner, and Lyness, 1981]. They must be related not only to the desired ultimate accuracy, EPS, and the characteristics of the individual iterated integrands but also to the routines QUADA and QUADB. If these routines are duplicates of each other, then an acceptable strategy is to pick EPSA = EPS/2 and EPSB = EPS/$(2(\beta-\alpha))$.

PURE MULTIDIMENSIONAL METHODS

The adaptive algorithms described previously can usually be applied in a multidimensional setting with an appropriate change in the definition of a "cell." However, in addition to various kinds of domains, any software that employs a subdivision strategy has one or more natural kinds of "cells" which ought to be understood by the end user. For example, a typical subdivision strategy might divide a square into four congruent subsquares. Such a strategy is unlikely to work efficiently on an integrand function $f(x,y)$ which has very different behavior along orthogonal coordinate lines, e.g., $f(x,y) = \sqrt{x}$. Similarly, if f has a singularity along the line $y=x$, triangular subdivision will work very differently from square subdivision. Also, an initial dissection of the unit square into two congruent triangles can be done in two distinct ways. Finally, within the triangular subdivision algorithms we can expect substantial variations in the effort required to solve a problem, depending on the starting dissection.

One idea that appears consistently in multidimensional quadrature is the abandonment of polynomial order as a useful measure of the potential effectiveness of a given quadrature rule. Thus, in the univariate case we saw that rules were classified according to the highest degree polynomial that they could integrate exactly, with the implication (usually borne out in practice) that higher degree rules were "better." This amounts to thinking in terms of expansion of the integrand locally in Taylor series. Alternatively, we could consider expansion in Fourier series, which proved fruitful for studying the trapezoidal rule.

In one dimension, the use of Fourier expansions, or at least expansions in orthogonal functions, has been popularized as Clenshaw-Curtis quadrature [Clenshaw and Curtis, 1960]. However, these expansions have never really been in the mainstream of either research or software. But in higher dimensions, low degree polynomials do not describe multivariate functions well, while high degree polynomials have a great many coefficients. Furthermore, a technique such as Gaussian quadrature (which is based on the zeros of orthogonal polynomials) does not generalize nicely to multivariate problems. Currently, polynomial-based algorithms are more common in lower dimensional programs, especially in two or three dimensions where they are also linked with a subdivision strategy. These algorithms are often extensions of one-dimensional schemes.

Software for very high dimensional integrals almost never uses polynomials. In those cases Korobov or Monte Carlo methods described below are more popular. These are characterized by low or no subdivision and little overhead other than function evaluation.

Korobov Quadrature

We have seen that the trapezoidal rule $T_n f$, when applied to a function which is periodic on $[0,1]$ and has k continuous derivatives, can be surprisingly accurate; in fact, $R_n f \sim 1/n^k$ [Lyness, 1970; Haber, 1977].

There is an analogous result if $f(x)$ is a multivariate function [Hsu, 1961], but we must be much more careful in counting to assess its true meaning. Let $f(x)$ be periodic in each variable (on the unit cube in s dimensions) and also sufficiently smooth, with k continuous partial derivatives in each variable. Then using T_n for each variable,

$$T_n f = \frac{1}{n^s} \sum_{r_1, r_2, \ldots, r_s = 1}^{n} f\left(\frac{r_1}{n}, \frac{r_2}{n}, \ldots, \frac{r_s}{n}\right), \tag{19}$$

we get $R_n f \sim 1/n^k$. However, the actual number of function evaluations is proportional to n^s. Calling this number (which measures the amount of work) N, we find that the error goes to zero only as $N^{-k/s}$ — not very rapidly for large s.

Korobov rules [Korobov, 1963; Conroy, 1967; Hlawka, 1964] may be thought of as generalizations of the trapezoidal rule which replace the factor $N^{-k/s}$ by

$$\frac{(\log N)^{k \cdot (s-1)}}{N^k}. \tag{20}$$

This is accomplished by using not the s-dimensional trapezoidal rule but a rule of the form

$$K_N^a f = \frac{1}{N} \sum_{r=1}^{N} f\left(\frac{ra_1}{N}, \frac{ra_2}{N}, \ldots, \frac{ra_s}{N}\right). \tag{21}$$

The point $a = (a_1, a_2, \ldots, a_s)$ is selected to give the rule its error term (20). The computation of a is tedious but needs to be done only once for each s; these points can then be tabulated in a program that actually evaluates the rule. The points $\frac{1}{N} a$, $\frac{2}{N} a, \ldots, a$ are equally spaced on a ray in s space; but since the function being integrated is periodic, the effect is to evaluate the integrand on a slanted mesh in the unit cube.

A difficulty with Korobov rules is that there is no natural error bound associated with the quadrature estimate. The error term (20) is asymptotic, and $K_N^a f$ has no direct means of estimating it in a practical manner without additional computing. Current practice is to compute $K_N^b f$ also (i.e., the Korobov estimate at a different point) and compare these. In fact, many points can be used to produce the same order error term (20). Patterson's program [Patterson, 1981] allows the user the option of choosing a variable number of these points at random from among the "good" ones and thereby obtaining an error estimate based on the variance of these different quadrature estimates.

Monte Carlo Quadrature

Monte Carlo methods were developed systematically beginning in the 1940's, although several isolated instances occur much earlier. They are still among the most common methods to evaluate multidimensional integrals. The basic idea is to replace the analytic problem "evaluate this integral" with a probabilistic problem "find the mean of a set of random numbers chosen according to a specific distribution."

There are a number of good expositions on the subject, including Hammersley [1960]; Hammersley and Handscomb [1964]; and Schreider [1966]. We refer the interested reader to these for more background.

The major practical interest in these methods stems from the fact that error estimates, in the form of standard errors, are known to have the form

$$R_n f \sim K N^{-\frac{1}{2}},$$

where K depends upon f, the region and the selection of evaluation points. According to Haber [1970], "This convergence does not seem to be very rapid until we note that neither the dimensionality s of the integration region, nor any specification of the degree of smoothness of f, entered into the determination of the error estimate .... f need not even be continuous."

The evaluation or sampling points play a crucial role in the constant K above. The strategies used (called variance reduction methods) attempt to reduce this constant by choosing the evaluation points so that their density is less uniform and more proportional to $|f|$. Sometimes this is done by dissecting the integration region R into smaller pieces and using a uniform density of points in each piece. The most efficient procedure is to make the same number of evaluations in each subregion but selectively adjust the subvolumes.

Early Monte Carlo programs used a very simple strategy. Currently developing software is beginning to use ideas from one-dimensional or low-dimensional adaptive quadrature. This suggests that future programs will contain both probabilistic sampling techniques and adaptive subdivision strategy.

SOFTWARE IN THE STANDARD LIBRARIES

The first place to look for programs for quadrature is in one of the widely available commercial software libraries. Of these, three seem most useful: IMSL,

NAG and PORT. Alternatively, the SLATEC Library contains routines of similar quality and is available without charge.

Table I describes the subprograms that are available from these sources. The routines are subdivided into categories by the type of problems and basic approach, e.g., automatic vs. non-automatic. The classification scheme is a portion of one used at the National Bureau of Standards to organize currently available mathematical software [Boisvert *et al.*, 1981]. The routine names listed under the NAG library all end with "E/F." This refers to a pair of routines, one ending in E and the other in F, for different precisions. Similarly, a name like (D)ODEQ refers to DODEQ and ODEQ in double and single precision, respectively.

Each of these libraries has numerous routines for numerical evaluation of integrals. These represent the state-of-the-art in portable software.

Table I

Standard Libraries or Subprograms
Available from NAG, PORT, SLATEC, and IMSL

CATEGORY: D1a   Compute weights and nodes for quadrature formulas.

LIBRARY	NAG	
	D01BBE/F:	Weights and abscissas for Gaussian quadrature rules, restricted choice of rule using precomputed weights and abscissas.
	D01BCE/F:	Weights and abscissas for Gaussian quadrature rules, more general choice of rule calculating weights and abscissas.
	PORT	
	(D)GAUSQ:	Weights and abscissas for Gaussian quadrature on (a,b) for a general weight function with known moments.
	(D)GQ0IN:	Weights and abscissas for Gauss-Laguerre quadrature on the interval $(0,\infty)$.
	(D)GQM11:	Weights and abscissas for Gauss-Legendre quadrature on the interval $[-1,1]$.

CATEGORY: D1b1a1   Automatic finite-interval quadrature for a general user-defined integrand.

LIBRARY   SLATEC
    (D)QAG: Automatic adaptive integrator, will handle many non-smooth integrands using Gauss-Kronrod formulas.

    (D)QAGE: Automatic adaptive integrator, will handle many non-smooth integrands, provides more information than QAG.

    (D)QAGS: Automatic adaptive integrator, will handle most non-smooth integrands including end-point singularities, uses extrapolation.

    (D)QAGSE: Automatic adaptive integrator, will handle most non-smooth integrands including end-point singularities, provides more information than QAGS.

    (D)QNG: Automatic non-adaptive integrator, will handle most smooth integrands.

    IMSL
    DCADRE: Automatic adaptive integrator using cautious Romberg extrapolation, will handle many non-smooth integrands.

    NAG
    D01ACE/F: Computes definite integral over finite range to specified relative or absolute accuracy using Patterson's non-adaptive method (in Mark 8 but deleted from Mark 9).

    D01AGE/F: Calculates definite integral over finite range to specified absolute accuracy by Oliver's Clenshaw-Curtis method (in Mark 8 but deleted from Mark 9).

    D01AHE/F: Automatic adaptive quadrature over finite range, suitable for well-behaved integrands.

D01AJE/F: Automatic adaptive quadrature over finite range, will handle badly-behaved integrands.

D01BDE/F: Automatic non-adaptive quadrature over finite range.

PORT
(D)ODEQ: Integrates a set of smooth functions over a finite range by using the initial-value solver 0DES1.

(D)QUAD: Automatic adaptive quadrature of function over a finite range to specified absolute accuracy.

(D)RQUAD: Automatic adaptive quadrature of function over a finite range to specified accuracy, combined absolute and relative error control.

CATEGORY: D1b1a2   Non-automatic finite interval quadrature for general user-defined integrand.

LIBRARY   SLATEC
(D)QK15:   Evaluates integral of a given function on a finite interval with a 15-point Gauss-Kronrod formula and also returns error estimate.

(D)QK21:   Evaluates integral of a given function on a finite interval with a 21-point Gauss-Kronrod formula and also returns error estimate.

(D)QK31:   Evaluates integral of a given function on a finite interval with a 31-point Gauss-Kronrod formula and also returns error estimate.

(D)QK41:   Evaluates integral of a given function on a finite interval with a 41-point Gauss-Kronrod formula and also returns error estimate.

(D)QK51:   Evaluates integral of a given function on a finite interval with a 51-point Gauss-Kronrod formula and also returns error estimate.

(D)QK1:   Evaluates integral of a given function on a finite interval with a 61-point Gauss-Kronrod formula and also returns error estimate.

NAG
D01BAE/F: Evaluates integral of a given function by evaluating a Gauss quadrature formula.

CATEGORY: D1b1b2   Non-automatic finite-interval quadrature for general grid-defined integrand.

LIBRARY   PORT
(D)CSPQU: Finds the integral of a function defined by pairs (x,y) of input points using spline interpolation. The x's can be unequally spaced.

NAG
D01GAE/F: Finds the one-dimensional integral of a function defined by data values only.

CATEGORY: D1b2a1   Automatic finite-interval quadrature for user-defined integrands of special forms, including weight functions, oscillating and singular integrands, Principal Value integrals, and splines.

LIBRARY   SLATEC
(D)BFQAD: Automatic adaptive integration of a general function times derivative of a B-spline on a finite interval. Spline is in B representation.

(D)PFQAD: Automatic adaptive integration of a general function times derivative of a B-spline on a finite interval. Spline is in piecewise polynomial representation.

(D)QAG0:   Automatic adaptive integration of a general function times oscillatory sine or cosine factor on a finite interval.

(D)QAG0E: Automatic adaptive integration of a general function times oscillatory sine or cosine factor on a finite interval, provides more information than QAG0.

(D)QAGP: Automatic adaptive integrator of function with known user-specified points of singularities or difficulties of integrand.

(D)QAGPE: Automatic adaptive integration of function with known user-specified points of singularities or difficulties of integrand, provides more information than QAGP.

(D)QAWC: Automatic adaptive integrator for Cauchy Principal Value integrals, uses Clenshaw-Curtis technique (real Hilbert transform).

(D)QAWCE: Automatic adaptive integrator for Cauchy Principal Value integrals, uses Clenshaw-Curtis technique, provides more information than QAWC.

(D)QAWS: Automatic adaptive integrator for functions with explicit algebraic and/or logarithmic end-point singularities.

(D)QAWSE: Automatic adaptive integrator for functions with explicit algebraic and/or logarithmic end-point singularities, provides more information than QAWS.

(D)QMOMO: Automatic integrator of Tchebycheff polynomial times weight function. A specific selection of weight functions with various singularities is permitted.

NAG

D01AKE/F: Automatic adaptive integrator of oscillating function on finite interval.

D01ALE/F: Automatic adaptive integrator of oscillating function on finite interval, allowing for singularities at user-specified points.

D01ANE/F: Automatic adaptive integrator of a function times weight function $\cos(\omega x)$ or $\sin(\omega x)$ over a finite interval.

D01APE/F: Automatic adaptive integrator of a function with algebraic-logarithmic endpoint singularities over a finite interval.

D01AQE/F: Automatic adaptive integrator for Cauchy Principal Value integrals (real Hilbert transform).

PORT
(D)BQUAD: Automatic adaptive integrator of a function with discontinuities in derivatives; user can specify locations of these points.

CATEGORY: D1b2a2:   Non-automatic finite-interval quadrature for special user-defined integrand, including weight functions, oscillating and singular integrands, Principal Value integrals, and splines.

LIBRARY   SLATEC
(D)QC25C: Clenshaw-Curtis 25-point formula used to estimate finite-interval integral of $F(x)/(x-c)$, also returns error estimate.

(D)QC25F: Either Clenshaw-Curtis formula or 15-point Gauss-Kronrod formula used to estimate integral of $F(x)$ times $\sin(\omega x)$ or $\cos(\omega x)$, also returns error estimate.

(D)QC25S: Clenshaw-Curtis 25-point formula used to estimate finite-interval integral of function with algebraic-logarithmic singularities, also returns error estimate.

(D)QK15W: 15-point Gauss-Kronrod formula used to estimate integral of $F(x)\,W(x)$ for arbitrary weight function W, also returns error estimate.

CATEGORY: D1b3a1   Automatic semi-infinite interval quadrature for user-defined integrand.

LIBRARY   SLATEC
(D)QAGI: Automatic adaptive integrator for infinite and semi-infinite intervals, uses nonlinear

transformation and extrapolation.

(D)QAGIE: Automatic adaptive integrator for infinite and semi-infinite intervals, uses nonlinear transformation and extrapolation, provides more information than QAGI.

(D)QAWF: Automatic integrator for Fourier integrals on (a,∞) with factors $\sin(\omega x)$ or $\cos(\omega x)$ by integrating between zeros.

(D)QAWFE: Automatic integrator for Fourier integrals on (a,∞) with factors $\sin(\omega x)$ or $\cos(\omega x)$ by integrating between zeros, provides more information than QAWF.

NAG
D01AME/F: Automatic adaptive integrator for infinite or semi-infinite intervals, uses nonlinear transformation and extrapolation.

CATEGORY: D1b3a2 Non-automatic semi-infinite-interval quadrature for user-defined integrand.

LIBRARY SLATEC
(D)QK15I: Evaluates integral of a given function on semi-infinite or infinite interval with transformed 15-point Gauss-Kronrod formula, also gives error estimate.

NAG
D01BAE/F: Evaluates integral of a given function by evaluating a Gauss quadrature formula.

CATEGORY: D1b5a1 Automatic infinite-interval quadrature for user-defined integrand.

LIBRARY SLATEC
(D)QAGI: Automatic adaptive integrator for infinite or semi-infinite intervals, uses nonlinear transformation and extrapolation.

(D)QAGIE: Automatic adaptive integrator for infinite or semi-infinite intervals, uses nonlinear transformation and extrapolation, provides more information than QAGI.

NAG
   D01AME/F: Automatic adaptive integrator for infinite or semi-infinite intervals, uses nonlinear transformation and extrapolation.

CATEGORY: D1b5a2   Non-automatic infinite-interval quadrature for user-defined integrand.

LIBRARY   SLATEC
   (D)QK15I: Evaluates integral of given function on semi-infinite or infinite interval with transformed 15-point Gauss-Kronrod formula, also gives error estimate.

NAG
   D01BAE/F: Evaluates integral of a given function by evaluating a Gauss quadrature formula.

CATEGORY: D1c1a1   Automatic hyperrectangle quadrature for user-defined integrand.

LIBRARY   IMSL
   DBLINT: Automatic numerical quadrature of a function of two variables over a rectangular region.

NAG
   D01FAE/F: Automatic numerical quadrature for multidimensional integrals over a hyperrectangle by Monte Carlo method.

   D01FCE/F: Adaptive automatic quadrature for multidimensional integrals over a hyperrectangle.

   D01DAE/F: Automatic quadrature for two-dimensional integrals over a finite region.

CATEGORY: D1c1a2   Non-automatic hyperrectangle quadrature for user-defined integrand.

LIBRARY   NAG
              D01FBE/F: Quadrature for multidimensional integrals
                          over a hyperrectangular region by Gaussian
                          rule evaluation.

CATEGORY: D1c1b2   Non-automatic hyperrectangle quadrature for grid-
                       defined integrand.

LIBRARY   IMSL
              DBCQDU: Bicubic spline quadrature.

REFERENCES

Boisvert, R., S. Howe, D. Kahaner, J. Knapp-Cordes, and M. Knapp-Cordes
[1981]. *Guide to Available Mathematical Software,* DoC, NBS, Center for
Applied Mathematics, Washington, D.C.

Clenshaw, C., and C. Curtis [1960]. "A method for numerical integration on an
automatic computer." *Numer. Math.* 2:197-205.

Conroy, J. [1967]. "Molecular Schrodinger equation VIII." *J. Chem. Phys.*
47:5307-5318.

Davis, P., and P. Rabinowitz [1975]. *Methods of Numerical Integration.*
Academic Press, New York.

de Boor, C., and J. Rice [1979]. "An adaptive algorithm for multivariate approximation giving optimal convergence rates." *J. Approx. Th.* 25:337-359.

de Doncker, E. [1978]. "An adaptive extrapolation algorithm for automatic
integration." *SIGNUM Newsletter* 13:12-18.

de Doncker, E., and R. Piessens [1976]. "A bibliography on automatic integration." *J. Comp. Appl. Math.* 2:273-280.

Dixon, V. [1974]. "Numerical quadrature. A survey of the available algorithms." *Software for Numerical Mathematics.* Ed. D. J. Evans. Academic
Press, New York.

Engels, H. [1980]. *Numerical Quadrature and Cubature.* Academic Press, New
York.

Forsythe, G., M. Malcolm, and C. Moler [1977]. *Computer Methods for Mathematical Computations.* Prentice Hall, Englewood Cliffs, New Jersey.

Friedman, J., and M. Wright [1981]. "A nested partitioning procedure for numerical multiple integration." *ACM Trans. on Math. Soft.* 7:76-92.

Fritsch, F., D. Kahaner, and J. Lyness [1981]. "Double integration using one-dimensional adaptive quadrature routines: a software interface problem." *ACM Trans. on Math. Soft.* 7:46-75.

Gautschi, W. [1982]. "A survey of Gauss-Christoffel quadrature formulae." *SIAM Rev.* To appear.

Genz, A., and A. Malik [1980]. "Remarks on Algorithm 006: an adaptive algorithm for numerical integration over an N-dimensional rectangular region." *J. Comp. Appl. Math.* 6:295-302.

Ghizzetti, A., and A. Ossicini [1970]. *Quadrature Formulas.* Academic Press, New York.

Haber, S. [1970]. "Numerical evaluation of multiple integrals." *SIAM Rev.* 12:481-526.

Haber, S. [1977]. "The $\tanh$ rule for numerical integration." *SIAM J. Numer. Anal.* 14:668-685.

Hammersley, J. [1960]. "Monte Carlo methods for solving multivariate problems." *Ann. N.Y. Acad. Sci.* 86:844-874.

Hammersley, J., and D. Handscomb [1964]. *Monte Carlo Methods.* Wiley, New York.

Haskell, K., W. Vandevender, and E. Walton [1980]. *The SLATEC Common Mathematical Subprogram Library: SNLA Implementation.* Sandia National Laboratories Report SAND80-2792.

Hlawka, E. [1964]. "Uniform distribution modulo 1 and numerical analysis." *Compositio Math.* 16:92-105.

Hsu, L. [1961]. "Note on the numerical integration of periodic functions and of partially periodic functions," *Numer. Math.* 3:169-173.

Kahaner, D. [1980]. "Algorithm 561, Fortran implementation of Heap programs for efficient table maintenance." *ACM Trans. on Math. Soft.* 6:444-449.

Kahaner, D., and J. Stoer [1982]. "Extrapolated adaptive quadrature." *SIAM J. of Sci. and Statis. Computing.* To appear.

Korobov, N. [1963]. *Number Theoretic Methods of Approximate Analysis.* Fizmatgiz, Moscow.

Kronrod, A. [1965]. *Notes and Weights of Quadrature Formulas.* Consultants Bureau, New York.

Lapidus, L., and J. Sinfeld [1971]. *Numerical Solution of Ordinary Differential Equations.* Academic Press, New York.

Lauri, D. [1982]. "CUBTRI: automatic cubature over a triangle," *ACM Trans. on Math. Soft.* 8(2):210-218.

Lyness, J. [1969]. "Notes on the adaptive Simpson quadrature routine." *JACM* 16:483-495.

Lyness, J. [1970a]. "Algorithm 379 - SQUANK." *CACM* 13:260-263.

Lyness, J. [1970b]. "The calculation of Fourier coefficients by the Mobius inversion of the Poisson summation formula, Pt. I." *Math. Comp.* 24:101-135.

Lyness, J. [1972]. "Guidelines for automatic quadrature routines." *Information Processing '71.* Ed. C. V. Freeman. North Holland, Amsterdam, pp. 1351-1355.

Lyness, J., and B. Ninham [1967]. "Numerical quadrature and asymptotic expansions." *Math. Comp.* 21:162-178.

Malcolm, M., and R. Simpson [1975]. "Local versus global strategies for adaptive quadrature." *ACM Trans. on Math. Soft.* 1:129-146.

McKeeman, W. [1963]. "Certification of Algorithm 145, adaptive numerical integration by Simpson's rule." *CACM* 6:167-168.

Patterson, T. [1968]. "The optimum addition of points to quadrature formulae." *Math. Comp.* 22:847-856.

Patterson, T. [1981]. Notes (unpublished) on Multiple Quadrature, Queens University, Belfast, BT7 1NN, Northern Ireland.

Rice, J. R. [1975]. "A metalgorithm for adaptive quadrature." *JACM* 22:61-82.

Robinson, I. [1975]. "A comparison of numerical integration programs." *J. Comp. Appl. Math.* 5:207-223.

Robinson, I., and E. de Doncker [1982]. "Automatic integration over a bounded or unbounded planar region." To appear.

Schreider, Y., ed. [1966]. *The Monte Carlo Method.* Pergamon Press, New York.

Shampine, L., and R. Allen [1973]. *Numerical Computing: An Introduction.* Saunders, Philadelphia.

Stroud, A. H., and D. Secrest [1966]. *Gaussian Quadrature Formulas.* Prentice Hall, Englewood Cliffs, New Jersey.

Stroud, A. H. [1971]. *Approximate Calculation of Multiple Integrals.* Prentice Hall, Englewood Cliffs, New Jersey.

Stroud, A. H. [1974]. *Numerical Quadrature and Solution of Ordinary Differential Equations.* Springer-Verlag, New York.

Szego, G. [1959]. *Orthogonal Polynomials.* American Mathematical Society, New York.

Chapter 8

A SURVEY OF SPARSE MATRIX SOFTWARE

Iain S. Duff
Computer Science and Systems Division
A.E.R.E. Harwell, Oxon
United Kingdom

INTRODUCTION

For this survey of sparse matrix software, we have concentrated on codes that are readily available, are well-documented and are easily transported to a wide range of computers. We make limited reference to uncoded algorithms and experimental versions of codes in order to highlight some areas of algorithmic research that are likely to produce high-quality software in the near future.

In the area of sparse matrix research, it is difficult to make a distinction between algorithm and code development since the coding itself often vitally affects the performance of the algorithm and since the developing and testing of an algorithm usually requires the writing of an experimental code. Thus, the development of sparse matrix software occurs in two phases: the design of an algorithm together with an experimental code, and the conversion of that code into high-quality software. The time lag between these two phases can be very great. Although we primarily discuss the final software product, we also consider some experimental codes, particularly when there is no more polished software in the same area or when the version for general release is imminent. In the instances when we have anticipated the appearance of a piece of software, we have reported in the best of faith the situation as described by the author; but we know from our own experience how easy it is to underestimate code development times. Thus readers should beware of staking their future on the prompt arrival of a promised code. If nothing else, however, the "to appears" do give pointers to the likely future development of sparse software.

We have intentionally avoided actual comparisons of code. In many areas only very few comparisons of software have been conducted; and in others (for example, Duff [1979], Duff and Reid [1979]) it is a highly specialized enterprise, with the difficulties only just being appreciated. We have, however, some familiarity with the codes described (or their authors) and believe that the software we discuss is of high quality. Perhaps one effect of this survey will be to develop an awareness of what software is available and to stimulate future development and further comparisons of existing codes.

In most instances when we describe software, we give its source, cost, and restrictions on use (when known). Often the source is simply the author or authors referenced, the cost is either low or nonexistent, and the restrictions are slight. Another principal source is the *Collected Algorithms of the ACM* (usually after the code has appeared in the *ACM Transactions on Mathematical Software*) when the software is in the public domain and the costs are essentially a handling charge. Codes from the Harwell Subroutine Library are similarly inexpensive, and the only real restrictions concern their commercial use (which is agreed on a one-to-one basis). The *Computer Physics Communications* works on a subscription system, although individual codes can also be requested; most other journals, for example, *Computing* and the *International Journal of Numerical Methods in Engineering,* publish listings only and recommend that the author be contacted for machine-readable versions. The two main proprietary libraries have little sparse software: IMSL has nothing in its mathematical library (except band solvers) but does act as a distribution agent for some codes (for example, ITPACK); and NAG has a few sparse codes (identified in the text), some of which come from its collaboration with Harwell. In conjunction with a Sparse Matrix Symposium which will be held at Fairfield Glade in Tennessee, a catalog on sparse matrix software has recently been produced. This is available from Heath or Ward of the Computer Science Division of Oak Ridge National Laboratory. Inclusion in the catalog was dependent on the author's or vendor's completing a catalog form; the information was thus provided solely by the author or distributor.

We also commented on the portability of the software described. When we speak explicitly of portable Fortran codes, we mean that they satisfy the PFORT verifier [Ryder, 1974]. However, unless a code is described as machine-dependent, we believe it should be a fairly easy matter to transport it to a reasonable set of machine ranges.

An area that we have intentionally avoided is sparse matrix techniques embedded in large packages. There are many such large applications codes — for example, for finite element calculations on structures (e.g., SAP, NASTRAN or STRUDL) and economics [Szyld, 1981b], but we feel these merit a separate study of their own and that the best route to familiarity with these packages is through the applications discipline involved. Here we are concerned primarily with sparse codes that can be accessed independently of such larger entities. A short discussion of sparse matrix packages is given in the last section.

We first discuss the solution of sparse linear systems, considering direct methods and iterative methods. Since much of the recent stimulus for research on sparse systems has come from the solution of elliptic partial differential equations, we examine software for these problems in a separate section, concentrating on that part of the codes for solving the sparse equations and stressing techniques not explicitly covered by the earlier sections. Next, we examine software for the solution of sparse eigensystems, sparse least-squares problems, nonlinear systems, and linear programming problems. The final section is devoted to remarks on sparse matrix packages and other applications.

We would like to reiterate that this survey primarily considers codes and not algorithms. Thus we have kept the discussion of underlying methods and theory to a bare minimum. Readers with no previous background are encouraged to consult an appropriate survey in this area (for example, Duff [1977b]). Details on surveys and books containing background material on each area can be found in the appropriate section.

DIRECT SOLUTION OF LINEAR EQUATIONS

In this section we consider software for the solution of the sparse linear equations

$$A\underline{x} = \underline{b} \tag{1}$$

based on Gaussian elimination. That is, we consider an implicit or explicit factorization of the form

$$PAQ = LU, \tag{2}$$

where P and Q are permutation matrices and L and U are lower and upper triangular matrices, respectively. When A is symmetric, we can use the decomposition

$$PAP^T = LDL^T, \tag{3}$$

where P and L are as before and D is a diagonal matrix (for indefinite matrices A, D may need to be block diagonal with blocks of order 1 and 2).

We divide our discussion of software for the direct solution of (1) according to the properties of the matrix A. First we examine codes applicable when A is of a special structure (for example, banded). We then describe software applicable when A is symmetric positive definite or when no numerical pivoting is required. This is followed by a discussion of codes available when A is a general matrix. We conclude by considering cases when A is not formed explicitly but is generated a row at a time or as a sum of submatrices, as is common in finite element applications. A critical analysis of several features available in direct codes and a comparison of some of the codes discussed here are given by Duff [1979].

Band matrices are a particularly simple example of sparse matrices, and codes for solving both symmetric and unsymmetric band systems can be found in almost all software libraries (NAG, IMSL, Harwell) and collections (LINPACK). The particular case of tridiagonal systems is often handled by a different code from that for matrices with a wider band.

A more general class of sparse matrices is that of variable band matrices [Jennings, 1966], where the first non-zero in each row and column need not be the same distance from the diagonal. Some of the standard orderings on grid-based problems give matrices suitable for variable band storage and elimination schemes. Although the number of non-zeros in the LU factors for such schemes is usually higher than for the more general methods we will discuss later, variable band methods have the merit of being very simple and requiring much less storage overhead than the general schemes. Another benefit of both the band and variable band matrices is that only a relatively small part of the matrix need be in main storage at any one time. The NAG routines F01MCF and F04MCF solve positive definite systems of variable band structure holding the matrix and factors in-core, while the Harwell Subroutine Library code MA15 has the option of using auxiliary storage. Perry [1981] is developing a collection of subroutines (BESS) for auxiliary storage band elimination on symmetric positive definite linear systems. The same type of structure is also termed "profile" or "skyline," and several codes implementing these schemes have appeared in journals (for example, Thompson and Shimazaki [1980], Hasbani and Engelman [1979], Felippa [1975]). The SPARSPAK package [George, Liu, and Ng, 1980]), which we will discuss shortly, also has an option for profile elimination.

When the matrix is symmetric positive definite, then the computation of the permutation matrix P in (3) can be completely decoupled from the numerical decomposition. This has led to the development of extremely efficient algorithms both for the determination of the ordering P and for its use in the subsequent creation of data structures to facilitate numerical factorization. The most common method of choosing an ordering is that of minimum degree. This algorithm chooses as pivot, at each stage, the diagonal entry from a row with the least number of non-zeros, where fill-ins from previous elimination steps are included in the non-zero count. Remarkable advances have been made in recent years in the implementation of the minimum degree ordering. For example, on problems arising from the finite element triangulation of the L-shaped region, the codes that we will discuss shortly run over 100 times faster than codes considered supreme twelve years ago.

The SPARSPAK package of George, Liu, and Ng [1980] has a particularly carefully designed user interface and offers several options for the ordering P:

1. Reverse Cuthill-McKee (profile elimination),
2. One-way dissection,
3. Refined quotient tree,
4. Nested dissection, and
5. Minimum degree.

A storage factorization and solution scheme appropriate to the particular ordering is used in each case. An excellent discussion of these orderings and their merits can be found in the book by George and Liu [1981]. Great care has been taken by the authors of SPARSPAK to screen the user from the internal data structures. Although SPARSPAK consequently has a commendably easy user interface, some efficiency has been sacrificed and the codes can be difficult to modify for use in application packages. SPARSPAK is available at fairly low cost (currently $600 Can.) from George at Waterloo, and the licensing agreement is not very restrictive. The Yale Sparse Matrix Package (YSMP) of Eisenstat *et al.* [1977 and 1982] uses a minimum degree ordering to compute P, although, in common with most of their codes, a user-specified ordering can be entered. YSMP can be ordered from Eisenstat at Yale for a nominal handling charge (currently $75) and has few restrictions on use. Eisenstat [Eisenstat, Schultz, and Sherman, 1979] is currently developing a code (MSSE) whose storage requirements are substantially reduced by discarding some of the factors and regenerating them when required. The MSSE code works on any ordering but performs best on nested dissection or minimum degree orderings. In an approach that is similarly based on viewing the decomposition as an elimination tree, Duff and Reid [1982b] have designed Harwell subroutine MA27 with a very fast minimum degree ordering and an option for obtaining a stable decomposition of A when the matrix is indefinite. Additionally, MA27 has been designed to take advantage of vectorization on machines like the CRAY-1. A comparison between SPARSPAK, YSMP, and MA27 is given by Duff and Reid [1982d].

Both SPARSPAK and the Yale package have an option for solving systems whose coefficient matrix is unsymmetric but has a symmetric structure. A version of MSSE has been developed for this case also. In these codes, pivots are chosen from the diagonal with no numerical restrictions, and so the decomposition could be unstable. MA27 has been designed so that routines for obtaining a stable decomposition in this structurally symmetric case could be incorporated in the package, and this work is currently in progress [Duff and Reid, 1982d].

The Yale package has a version that factors unsymmetric matrices using diagonal pivoting without numerical control, employing an ordering supplied by the user or generated from applying the minimum degree ordering to the symmetric pattern whose upper triangle has the same structure as the upper triangle of the coefficient matrix. In order, however, to obtain a stable decomposition for general unsymmetric A, numerical pivoting must be used, and it is normal to combine the ordering and factorization phases. The NSPIV code of Sherman [1978] performs partial pivoting by columns within a given row ordering. His code has the merit of being short and produces a very stable decomposition, but it is difficult to choose a good ordering for the rows. For many orderings there can be considerable fill-in, and the computation times can be high. The code

MA28 in the Harwell Subroutine Library [Duff, 1977a] selects its pivots by combining the Markowitz [1957] sparsity criterion with threshold pivoting, where a_{lk} can be used as a pivot only if

$$|a_{lk}| \geqslant u.\max_i |a_{ik}|$$

where u is a user-set parameter in the range (0,1]. This code also incorporates an optional preordering to block triangular form. A version of this code is in the NAG library as subroutines F01BRF, F01BSF and F04AXF. There is a corresponding code, ME28, in the Harwell Subroutine Library for sparse complex equations. The current release of NSPIV (NSPFAC, available directly from the author) also employs threshold pivoting to gain some sparsity over the original NSPIV code. Zlatev, Barker, and Thomsen [1978] use threshold pivoting with Markowitz's criterion in their code SSLEST. They additionally include options for restricting the pivot search to a specified number of rows and for dropping non-zeros below a specified value from the sparsity structure. The use of drop tolerances coupled with iterative refinement can greatly reduce storage requirements for the factors L and U when solving (1). This approach is used in the codes SIRSM [Zlatev and Nielsen, 1977] and Y12M [Zlatev, Barker, and Thomsen, 1978], and work is under way to incorporate such a feature in the MA28 package. This use of drop tolerances is rather similar to the incomplete factorization semi-direct methods that we will discuss in the next section, although there the iterative scheme normally used is conjugate gradients rather than iterative refinement. The SLMATH code [IBM, 1976] based on work of Fred Gustavson uses full code when the reduced matrix becomes full and has a fast factorization entry (see, for example, Duff [1979]). An experimental version of MA28 [Duff, 1982], which switches to full code before the reduced matrix becomes completely full, has proved very successful, particularly on the CRAY-1 where the innermost loop of the full code vectorizes. It is planned to include a switch to full code in a future release of the MA28 package. Kaufman [1982] has modified both YSMP and NSPFAC and will be incorporating these together with some utility routines in a future release of the PORT library (see Chapter 13).

The last class of matrices we consider are those that are not formed explicitly but are entered row by row or element by element in a finite element calculation. As we mentioned earlier, the band and variable band matrices are well suited to row-by-row input, but the first generally available code for element input was that of Irons [1970] for the frontal solution of positive definite element matrices. This was extended by Hood [1976], who allowed the assembled matrix to be unsymmetric. The MA32 package [Duff, 1981] from the Harwell Subroutine Library extends a robust and stable version of an unsymmetric frontal scheme to allow the input of a general unsymmetric matrix by rows. This latter mode of input is particularly designed for the direct solution of unsymmetric problems arising from finite difference discretizations of partial differential equations. Since these frontal codes use full matrix indexing in their innermost loops, it is possible to make them perform well on machines capable of vectorization. For example, the MA32 code running on the CRAY-1 at Harwell has solved equations from discretizations of partial differential equations at a rate (in millions of floating point operations per second) that, even including I/O overhead, is not far short of the rate for solving full systems on the CRAY-1 using Fortran code [Duff and Reid, 1982a]. Sherman (1982) is tailoring a version of MSSE to run on the CRAY-1.

ITERATIVE METHODS

The history and current development of iterative methods are tied strongly to the solution of elliptic partial differential equations. We consider in this section techniques that have more general applicability; discussion of the more specialized applications of iterative methods is deferred to the next section. Recently, there has been a rapid growth in algorithms that accelerate iterative methods by preconditioning the problem. Since effective preconditionings often employ partial factorizations using direct methods, this class of techniques lies midway between direct and iterative methods and is often termed semi-direct. We choose to discuss software based on these techniques in this present section. An excellent description of the theory and algorithms for iterative and semi-direct methods is given in the book by Hageman and Young [1981].

The beauty of iterative methods for solving sparse systems lies in the fact that the matrix factorization of direct methods is avoided, with a consequent saving in storage. Indeed, in some applications, the matrix itself need not be stored; it is sufficient to generate its product with any vector. The problem with basic iterative methods is that convergence is often slow on all but the simplest problems. This is one area in which there is a dearth of generally available high-quality software. One reason is that it is often easier to embed an iterative solution within a larger code. Another is that there are severe difficulties with construction of a user interface and with automatic selection of iteration parameters.

A very ambitious software project for iterative methods is ITPACK [Kincaid *et al.*, 1982; Young and Kincaid, 1981]. The ITPACK codes are written in a portable subset of Fortran and currently offer a unified interface to seven methods for solving linear equations with a sparse symmetric positive definite matrix. The methods available are listed below:

1. Jacobi iteration with Chebyshev acceleration;

2. Jacobi iteration with conjugate gradient acceleration;

3.   Reduced system with Chebyshev acceleration;

4.   Reduced system with conjugate gradient acceleration;

5.   Successive overrelaxation;

6.   Symmetric successive overrelaxation with Chebyshev acceleration; and

7.   Symmetric successive overrelaxation with conjugate gradient acceleration.

The reduced system corresponds to implementing Richardson's method on the system obtained after eliminating all the red nodes of a red-black ordering.

A principal feature of the package is the automatic selection of acceleration parameters and stopping criteria. The software, coded in a portable subset of Fortran, is in the public domain, is available through IMSL for a handling charge of $75, and has appeared in the *ACM Transactions on Mathematical Software*. The package is oriented towards teaching and research, with the emphasis on flexibility rather than maximum efficiency, and so may not be suitable for use in a production environment. Although the routines are designed primarily to solve symmetric positive definite systems and are guaranteed only theoretically to converge on such problems, the interface allows matrices that are not symmetric and the experience of the developers of the package has been that the algorithms work well when the coefficient matrix is nearly symmetric. These routines have been incorporated into the ELLPACK software package [Rice, 1978] for the solution of elliptic partial differential equations in a teaching and research environment. The authors plan to extend the ITPACK package by including robust algorithms for the solution of unsymmetric problems [Young and Jea, 1980].

The conjugate-gradient algorithm for the solution of positive definite systems has a long history [Hestenes and Stiefel, 1952]; but although Reid wrote the code MA16 for the Harwell Subroutine Library in 1971, the method has not been greatly favored on systems arising from the discretization of partial differential equations because it uses more storage and is not significantly faster than traditional methods such as SOR.

Attempts to improve the convergence of the conjugate-gradient method have led to a veritable growth industry in the development of techniques for preconditioning the coefficient matrix to give one whose spectrum is more favourable. The normal effect of preconditioning is to bunch the eigenvalues. A common preconditioning technique is to perform a partial decomposition

$$A = LL^T + C, \tag{4}$$

where the matrix C is not at round-off level and we have omitted permutations for the sake of clarity. The preconditioned matrix

$$L^{-1}AL^{-T} \tag{5}$$

is then used in the conjugate-gradient iteration. It is because of the partial factorization (4) that such methods are often termed semi-direct.

The concept of preconditioning to accelerate iterative techniques has been around for some time (for example, Evans [1967]), but it was not until Meijerink and van der Vorst [1977] demonstrated the stability of simple preconditioning techniques when using conjugate gradients on M-matrices that the current growth industry was established. Van der Vorst and his colleagues [van Kats, Rusman, and van der Vorst, 1980] have developed software in this area which can be obtained free of charge (apart from supplying a tape) from ACCU, Utrecht. Included are codes in portable Fortran for conjugate gradients with or without a user-supplied preconditioning, together with subroutines for use when the matrix is obtained from the five-point formula with possible periodic boundary conditions (the ICCG methods). When the structure of the matrix is more general or when diagonal pivoting is inadvisable, the automatic calculation of preconditioning requires stabilization and the codes are normally much more complicated. The software of Ajiz and Jennings [1981] and the code MA31 of Munksgaard [1980] in the Harwell Subroutine Library stabilize the partial factorization by modifying diagonal entries during the decomposition.

In the symmetric case, the conjugate-gradient method (with or without preconditioning) can break down, and biconjugate gradients [Fletcher, 1976] or a variant of the Lanczos algorithm [Parlett, 1978, or Paige and Saunders, 1975] are commonly used. Code implementing the SYMMLQ algorithm of Paige and Saunders with optional preconditioning is available from Saunders at Stanford. The code of Parlett should be more efficient when the matrix is positive definite (it reduces to the conjugate-gradient algorithm in this case) but does not include any explicit provision for preconditioning. Additionally, Parlett has not yet established a distribution system for his software. Saad [1981a,1981b] has recently developed codes employing a similar approach to SYMMLQ but using an LU factorization, with partial pivoting, of the projection of A in the Krylov subspace. The IOMDRV code of Saad will soon be available directly from the author at Yale University. It should prove more economical than SYMMLQ and have the added advantage that the same code can be used to solve unsymmetric systems. There is, however, no explicit provision for preconditioning in Saad's code.

Although many of the methods used in the symmetric case can be extended to solve unsymmetric systems, the theory and practice of preconditioning is not as well understood. For unsymmetric systems it is possible to form the normal equations (usually implicitly) and solve the resulting symmetric positive definite system, but this is not generally satisfactory for two reasons. First, the amount of computation for each iteration is doubled; and second, the convergence of the iterative method on the normal equations is often very slow.

There are two main approaches on which most software for unsymmetric systems is based; extensions of conjugate gradients is one avenue and the use of Chebyshev iterations another. Many unsymmetric systems can be split,

$$A = M - N,   \tag{6}$$

where the matrix M is symmetric positive definite and N is skew symmetric. A version of the conjugate-gradient algorithm [Concus and Golub, 1975] can then be applied to solve

$$M\underline{x}^{k+1} = N\underline{x}^k + \underline{b},   \tag{7}$$

but we do not know of any easily available software that implements this approach. Another way in which conjugate gradients can be extended is to ensure local conjugacy between recent search directions by explicit orthogonalization. This approach, usually coupled with an incomplete LU factorization as a preconditioning, is used in several large oil reservoir codes (for example, Vinsome [1976]), but we know of no readily available software that implements it. The main problems with this approach lie in the lack of good theoretical results, the difficulties in choosing a good preconditioning, and the amount of reorthogonalization required on realistic problems. Detailed discussions on these topics are presented in recent theses of Jea [1982] and Elman [1982a]. The biconjugate gradient algorithm can be extended to general systems, and Jacobs [1982a and 1982b] has produced software based on this approach. Paige and Saunders [1982] have written a code based on the bidiagonalization procedure of Golub and Kahan [1965] for solving either unsymmetric equations or the least-squares problem. Their code, LSQR, is equivalent to the unsymmetric conjugate-gradient method of Hestenes and Stiefel but is numerically more reliable on ill-conditioned systems. It is possible to use any nonsingular preconditioning matrix C, say, with LSQR (as long as the equations $C\underline{x} = \underline{b}$ and $C^T\underline{x} = \underline{b}$ can be solved); but again the main problem is in calculating a good preconditioning, without which convergence can be slow. The IOMDRV code of Saad [1981a,1981b], which we discussed earlier, can also be used to solve unsymmetric equations, although no explicit provision is made for preconditioning.

Manteuffel [1977] has written a package, TCHLIB, consisting of five subroutines that solve both symmetric and unsymmetric systems using Chebyshev iteration:

TCHSYM: Matrix symmetric positive definite.

TCHSYP: Matrix symmetric; symmetric preconditioning matrix.

TCHFPS: Matrix unsymmetric; symmetric part used as preconditioner.

TCHEB: Matrix unsymmetric.

TCHEBP: Matrix unsymmetric; preconditioning used.

These routines, written in portable Fortran, are in the public domain and will be distributed by the National Energy Software Center at Argonne. An important feature of the routines is that the optimal iteration parameters are determined automatically. However, the user must supply the preconditioning matrix C ($C=I$ when no preconditioning is used), which must be such that the eigenvalues of $C^{-1}A$ lie in the right-hand complex plane. The ADACHE code of van der Vorst and van Kats [1979] with an optional stabilized partial LU preconditioning for five-diagonal systems also implements Manteuffel's algorithms in the case where all the eigenvalues of A lie in the right half-plane.

Both Elman [1982b] at Yale and Markham [1982] at CERL, Leatherhead, England, are working on ambitious packages employing preconditioned conjugate-gradient-like methods on a large range of matrices. Other active workers in this area include Widlund [1978], O'Leary [1979], Axelsson [1980] and Gustafsson [1979], although their codes are largely experimental rather than for general release.

ELLIPTIC PARTIAL DIFFERENTIAL EQUATIONS

The solution of elliptic partial differential equations is a vast subject. We refer the reader particularly interested in this area to Machura and Sweet [1980], who have written a recent survey of software for partial differential equations. In this section, we concentrate on the solution of the sparse linear equations that arise after the discretization of such problems. There will, of course, be some overlap with the material in the previous two sections; but our emphasis here is quite different and some specialized methods not applicable to the general case will be discussed.

Although direct methods by themselves are not practical for really large systems on general geometries, nearly all large production codes (particularly in the oil industry) have a direct solver option for use as a benchmarking tool, because of its robustness, and as the preferred method for small to medium two-dimensional problems. Most of the codes described above for linear equations are suitable, with the out-of-core frontal (e.g., Harwell's MA32) and variable band methods (e.g., Harwell's MA15) applicable even for quite large systems, especially if the geometry is long and narrow. Direct methods are particularly useful when the coefficient matrix is not symmetric or diagonally dominant, an uncommon occurrence in model problems but very common in many practical situations (for example, in fluid-flow calculations).

For many years classical iterative techniques like Jacobi, SOR, and their extensions were used to solve the discrete problem; but no codes implementing them were published, and researchers wrote their own independent versions. Additionally, problems of parameter selection and user interface made the production of high-quality software difficult. It is only recently that such methods have been readily available in a software package. We discussed such a package, called ITPACK, in the previous section and need not comment further except to note that it is designed as a teaching and research tool and thus might be very useful for the purposes of comparing algorithms.

The strongly implicit procedure of Stone [1968] is geared to the structure obtained from finite difference discretizations of elliptic equations. Jacobs [1980] has developed a series of routines DI205, DI209, DI213 and DI307 implementing this method for 5-, 9-, and 13-diagonal systems derived from 2-dimensional models and for 7-diagonal systems and derived from 3-dimensional models, respectively. The subroutines DI205 and DI307 are also available from NAG as subroutines D03EBF and D03ECF, respectively. Jacobs has also developed SIP codes for use when, in addition to the diagonal structure, a few extra non-zeros or a number of small non-zeros are present; however, these algorithms exhibit very slow convergence on large problems (see, for example, Jacobs [1981]). Recently, Jacobs [1982a and 1982b] has written preconditioned conjugate-gradient codes for symmetric (CG25, CG37), non-symmetric non-positive definite (BCG25, BCG37) and complex systems (CCG25, CCG37). These latter routines use variants of the biconjugate gradient algorithm and rely on the diagonal structure arising from the 5- and 7-point formulae in 2 and 3 dimensions, respectively. Another code which uses the regular structure from finite difference discretizations to generate preconditioners for conjugate gradients is that of Meijerink and van der Vorst [1977], which was mentioned earlier. The subroutines ICCG13, ICCG11, ICCG1P of van der Vorst employ $ICCG(1,3)$, $ICCG(1,1)$ preconditionings and $ICCG(1,1)$ applied to problems with periodic boundary conditions, respectively. Kuo-Petravic and Petravic [1981] have a similar ICCG code in the CPC Program Library and have a version [Kuo-Petravic and Petravic, 1979] implementing an unsymmetric preconditioning. The codes of Ajiz and Jennings [1981] and Munksgaard [1980] do not rely on the diagonal structure of the matrix but generate a stable preconditioning automatically. Their codes, however, still require the matrix to be symmetric positive definite. For nonsymmetric systems Manteuffel's Chebyshev codes discussed in the previous section may be used. The ADACHE code of van der Vorst and van Kats [1979] with an optional stable $LU(1,1)$ preconditioning for five-diagonal systems (LUEQ2D and ASPC11) also implements Manteuffel's algorithms in the case where all the eigenvalues of A lie in the right half-plane. Finally, Markham (CERL, Leatherhead) and Elman (Yale) are both developing codes for applying preconditioned conjugate-gradient-like methods to quite general systems.

The very regular structure of Laplace, Poisson or Helmholtz equations give rise to a number of methods termed fast or fast direct methods. They are based on a variety of techniques including marching methods, cyclic and total reduction, and the use of the fast Fourier transform [Swarztrauber, 1977]. A recent survey of Detyna [1981] discusses many of the current approaches. They are termed fast since their computation time varies almost linearly with the number of grid points. An idea of their actual speed on model problems can be gleaned from the GAMM report [Schumann, 1978] on a contest between such solvers. This is an area in which fine tuning has resulted in many experimental codes that are unattractive pieces of software and are difficult to use. However, we can recommend the FISHPAK package [Swarztrauber and Sweet, 1979] which is well-designed and easy to use. It solves Laplace, Poisson or Helmholtz equations in Cartesian and polar coordinates using a combination of cyclic reduction and fast Fourier transforms. It is written in portable Fortran and has appeared in the *ACM Transactions on Mathematical Software*. Software based on the more general and very stable method of total reduction, where unlike cyclic reduction the reduction is in both directions [Schröder, Trottenberg, and Witsch, 1978], has been developed at Bonn and is incorporated in the software of the club MODULEF (INRIA, France). The group from GMD-IMA at Bonn have codes for Helmholtz's equation on a rectangle with Dirichlet [Foerster and Förster, 1978] or Neumann or periodic [Förster, 1980] boundary conditions and for the modified biharmonic equation

$$\Delta \Delta u - b \Delta u + c u = s$$

with boundary conditions

$$-\Delta u = f$$

and

$$u = g$$

over a rectangle [Förster and Reutersberg, 1979]. Another package by Swarztrauber [1981], FFTPACK, contains routines for computing periodic as well as symmetric discrete Fourier transforms and can be used to solve Poisson's equation subject to the standard boundary conditions including Dirichlet and Neumann conditions on opposite boundaries. Van Kats and van der Vorst [1979] also have a code (TBPSXY) implementing cyclic reduction with Buneman stabilization [Buzbee, Golub, and Nielsen, 1970] which can be used to solve Poisson's or Helmholtz's equation on a rectangular domain with a uniform grid and with Dirichlet boundary conditions on two opposite sides and Dirichlet or Neumann conditions on the other sides. Although these fast methods are quite limited in the geometries to which they are applicable, it is sometimes possible to divide the region so that fast methods will work on each part and then use some correction technique to obtain the solution for the whole region. The capacitance matrix method is one way of doing this, and O'Leary and Widlund [1981] have an algorithm (Algorithm 572) in the *ACM Transactions on Mathematical Software* that uses this approach to solve the Helmholtz equation with Dirichlet boundary conditions on general bounded three-dimensional regions.

An approach that is at least as active as research into preconditionings for conjugate gradients is the multigrid approach. Here, an approximation to the solution of the differential equation on a fine grid is improved by a few sweeps (perhaps one) of an iterative technique (called smoothing) and then improved by calculating a correction on a coarser grid after which further smoothing may be performed. The procedure can be nested and a variety of methods used for the smoothing, the restriction and prolongation between grids, and the solution on the coarsest grid. As in any highly active area, many ideas and experimental codes abound, but there is relatively little implementation of efficient and reliable software. Below we report on the codes we know of and suggest that the reader interested in this area subscribe to the multigrid newsletter (free, at present, to Europeans) edited jointly by McCormick (Fort Collins, Colo.) and Trottenberg (Bonn). A report on a recent conference on multigrids is also recommended [Hackbusch and Trottenberg, 1982].

MLAT, an early code of Brandt [1977], is based on finite difference discretizations and can be used to solve general second-order equations in rectangular domains with general boundary conditions. Bank and Sherman [1981] use a triangulation of linear finite elements in their PLTMG code for solving general variable-coefficient elliptic second-order equations on arbitrary bounded planar domains. Their package can be used interactively on many computing systems and has a good graphics interface. The MGOO package of Foerster and Witsch [1982], some of which is incorporated in the MODULEF package, is an ongoing software project that currently solves the problems

$$-\Delta u = f$$

$$-\Delta u + cu = f$$

$$-au_{xx} - bu_{yy} + cu = f$$

$$-\text{div}(\lambda \text{ grad } u) + cu = f,$$

where a, b, c are variable coefficients on rectangular domains with general boundary conditions. Hackbusch [1976] has written a code, HELM, for the solution of the Helmholtz equation with Dirichlet boundary conditions on an arbitrary bounded domain. In addition, he has a code RECTCF written in Fortran IV (there is an Algol 60 version RECTC) for the solution of a general non-constant coefficient equation with general boundary conditions on a rectangular domain [Hackbusch, 1977]; RECTCF and RECTC are available from the OECD NEA Data Bank. A package, KLEV, with a similar capability is under development at Yale [Douglas, 1982]. There is likely to be a considerable growth of multigrid software in the near future, as is evident from the conference proceedings by Hackbusch and Trottenberg [1982].

ELLPACK [Rice, 1978] represents a major vehicle for comparing codes that solve partial differential equations. Its modular structure allows an easy interface for different methods of solving the discrete equations; indeed, one of the ELLPACK design criteria was to facilitate such comparisons. Currently there are direct, iterative, and semi-direct solvers in the package; and some codes based on multigrids methods are likely to be incorporated in the near future.

EIGENSYSTEMS

In this section we discuss software for the solution of the standard eigen-problem

$$A\underline{x} = \lambda \underline{x} \tag{8}$$

and the generalized eigenproblem

$$A\underline{x} = \lambda B\underline{x}, \tag{9}$$

where A and B are sparse. A number of methods for solving these sparse eigen-problems are given in the review articles by Stewart [1976] and Jennings [1981]. In this paper, we concentrate on the two main approaches: methods based on simultaneous iteration and the use of Lanczos recurrences. It is interesting to reflect that, although excellent methods and codes have been available for many years for eigenproblems where the matrices are full or banded [Peters and Wilkinson, 1969], it is really only very recently that any software for sparse systems has existed.

If B is symmetric positive definite with Cholesky factors LL^T, then (9) can thus be reformulated as

$$L^{-1}AL^{-T}L^T\underline{x} = \lambda L^T\underline{x}.$$

When B can be so factored, any code for the standard eigenproblem (8) can be used to solve the generalized eigenproblem. When we say a code is designed for the generalized eigenproblem, we mean that it can solve (9) without requiring the Cholesky factorization of B. Unless we explicitly state otherwise, A will be symmetric and B symmetric positive definite. Later in this section we will discuss software and algorithms for the unsymmetric problem.

The code EA12 in the Harwell Subroutine Library was developed from the work of Rutishauser [1970] on simultaneous iteration methods for the solution of the standard eigenproblem. This code has a particularly flexible interface so that the user is free to use inverse or direct iteration with or without shifts. A Harwell subroutine for the Hermitian-definite generalized eigenproblem is under development. Nikolai [1979] has published a code in the *ACM Transactions on Mathematical Software* based on Rutishauser's but has generalized it slightly so that it can solve real symmetric-definite generalized eigenvalue problems. Although there is no need for a Cholesky factorization of B, the user must solve linear systems with B as coefficient matrix.

Techniques based on the theoretically more powerful Lanczos method [Parlett, 1980] have been developed even more recently than those described above. Van Kats and van der Vorst [1977] divide the computation into two steps. They first generate the Lanczos tridiagonal matrix for the standard (LSVLAN) or the generalized problem (GENLAN). As with the Nikolai code, only a routine for solving $Bx = c$ is required for the generalized problem. A subsequent call to EVSCAN determines which eigenvalues of this tridiagonal matrix accurately represent eigenvalues of the original system. Although this approach works best when extreme eigenvalues are required, the whole spectrum can in practice be computed if the Lanczos recurrence is allowed to run for long enough, although there is currently no theoretical proof of this assertion. Cullum and Willoughby [1981] use similar techniques to compute the spectrum for the standard problem; their codes in IBM Fortran have recently been made available [Cullum and Willoughby, 1980]. Parlett and Reid [1981] have written a code, EA14, for the Harwell Subroutine Library that finds all or part of the spectrum for the standard eigenproblem using a carefully designed monitoring technique to decide how many Lanczos iterations are required. Gladwell [1982] has modified EA14 to return the eigenvectors also. The above three codes do not orthogonalize with respect to any earlier vectors and so will often require many more iterations than the order of the system if the whole spectrum is desired. A Lanczos code LASO, which employs selective orthogonalization, has been designed and written by Scott [1981a] and is suitable for finding a few eigenpairs of the standard problem. It can be obtained from the Argonne National Energy Software Center as code NESC 918.

The above codes are not able to determine the multiplicities of the eigenvalues unless one determines all of the eigenvectors. There are two approaches that perform this task more efficiently. The first is to use block Lanczos, and the second is to use inverse iteration coupled with an inertia count.

The block Lanczos methods will compute multiplicities up to the size of the block. The block Lanczos code of Underwood [1975; cf. Golub, Underwood, and Wilkinson, 1970] uses full orthogonalization and would be efficient if only extreme eigenvalues are required. This code also provides eigenvectors. Scott [1979] has block versions of his LASO2 code which, like the original code, use selective orthogonalization and hence should be less costly than Underwood's code, particularly if several eigenvalues are required. There are codes in LASO2 for computing a given number of eigenvalues at one end of the spectrum or for computing all those outside a given interval.

If it is possible to factorize $A - \sigma B$, then one can compute multiplicities efficiently, and faster algorithms for obtaining interior eigenpairs can be developed. Ericsson and Ruhe [1980a and 1980b] have written code that uses a sequence of shifted factorizations with the Lanczos algorithm to compute any number of eigenpairs of the symmetric-definite generalized eigenproblem in a user-specified interval; the necessary shifts are calculated by the code itself. Their software will appear in a forthcoming issue of the *Transactions on Mathematical Software*. Scott is also working on a code, SILA, based on a similar approach. Not only are the inertia counts (obtained by counting the number of negative pivots in Gaussian elimination on $A - \sigma B$) important for determining multiplicities, but also they can be used to detect if any eigenvalue has been missed. This might happen if the starting vector(s) are in a subspace orthogonal to the eigenvector of the desired eigenvalue. The normal cure is to restart with different vectors. This is done automatically in the programs of Ericsson-Ruhe and Scott which we have previously discussed.

Scott [1981c] is designing a code NOFAC that can be used to find a few eigenpairs of the symmetric-definite generalized eigenproblem without requiring any matrix factorizations or the solution of linear equations. The code, which will be distributed by the NESC, uses an algorithm that combines Rayleigh quotient estimates for the eigenvalues with the Lanczos algorithms on $A - \sigma B$.

Scott and Ward are currently developing software to find a few eigenpairs of the quadratic λ-matrix problem

$$(M\lambda^2 + C\lambda + K)\underline{x} = \underline{0}, \tag{10}$$

where M, C, K are symmetric, M is negative definite and K is positive definite. It is, of course, possible to transform λ-matrix problems to generalized eigenproblems of higher order, but Scott and Ward do not do this explicitly. They have two approaches depending on whether $(M\sigma^2 + C\sigma + K)$ can be factored [Scott, 1981b] or not [Scott and Ward, 1982]. The NOFAC package, mentioned earlier, will also include a capability for using the algorithm of Scott and Ward [1982] to obtain a few eigenpairs of (10) of smallest or largest magnitude without factoring.

There is much less software available for the unsymmetric eigenproblem. Stewart and Jennings [1981] have written a simultaneous iteration algorithm for real matrices; the code, which is called LOPSI, treats the standard unsymmetric eigenproblem obtaining either the right or left eigenvectors. Parlett and Taylor [1982] are developing a Lanczos method for the simple unsymmetric eigenproblem and have an experimental working code. Saad [1980] has developed theory and algorithms based on the method of Arnoldi, but again his codes are experimental.

The traditional method for obtaining the singular value decomposition destroys sparsity in a similar way to the classical reduction of matrices to condensed forms in eigensystem analysis. However, Golub, Luk, and Overton [1981] have suggested a technique using block Lanczos on an augmented system which will produce the singular values and corresponding vectors without destroying sparsity. The technical report by the authors [Golub, Luk, and Overton, 1977] has a code implementing the algorithm. The user need only form matrix-vector products using the matrix and its transpose. The code is recommended for computing a few greatest singular values of a matrix.

Finally, many large packages have sparse eigenvalue routines embedded within them. For example, Lewis has modified some of the large structures packages to incorporate his early work [Lewis, 1977] on Lanczos methods, and Scott has implemented similar embeddings with some of his codes which we discussed earlier.

LEAST-SQUARES PROBLEMS

There is currently much interest in developing algorithms to solve

$$\min_{\underline{x}}||\underline{b}-A\underline{x}||_2$$

subject to $C^T\underline{x} = \underline{d},$ (11)

where A and C are sparse. However, the time delay which we discussed in the introduction means that very few codes are available in this area.

Although techniques based on the orthogonal reduction of A to triangular form suffer very heavily from fill-in [Duff, 1974], there are two main approaches that allow such methods to be used on sparse systems. Both of these normally discard the orthogonal factors, Q, of the decomposition

$$A = QU$$ (12)

and solve further systems, if required, through

$$U^T U\underline{x} = A^T\underline{b}.$$ (13)

Zlatev and Nielsen [1979] have a code, LLSS01, for solving the unconstrained least-squares problem using square-root-free Givens for the orthogonal transformations, operating on the system an equation at a time, and optionally performing iterative refinement. George and Heath [1980] greatly facilitate the factorization (12) by determining the structure of U from a symbolic factorization of the normal equations matrix. This factorization also gives a good sparsity ordering for the columns of A. They have continued to work on this approach developing row orderings to reduce work [George and Ng, 1981], extending to the constrained problem (11) [Heath, 1981], and applying their technique to very large and structured systems [George, Heath, and Plemmons, 1981; Golub and Plemmons, 1980]. They are currently incorporating some of this work in an extension of the SPARSPAK package (mentioned earlier).

Paige and Saunders [1982] have written a code based on the bidiagonalization procedure of Golub and Kahan [1965] for solving either unsymmetric equations or the least-squares problem. (We discussed the use of this code in the context of unsymmetric equations earlier.) Their code, LSQR, is equivalent to the unsymmetric conjugate-gradient method of Hestenes and Stiefel but is numerically more reliable on ill-conditioned systems. LSQR solves the damped least-squares problem and provides standard error estimates for the components of $\underline{x}$; it has recently appeared in the *Transactions on Mathematical Software*.

Finally, probably the most commonly used technique in large production codes is the method of normal equations

$$A^T A \underline{x} = A^T \underline{b}, \tag{14}$$

where the matrix $A^T A$ may or may not be formed explicitly. Clearly many of the iterative methods discussed earlier could be used on (14), but only experimental or embedded codes are readily available in this instance. Björck and Elfving [1978] have code for solving (14) using conjugate gradients preconditioned by SSOR, which avoids forming $A^T A$ explicitly. Björck and Duff [1980] avoid the ill-conditioning of the normal equations by first performing an LU factorization of A as suggested by Peters and Wilkinson [1970]. They have also extended it to handle the constrained problem and to deal with non-sparse equations efficiently. They intend developing their experimental code for incorporation in the Harwell Subroutine Library.

NONLINEAR SYSTEMS

Many research workers in nonlinear programming have recently been considering large sparse systems. Although there has been much research on sparse quasi-Newton updates, experience with them has been mixed [Thapa, 1981; Dennis and Gay, 1982]. There are now established techniques for calculating sparse Hessians or Jacobians (see below), and many of the methods discussed earlier can be used to solve the linear problem. It is not clear, however, what the best way is of computing an acceptable step or of updating the linear problem [Dennis, 1982]. As a result of this uncertainty, little effort has been spent on writing sparse nonlinear software. We give details of the software we know to be available but intentionally present no details on current algorithmic research because of the uncertainty as to which work will result in useful software.

Subroutine VE05A in the Harwell Subroutine Library [Buckley, 1975] minimizes a general function subject to bounds and linear constraints taking advantage of sparsity in the constraints. Subroutine NS03 [Reid, 1972], also in the Harwell Subroutine Library, uses a Levenberg-Marquardt algorithm to solve sparse nonlinear least-squares problems; and Harwell Subroutine Library routines TD02 [Curtis, Powell, and Reid, 1974] and TD03 [Powell and Toint, 1979] calculate finite difference approximations to Jacobians and Hessians respectively, evaluating several columns at once to reduce the number of function evaluations. Coleman and Moré [1982] have developed ordering techniques to reduce the number of evaluations even further and have written software to implement them.

Naturally, any conjugate-gradient code for nonlinear systems is well suited to the solution of sparse systems. Examples of such software are subroutines VA08A and VA14A in the Harwell Subroutine Library and E04DBF in NAG.

The MINOS package of Murtagh and Saunders, which solves a class of constrained nonlinear problems, is discussed in the next section.

Possible software developments in the near future include sparse quasi-Newton algorithms using updates that preserve sparsity, coupled with occasional re-evaluations of the Hessian using finite differences and trust region methods for the solution of sparse nonlinear equations or sparse nonlinear least squares.

LINEAR PROGRAMMING

Some of the earliest research in sparse matrix techniques, particularly the development of direct methods, was in the area of linear programming. However, because of the highly commercial nature of the applications of these techniques, there is very little software cheaply available without major restrictions. Indeed, in this area, it is often consultancy rather than codes that is marketed.

The MINOS package of Murtagh and Saunders [1977,1980] distributed by Saunders is not free (currently $300 for research or educational centers and $3000 for profit-making institutions) and does require a license agreement, but the restrictions are not too onerous. The package solves more general problems than linear programming. MINOS solves large-scale nonlinear programming problems expressed in the form

minimize $f^0(\underline{x}) + \underline{c}^T \underline{x} + \underline{d}^T \underline{y}$

subject to $\underline{f}(\underline{x}) + A_1 \underline{y} = \underline{b}_1$

$$A_2\underline{x}+A_3\underline{y} = \underline{b}_2$$

$$\underline{l} \leqslant \begin{pmatrix} x \\ y \end{pmatrix} \leqslant \underline{u},$$

where f^0 represents nonlinear terms in the objective function, and $\underline{f}(\underline{x})$ is a vector of nonlinear functions in the constraints. Advantage is taken of sparsity in both the Jacobian of $\underline{f}$ and the matrices A_1, A_2 and A_3. All of the nonlinear functions must be smooth and have known gradients. Most of the code is portable, and the distribution tape contains source code for all of the common large computers.

Two other noncommercial linear programming codes that are readily available are not complete by themselves but are useful building blocks for a general code. Gay [1979] has coded a set of subroutines implementing the P^4 procedure of Hellerman and Rarick [1971] which partitions the matrix into bordered block triangular form; and the LA05 code in the Harwell Subroutine Library [Reid, 1976] handles linear programming bases by using an LU factorization (elimination form of the inverse) and updating it on change of basis so that sparsity is maintained.

Marsten [1981] has designed a linear programming package, XMP, as a tool for algorithmic research and development and model development in operations research and related disciplines. This package, which uses LA05 as a basis handler, is aimed at the experienced researcher or user who is able and willing to use a rather complicated system to obtain much greater freedom and flexibility than is allowed by standard commercial packages. XMP is portable and is available from Marsten at Arizona. The SPLP package of Hanson and Hiebert [1981] also incorporates LA05 and offers comprehensive facilities for solving LP problems. The authors of SPLP, which is also in the public domain, claim that their package is more flexible and easier to use than XMP.

We should stress again that there are commercial packages for sparse LP — for example, from most mathematical software houses and mainframe manufacturers, but we do not discuss them here.

SPARSE MATRIX PACKAGES

In the preceding discussion, we have tried to identify particular problem areas where there is significant sparse software. As we suggested in the introduction, there are many large codes in which sparse software is embedded. We intentionally do not consider them because the users of these large codes are generally less concerned with sparsity than with the problem class they are trying to solve and it is to surveys peculiar to that class that they will turn.

Additionally, the embedded nature of software in such packages usually makes it impossible to extract the sparse code for more general use. However, in this section we tie up a few loose ends not covered by the other sections.

Much of the early sparse matrix research in direct systems was motivated by the need to solve large linear systems arising from implicit methods for solving stiff ordinary differential equations. There are several packages for solving such systems, including FACSIMILE [Curtis, 1980], GEAR in various versions [Hindmarsh, 1974], LARKIN [Deuflhard, Bader, and Nowak, 1980], and a semi-implicit Runge Kutta code SPARKS [Houbak and Thomsen, 1979]. The solution of two-point boundary-value problems using finite differences gives rise to block bidiagonal sparse matrices — for example, the PASVA codes of Lentini and Pereyra [1977], a version of which (PASVA3) is available as DD04 in the Harwell Subroutine Library and D02RAF in NAG.

Finally, there have been a few attempts to create a sparse matrix environment where the user can easily change data formats and multiply, add, and transpose matrices as well as solve sparse linear systems. One such code was presented in the *Communications of the ACM* a decade ago [McNamee, 1971], and Harms [1981] and his colleagues have attempted to combine this work with Harwell and NAG routines to create a more versatile environment. Another such package, RALAPACK, has been developed by Pissanetzky [1981b]; intended as a building block for general-purpose programs, it has been implemented on several installations as part of a finite element program or preprocessor [Pissanetzky, 1981a]. Szyld [1981a] also has a set of subroutines for data management of sparse matrices. Lest the casual reader imagine that software for multiplying or transposing sparse matrices is entirely trivial, a reading of Gustavson [1978] would be in order. He indicates well some of the tricks required for an efficient code in these cases.

ACKNOWLEDGMENTS

During the writing of this report, I consulted several people concerning their sparse software. I would like to record my thanks to them for sending me letters, reports and listings. I am also grateful to John Reid for his comments on a draft version of this report and to Gene Golub, Axel Ruhe, Mike Saunders, Andy Sherman, and Bob Ward for their helpful comments on particular sections.

REFERENCES

Ajiz, M. A., and A. Jennings [1981]. *A Robust ICCG Algorithm.* Queen's University Civil Engineering Department Report, Belfast.

Axelsson, O. [1980]. "Conjugate gradient type methods for unsymmetric and inconsistent systems of linear equations." *Lin. Alg. and its Appl.* 29:1-16.

Bank, R. E., and A. H. Sherman [1981]. "An adapive, multi-level method for elliptic boundary value problems." *Computing* 26:91-105.

Björck, A., and I. S. Duff [1980]. "A direct method for the solution of sparse linear least squares problems." *Lin. Alg. and its Appl.* 34:43-67.

Björck, A., and T. Elfving [1978]. *Accelerated Projection Methods for Computing Pseudoinverse Solutions of Systems of Linear Equations.* Department of Mathematics, Linköping University, Sweden, Report LiTH-MAT-R-1978-5 (paper without program appeared in *BIT* 9:145-163).

Brandt, A. [1977]. "Multi-level adaptive techniques (MLAT) for partial differential equations: ideas and software." *Numerical Software III.* Ed. J. Rice. Academic Press, New York, pp. 277-318.

Buckley, A. [1975]. "An alternate implementation of Goldfarb's minimization algorithm." *Math. Prog.* 8:207-231.

Buzbee, B. L., G. H. Golub, and C. W. Nielsen [1970]. "On direct methods for solving Poisson's equations." *SIAM J. Numer. Anal.* 7:627-656.

Coleman. T. F., and J. J. Moré [1982]. *Software for Estimating Sparse Jacobian Matrices.* Argonne National Laboratory Report ANL-82-37.

Concus, P., and G. H. Golub [1975]. *A Generalized Conjugate Gradient Method for Nonsymmetric Systems of Linear Equations.* Stanford University Computer Science Department Report STAN-CS-75-535.

Cullum, J., and R. A. Willoughby [1980]. *Computing Eigenvalues of Large Symmetric Matrices — An Implementation of a Lanczos Algorithm without Reorthogonalization, Computer Programs.* IBM Research Center Report RC8298, Yorktown Heights, New York.

Cullum, J., and R. A. Willoughby [1981]. "Computing eigenvalues of very large symmetric matrices — an implementation of a Lanczos algorithm with no reorthogonalization." *J. Comp. Phys.* 44:329-358.

Curtis, A. R. [1980]. "The FACSIMILE numerical integrator for stiff initial value problems." *Computational Techniques for Ordinary Differential Equations.* Ed. I. Gladwell and D. K. Sayers. Academic Press, New York, pp. 47-82.

Curtis, A. R., M. J. D. Powell, and J. K. Reid [1974]. "On the estimation of sparse Jacobian matrices." *J. Inst. Math. Appl.* 13:117-120.

Dennis, J. [1982]. Private communication.

Dennis, J., and D. M. Gay [1982]. Private communication.

Deytna, E. [1981]. "Rapid elliptic solvers." *Sparse Matrices and Their Uses.* Ed. I. S. Duff. Academic Press, New York, pp. 245-264.

Deuflhard, P., G. Bader, and U. Nowak [1980]. *LARKIN — A Software Package for the Numerical Simulation of LARge Systems Arising in Chemical Reaction KINetics.* Inst. für Angewandte Mathematik, University of Heidelberg, Report Nr. 100.

Douglas, C. C. [1982]. *A Multi-grid Optimal Order Elliptic Partial Differential Equation Solver.* Research Center for Scientific Computation, Yale University Report. To appear.

Duff, I. S. [1974]. "Pivot selection and row ordering in Givens reduction on sparse matrices." *Computing* 13:239-248.

Duff, I. S. [1977a]. *MA28 — A Set of Fortran Subroutines for Sparse Unsymmetric Linear Equations.* AERE Harwell Report R.8730, HMSO, London.

Duff, I. S. [1977b]. "A Survey of Sparse Matrix Research." *Proc. IEEE* 65:500-535.

Duff, I. S. [1979]. "Practical Comparisons of Codes for the Solution of Sparse Linear Systems." *Sparse Matrix Proceedings 1978.* Ed. I. S. Duff and G. W. Stewart. SIAM, pp. 107-134.

Duff, I. S. [1981]. *MA32 — A Package for Solving Sparse Unsymmetric Systems Using the Frontal Method.* AERE Harwell Report R.10079, HMSO, London.

Duff, I. S. [1982]. *The Solution of Sparse Linear Equations on the CRAY-1.* AERE Harwell Report CSS 125. Presented at the Cray Research, Inc., Symposium "Science, Engineering, and the CRAY-1," Minneapolis, April 5-7, 1982.

Duff, I. S., and J. K. Reid [1979]. "Performance evaluation of codes for sparse matrix problems." *Performance Evaluation of Numerical Software.* Ed. L. Fosdick. North Holland, pp. 121-135.

Duff, I. S., and J. K. Reid [1982a]. "Experience of sparse matrix codes on the CRAY-1." *Comput. Phys. Comm.*. 26:293-302.

Duff, I. S., and J. K. Reid [1982b]. *MA27 — A Set of Fortran Subroutines for Solving Sparse Symmetric Sets of Linear Equations.* AERE Harwell Report R10533, HMSO, London.

Duff, I. S., and J. K. Reid [1982c]. *The Multi-Frontal Solution of Indefinite Sparse Symmetric Linear Systems.* AERE Harwell Report CSS 122.

Duff, I. S., and J. K. Reid [1982d]. *The Multifrontal Solution of Sparse Unsymmetric Equations.* AERE Harwell Report. To appear.

Eisenstat, S. C., M. C. Gursky, M. H. Schultz, and A. H. Sherman [1977]. *Yale Sparse Matrix Package. I. The Symmetric Codes. II. The Nonsymmetric Codes.* Yale University Department of Computer Science Reports 112 and 114.

Eisenstat, S. C., M. C. Gursky, M. H. Schultz, and A. H. Sherman [1982]. "Yale sparse matrix package. I: The symmetric codes." *Int. J. Numer. Meth. Engng.* 18:1145-1151.

Eisenstat, S. C., M. H. Schultz, and A. H. Sherman [1979]. "Software for sparse Gaussian elimination with limited core storage." *Sparse Matrix Proceedings 1978.* Ed. I. S. Duff and G. W. Stewart. SIAM, pp. 135-153.

Elman, H. C. [1982a]. *Iteration Methods for Large, Sparse, Nonsymmetric Systems of Linear Equations.* Yale University Department of Computer Science Report 229.

Elman, H. C. [1982b]. *PCGPAK: A Package for Preconditioned Conjugate Gradient Routines for Symmetric and Nonsymmetric Problems.* Research Center for Scientific Computation, Yale University Report. To appear.

Ericsson, T., and A. Ruhe [1980a]. "The spectral transformation Lanczos method for the numerical solution of large sparse generalized symmetric eigenvalue problems." *Math. Comp.* 35:1251-1268.

Ericsson, T., and A. Ruhe [1980b]. *A User Manual for One Implementation of the Spectral Transformation Lanczos Method.* University of Umea, Sweden, Report UMINF-80.80.

Evans, D. J. [1967]. "The use of preconditioning in iterative methods for solving linear equations with symmetric positive definite matrices." *J. Inst. Math. Appl.* 4:295-314.

Felippa, C. A. [1975]. "Solution of linear equations with skyline-stored symmetric matrix." *Computers and Structures* 5:13-29.

Fletcher, R. [1976]. "Conjugate gradient methods for indefinite systems." *Numerical Analysis* (Dundee, 1975). Lecture Note 506. Ed. G. A. Watson. Springer-Verlag, Berlin, pp. 73-89.

Foerster, H., and H. Förster [1978]. *Modular programs for the Fast Solving of Linear Elliptic Equations by Reduction Methods: TR2D01, TR2D02 to Solve the Dirichlet problem for Helmholtz's Equation.* GMD-IMA, Bonn, GMD-Studie 47.

Foerster, H., and K. Witsch [1982]. "On multigrid software for the solution of elliptic problems on rectangular domains." *Multigrid Methods.* Ed. W. Hackbusch and U. Trottenberg. *Lecture Notes in Mathematics 960.* Springer-Verlag, Berlin. To appear.

Förster, H. [1980]. *Modular programs for the Fast Solving of Linear Elliptic Equations by Reduction Methods: TR2D04, TR2D05 to Solve the Neumann and Periodic Boundary Value Problems for Helmholtz's Equation.* GMD-IMA, Bonn, GMD-Studie 53.

Förster, H., and H. Reutersberg [1979]. *Modular Programs for the Fast Solving of Linear Elliptic Equations by Reduction Methods: TR2D11 to Solve the Modified Biharmonic Boundary Value problem.* GMD-IMA, Bonn, GMD-Studie 50.

Gay, D. M. [1979]. "On combining the schemes of Reid and Saunders for sparse LP bases." *Sparse Matrix Proceedings 1978.* Ed. I. S. Duff and G. W. Stewart. SIAM, pp. 313-334.

George, A., and M. T. Heath [1980]. "Solution of sparse linear least squares problems using Givens rotations." *Lin. Alg. and its Appl.* 34:69-83.

George, A., and J. Liu [1981]. *Computer Solution of Large Sparse Positive Definite Systems.* Prentice-Hall, Englewood Cliffs, New Jersey.

George, A., and E. Ng [1981]. *On Row and Column Orderings for Sparse Least Squares Problems.* University of Waterloo Computer Science Department Report CS-81-09.

George, A., M. T. Heath, and R. J. Plemmons [1981]. "Solution of large-scale sparse least squares problems using auxiliary storage." *SIAM J. Sci. Stat. Comput.* 2:416-429.

George, A., J. Liu, and E. Ng [1980]. *User Guide for SPARSPAK: Waterloo Sparse Linear Equations Package*. Department of Computer Science, University of Waterloo, Report CS-78-30 (Revised Jan. 1980).

Gladwell, I. [1982] Private communication.

Golub, G., and W. Kahan [1965]. "Calculating the singular values and pseudo-inverse of a matrix." *J. SIAM Numer. Anal.* 2:205-224.

Golub, G. H., and R. J. Plemmons [1980]. "Large-scale geodetic least-squares adjustment by dissection and orthogonal decomposition." *Lin. Alg. and its Appl.* 34:3-28.

Golub, G. H., F. T. Luk, and M. L. Overton [1977]. *A Block Lanczos Method to Compute the Singular Values and Corresponding Singular Vectors of a Matrix*. Stanford University Computer Science Department Report STAN-CS-77-635.

Golub, G. H., F. T. Luk, and M. L. Overton [1981]. "A block Lanczos method for computing the singular values and corresponding singular vectors of a matrix." *ACM Trans. on Math. Soft.* 7:149-169.

Golub, G. H., R. C. Underwood, and J. H. Wilkinson [1970]. *The Lanczos Algorithm for the Symmetric $A\underline{x}=\lambda B\underline{x}$ Problem*. Stanford Computer Science Department Report STAN-CS-142.

Gustafsson, I. [1979]. "On modified incomplete Cholesky factorization methods for the solution of problems with mixed boundary conditions and problems with discontinuous material coefficients." *Int. J. Numer. Meth. Engng.* 14:1127-1140.

Gustavson, F. G. [1978]. "Two fast algorithms for sparse matrices: multiplication and permuted transposition." *ACM Trans. on Math. Soft.* 4:250-269.

Hackbusch, W. [1976]. *Ein iteratives Verfahren zur schnellen Auflosung elliptischer Randwertprobleme*. Math. Inst., Univ. of Köln, Report 76-12.

Hackbusch, W. [1977]. *A Multi-grid Method Applied to a Boundary Value Problem with Variable Coefficients in a Rectangle*. Math. Inst., Univ. of Köln, Report 77-17.

Hackbusch, W., and U. Trottenberg, eds. [1982]. "Multigrid methods." *Proceedings of Conference at Köln-Porz* (November 23-27, 1981). *Lecture Notes in Mathematics 960*. Springer-Verlag. Berlin.

Hageman, L. A., and D. M. Young [1981]. *Applied Iterative Methods.* Academic Press, New York.

Hanson, R. J., and K. L. Hiebert [1981]. *A Sparse Linear Programming Subprogram.* Sandia National Laboratory, Albuquerque, Report SAND 81-0297.

Harms, U. [1981]. "The use of integer packing techniques at the RRZN." *Sparse Matrices and Their Uses.* Ed. I. S. Duff. Academic Press, New York, pp. 335-341.

Hasbani, Y., and M. Engleman [1979]. "Out-of-core solution of linear equations with non-symmetric coefficient matrix." *Computers and Fluids* 7:13-31.

Heath, M. T. [1982]. "Some extensions of an algorithm for sparse linear least squares problems." *SIAM J. Sci. Stat. Comput.* 3:223-237.

Hellerman, E., and D. Rarick [1971]. "Reinversion with the preassigned pivot procedure." *Math. Prog.* 1:195-216.

Hestenes, M. R., and E. Stiefel [1952]. "Methods of conjugate gradients for solving linear systems." *J. Res. Nat. Bur. Standards,* 49:409-436.

Hindmarsh, A. C. [1974]. *GEAR: Ordinary Differential Equation System Solver.* Lawrence Livermore National Laboratory Report UCID-30001.

Hood, P. [1976]. "Frontal solution program for unsymmetric matrices." *Int. J. Numer. Meth. Engng.* 10:379-399.

Houbak, N., and P. G. Thomsen [1979]. *SPARKS. A Fortran Subroutine for the Solution of Large Systems of Stiff ODE's with Sparse Jacobians.* Numerisk Institut, Lyngby, Denmark, Report NI-79-02.

IBM [1976]. "IBM System/360 and System/370 IBM 1130 and IBM 1800. Subroutine Library - Mathematics. User's Guide." Program Product 5736-XM7. IBM Catalogue SH12-5300-1.

Irons, B. M. [1970]. "A frontal solution program for finite element analysis." *Int. J. Numer. Meth. Engng.* 2:5-32.

Jacobs, D. A. H. [1980]. *A Summary of Subroutines and Packages (employing the strongly implicit procedure) for Solving Elliptic and Parabolic Partial Differential Equations.* Central Electricity Research Laboratories (Leatherhead, England) Report RD/L/N 55/80.

Jacobs, D. A. H. [1981]. "The exploitation of sparsity by iterative methods." *Sparse Matrices and Their Uses.* Ed. I. S. Duff. Academic Press, New York, pp. 191-222.

Jacobs, D. A. H. [1982a]. *CG37, BCG37 and CCG37: Packages for Solving Seven Diagonal Systems of Algebraic Equations Using Preconditioned Conjugate Gradient Methods.* Central Electricity Research Laboratories (Leatherhead, England) Report RD/L/2235P81.

Jacobs, D. A. H. [1982b]. *CG25, BCG25 and CCG25: Packages for Solving Five Diagonal Systems of Algebraic Equations Using Fundamental Conjugate Gradient Methods.* Central Electricity Research Laboratories (Leatherhead, England) Report RD/L/2234P81.

Jea, K. C. [1982]. *Generalized Conjugate Gradient Acceleration of Iterative Methods.* University of Texas Center for Numerical Analysis Report CNA-176.

Jennings, A. [1966]. "A compact storage scheme for the solution of symmetric linear simultaneous equations." *Comput. J.* 9:281-285.

Jennings, A. [1981]. "Eigenvalue methods and the analysis of structural vibration." *Sparse Matrices and Their Uses.* Ed. I. S. Duff. Academic Press, New York, pp. 109-138.

Kaufman, L. [1982]. *Usage of the Sparse Matrix Programs in the PORT Library.* Bell Laboratories Report.

Kincaid, D. R., J. R. Respess, D. M. Young, and R. G. Grimes [1982]. "Algorithm 586. ITPACK 2C: A Fortran package for solving large sparse linear systems by adaptive accelerated iterative methods." *ACM Trans. on Math. Soft.* 8 (3) 302-322.

Kuo-Petravic, G., and M. Petravic [1979]. "A program generator for the incomplete *LU* decomposition-conjugate gradient (ILUCG) method." *Comput. Phys. Comm.* 18:13-25.

Kuo-Petravic, G., and M. Petravic [1981]. "A program generator for the incomplete Cholesky conjugate gradient (ICCG) method with a symmetrizing preprocessor." *Comput. Phys. Comm.* 22:33-48.

Lentini, M., and V. Pereyra [1977]. "An adaptive finite difference solver for nonlinear two point boundary problems with mild boundary layers." *SIAM J. Numer. Anal.* 14:91-111.

Lewis, J. G. [1977]. *Algorithms for Sparse Matrix Eigenvalue Problems.* Stanford University Computer Science Department Report STAN-CS-77-595.

Machura, M., and R. A. Sweet [1980]. "A survey of software for partial differential equations." *ACM Trans. on Math. Soft.* 6:461-488.

McNamee, J. M. [1971]. "A sparse matrix package. Part 1. Algorithm 408." *Comm. ACM* 14:265-273.

Manteuffel, T. A. [1977]. "The Tchebychev iteration for nonsymmetric linear systems." *Numer. Math.* 28:307-327.

Markham, G. [1982]. *A User's Guide to CGS, BCGS and CCGS: Sparse Code Packages Using Preconditioned Conjugate Gradient Methods for Solving Systems of Algebraic Equations.* Central Electricity Research Laboratories (Leatherhead, England) Report. To appear.

Markowitz, H. M. [1957]. "The elimination form of the inverse and its application to linear programming." *Management Science* 3:255-269.

Marsten, R. E. [1981]. "The design of the XMP linear programming library." *ACM Trans. on Math. Soft.* 7:481-497.

Meijerink, J. A., and H. A. van der Vorst [1977]. "An iterative solution method for linear systems of which the coefficient matrix is a symmetric M-matrix." *Math. Comp.* 31:148-162.

Munksgaard, N. [1980]. "Solving sparse symmetric sets of linear equations by preconditioned conjugate gradients." *ACM Trans. on Math. Soft.* 6:206-219.

Murtagh, B. A., and M. A. Saunders [1977]. *MINOS User's Guide.* Stanford University Department of Operations Research Report SOL 77-9.

Murtagh, B. A., and M. A. Saunders [1980]. *MINOS/AUGMENTED User's Manual.* Stanford University Department of Operations Research Report SOL 80-14.

Nikolai, P. J. [1979]. "Algorithm 538. Eigenvectors and eigenvalues of real generalized symmetric matrices by simultaneous iteration." *ACM Trans. on Math. Soft.* 5:118-125.

O'Leary, D. P. [1979]. "Conjugate gradient algorithms in the solution of optimization problems for nonlinear elliptic partial differential equations." *Computing* 22:59-77.

O'Leary, D. P., and O. Widlund [1981]. "Algorithm 572. Solution of the Helmholtz equation for the Dirichlet problem on general bounded three-dimensional regions." *ACM Trans. on Math. Soft.* 7:239-246.

Paige, C. C., and M. A. Saunders [1975]. "Solution of sparse indefinite systems of linear equations." *SIAM J. Numer. Anal.* 12:617-629.

Paige, C. C., and M. A. Saunders [1982]. "LSQR: An algorithm for sparse linear equations and sparse least squares." *ACM Trans. on Math. Soft.* 8:43-71 and 195-209.

Parlett, B. N. [1978]. "A new look at the Lanczos algorithm for solving symmetric systems of linear equations." *Lin. Alg. and its Appl.* 29:323-346.

Parlett, B. N. [1980]. *The Symmetric Eigenvalue Problem.* Prentice-Hall, Englewood Cliffs, New Jersey.

Parlett, B. N., and J. K. Reid [1981]. "Tracking the progress of the Lanczos algorithm for large symmetric eigenproblems." *IMA J. Numer. Anal.* 1:135-155.

Parlett, B. N., and D. Taylor [1982]. "Estimation of the eigensystem of unsymmetric matrices using Lanczos with look ahead." To appear.

Perry, J. R. [1981]. *Secondary Storage Methods for Solving Symmetric, Positive Definite, Banded Linear Systems.* Yale University Department of Computer Science Report 201.

Peters, G., and J. H. Wilkinson [1969]. "Eigenvalues of $A\underline{x}=\lambda B\underline{x}$ with band symmetric A and B." *Comp. J.* 12:398-404.

Peters, G., and J. H. Wilkinson [1970]. "The least squares problem and pseudo-inverses." *Comput. J.* 13:309-316.

Pissanetzky, S. [1981a]. "An automatic three-dimensional finite element mesh generator." *Int. J. Numer. Meth. Engng.* 17:255-269.

Pissanetzky, S. [1981b]. *Listing of a Sparse Matrix Package.* Atomic Energy Commission of Argentina Report CNEA-NT 20/81.

Powell, M. J. D., and Ph. L. Toint [1979]. "On the estimation of sparse Hessian matrices." *SIAM J. Numer. Anal.* 16:1060-1074.

Reid, J. K. [1972]. *Fortran Subroutines for the Solution of Sparse Systems of*

Nonlinear Equations. AERE Harwell Report R.7293, HMSO, London.

Reid, J. K. [1976]. *Fortran Subroutines for Handling Sparse Linear Programming Bases.* AERE Harwell Report R.8269, HMSO, London.

Rice, J. R. [1978]. *ELLPACK 77. User's Guide.* Purdue University Computer Science Department Report CSD-TR-289.

Rutishauser, H. [1970]. "Simultaneous iteration method for symmetric matrices." *Numer. Math.* 16:205-223.

Ryder, B. G. [1974]. "The PFORT verifier." *Software Practice and Experience* 4:359-377.

Saad, Y. [1980]. "Variations on Arnoldi's method for computing eigenelements of large unsymmetric matrices." *Lin. Alg. and its Appl.* 34:269-295.

Saad, Y. [1981a]. "Krylov subspace methods for solving large unsymmetric linear systems." *Math. Comp.* 37:105-126.

Saad, Y. [1981b]. *Practical Use of Some Krylov Subspace Methods for Solving Indefinite and Unsymmetric Linear Systems.* Yale University Department of Computer Science Technical Report.

Schröder, J., U. Trottenberg, and K. Witsch [1978]. "On fast Poisson solvers and applications." *Numerical Treatment of Differential Equations.* Ed. R. Bulirsch, R. D. Grigorieff, and J. Schröder. *Lecture Notes in Mathematics 631.* Springer-Verlag, pp. 153-187.

Schumann, U., ed. [1978]. *Computers, Fast Elliptic Solvers, and Applications.* Advance Publications.

Scott, D. S. [1979]. *Block Lanczos Software for Symmetric Eigenvalue Problems.* Oak Ridge National Laboratory Computer Sciences Division Report ORNL/CSD-48.

Scott, D. S. [1981a]. "The Lanczos algorithm." *Sparse Matrices and Their Uses.* Ed. I. S. Duff. Academic Press, New York, pp. 139-159.

Scott, D. S. [1981b]. "Solving sparse symmetric definite quadratic λ-matrix problems." *BIT* 21:475-480.

Scott, D. S. [1981c]. "Solving sparse symmetric generalized eigenvalue problems without factorization." *SIAM J. Numer. Anal.* 18:102-110.

Scott, D. S., and R. C. Ward [1982]. "Solving quadratic λ-matrix problems without factorization." *SIAM J. Sci. Stat. Comput.* 3:58-67.

Sherman, A. H. [1978]. "Algorithm 533: NSPIV, a Fortran subroutine for sparse Gaussian elimination with partial pivoting." *ACM Trans. on Math. Soft.* 4:391-398.

Sherman, A. H. [1982]. Private communication.

Stewart, G. W. [1976]. "A bibliographical tour of the large sparse generalized eigenvalue problem." *Sparse Matrix Computations.* Ed. J. R. Bunch and D. J. Rose. Academic Press, New York, pp. 113-130.

Stewart, W. J., and A. Jennings [1981]. "Algorithm 570. LOPSI: A simultaneous iteration algorithm for real matrices." *ACM Trans. on Math. Soft.* 7:230-232.

Stone, H. L. [1968]. "Iterative solution of implicit approximations of multidimensional partial differential equations." *SIAM J. Numer. Anal.* 5:530-558.

Swarztrauber, P. N. [1977]. "The methods of cyclic reduction, Fourier analysis and the FACR algorithm for the discrete solution of Poisson's equation on a rectangle." *SIAM Rev.* 19:490-501.

Swarztrauber, P. N. [1981]. "Vectorizing the FFT's." *Methods in Computational Physics,* Volume on Parallel Algorithms. Lawrence Livermore Series. To appear.

Swarztrauber, P. N., and R. A. Sweet [1979]. "Algorithm 541. Efficient Fortran subprograms for the solution of separable elliptic partial differential equations." *ACM Trans. on Math. Soft.* 5:352-364.

Szyld, D. B. [1981a]. *A Sparse Matrix Package.* New York University Institute of Economic Analysis.

Szyld, D. B. [1981b]. "Using sparse matrix techniques to solve a model of the world economy." *Sparse Matrices and Their Uses.* Ed. I. S. Duff. Academic Press, New York, pp. 357-365.

Thapa, M. K. [1981]. *Optimization of Unconstrained Functions with Sparse Hessian Matrices — Quasi-Newton Methods.* Stanford University Department of Operations Research Report SOL-12.

Thompson, E., and Y. Shimazaki [1980]. "A frontal procedure using skyline storage." *Int. J. Numer. Meth. Engng.* 15:889-910.

Underwood, R. [1975]. *An Iterative Block Lanczos Method for the Solution of Large Sparse Symmetric Eigenproblems.* Stanford University Computer Science Department Report STAN-CS-75-496.

van der Vorst, H. A., and J. M. van Kats [1979]. *Manteuffel's Algorithm with Preconditioning for the Iterative Solution of Certain Sparse Linear Systems with a Non-symmetric Matrix.* ACCU, Utrecht, Report TR-11.

van Kats, J. M., and H. A. van der Vorst [1977]. *Automatic Monitoring of Lanczos Schemes for Symmetric and Skew-symmetric Generalized Eigenvalue Problems.* ACCU, Utrecht, Report TR-7.

van Kats, J. M., and H. A. van der Vorst [1979]. *Software for the Discretization and Solution of Second Order Self-adjoint Elliptic Partial Differential Equations in Two Dimensions.* ACCU, Utrecht, Report TR-10.

van Kats, J. M., C. J. Rusman, and H. A. van der Vorst [1980]. *ACCULIB Documentation - Minimanual.* ACCU, Utrecht., Report 9.

Vinsome, P. K. W. [1976]. *Orthomin, An Iterative Method for Solving Sparse Sets of Simultaneous Linear Equations.* 4th SPE Symposium on Numerical Simulation of Reservoir Performance. Paper number SPE 5729.

Widlund, O. [1978]. "A Lanczos method for a class of nonsymmetric systems of linear equations." *SIAM J. Numer. Anal.* 15:801-812.

Young, D. M., and K. C. Jea [1980]. *Generalized Conjugate Gradient Acceleration of Iterative Methods. Part II: The Nonsymmetrizable Case.* University of Texas at Austin Center for Numerical Analysis Report CNA-163 (Revised September 1981).

Young, D. M., and D. R. Kincaid [1981]. "The ITPACK package for large sparse linear systems." *Elliptic Problem Solvers.* Ed. M. H. Schultz. Academic Press, New York, pp. 163-185.

Zlatev, Z., and H. B. Nielsen [1977]. *SIRSM: Fortran Package for the Solution of Sparse Systems by Iterative Refinement.* Numerisk Institut, Lyngby, Denmark, Report NI-77-13.

Zlatev, Z., and H. B. Nielsen [1979]. *LLSS01: Fortran Subroutine for Solving Linear Least Squares Problems.* Numerisk Institut, Lyngby, Denmark, Report NI-79-07.

Zlatev, Z., V. A. Barker, and P. G. Thomsen [1978]. *SSLEST: A Fortran IV Subroutine for Solving Sparse Systems of Linear Equations. User's Guide.* Numerisk Institut, Lyngby, Denmark, Report NI-78-01.

Zlatev, Z., J. Wasniewski, and K. Schaumburg [1981]. "Y12M. Solution of large and sparse systems of linear algebraic equations." *Lecture Notes in Computer Science 121.* Springer-Verlag, Berlin.

Chapter 9

MATHEMATICAL SOFTWARE
FOR ELLIPTIC BOUNDARY VALUE PROBLEMS

Ronald F. Boisvert and Roland A. Sweet
Scientific Computing Division
National Bureau of Standards *
Washington, D.C. 20234

The solution of problems involving partial differential equations (PDEs) was one of the earliest applications of the automatic computing devices that became available during the 1950s. The fact that such problems still consume a substantial portion of budgets for scientific computing attests to the importance of these problems in many areas of science and engineering.

Because of the nature of the computing environment in those early years (slow CPUs, small memories, and the absence of standardized high-level programming languages), most programs were highly applications-oriented and machine-dependent. The result was much duplication of effort, and as early as 1960 the need for portable general-purpose programs was recognized (see the description of the SCOPE project in Forsythe and Wasow [1960]). Although many PDE problems still have special features that defy generalization, advances in numerical analysis and machine architecture have vastly increased the class of problems that can be handled with the aid of general-purpose software.

In this chapter we survey recent advances in software for one such class — elliptic boundary value problems (EBVPs). Such problems are fundamental to the modelling of static physical phenomena such as electromagnetic fields, steady-state diffusion, and structural stress and strain. In addition, elliptic problems often arise as intermediate steps in the numerical modelling of dynamic processes such as fluid flow. In preparing for this survey we have focused our attention on general-purpose mathematical software, and consequently much useful applications-oriented software is necessarily omitted. Interested readers are referred to the listing of sources of software for partial differential equations given later in this chapter.

We begin with a working definition of an elliptic boundary value problem, indicating the problem features of most concern in software development. Two examples that demonstrate how such computational problems may arise in practice are also included. Next we review the major numerical methods employed in existing software, pointing out situations where each is most appropriate. We then turn to software engineering issues. Various considerations relating to the

---

* Contribution of the National Bureau of Standards, not subject to copyright in the United States.

design of the user interface to software for elliptic problems are discussed, followed by case studies of two packages with fundamentally different designs — FISHPAK and ELLPACK. In each case study we discuss the philosophy underlying the development, describe the contents, and present examples of use. We next present a list of sources of software for partial differential equations, including commercial libraries, software clearinghouses, and journals that distribute software. We close with a detailed catalog of mathematical software for EBVPs. In each case the problem class, numerical methods, portability, and distribution are described.

MATHEMATICAL PRELIMINARIES

The specification of a second order elliptic boundary value problem requires three essential quantities: a partial differential operator L, a domain Ω over which the equation holds (typically two or three dimensional), and conditions B that hold at the boundary $\partial\Omega$ of the domain. We seek a function u, the solution of the problem, that satisfies

$$Lu=f \quad \text{in} \quad \Omega \tag{1a}$$

$$Bu=g \quad \text{on} \quad \partial\Omega \tag{1b}$$

for two given functions f and g. A statement of the precise conditions on L, Ω, and B required to guarantee ellipticity of the problem and the existence of a solution is beyond the scope of the chapter. (Interested readers may find conditions by referring to any standard textbook on the theory of PDEs.) Many special-purpose algorithms have been constructed for EBVPs with special properties. When applicable, these algorithms can often yield substantial savings in computational effort and should affect one's choice of software. We present here a brief description of these features and some associated terminology.

The Operator

The foremost property of the operator L is linearity. The several codes (B2DE, SLDGL, TWODEPEP)* designed for nonlinear operators reduce the nonlinearity to a sequence of linear problems through the use of a Newton iteration. We shall, henceforth, concentrate on linear operators. Furthermore, even though the three codes mentioned above treat systems of EBVPs, our remarks shall be directed to the single equation. The form of the operator is used extensively to characterize important properties.

The simplest and most frequently encountered form of L is the well-known Laplacian operator ∇^2 which in Cartesian coordinates (x,y) is

---

*Throughout this chapter we shall note examples of software that contain the features under discussion. All such software is described in detail in the catalog at the end of this chapter.

$$\nabla^2 u = \frac{\partial^2 u}{\partial x^2} + \frac{\partial^2 u}{\partial y^2}. \tag{2}$$

(We illustrate forms of L with two-dimensional operators; most software is restricted to this dimensionality.) The equation $\nabla^2 u = f$ is known as Poisson's equation. The Laplacian in cylindrical coordinates (assuming circular symmetry) (r,z) is

$$\nabla^2 u = \frac{1}{r} \frac{\partial}{\partial r} (r\frac{\partial u}{\partial r}) + \frac{\partial^2 u}{\partial z^2}. \tag{3}$$

It illustrates a slightly more general, important form of L, namely,

$$Lu = D_x u + \frac{\partial^2 u}{\partial y^2}, \tag{4}$$

where D_x is the general second-order linear operator with (possibly) variable coefficients depending only on the independent variable x, i.e.,

$$D_x u = a(x)\frac{\partial^2 u}{\partial x^2} + b(x)\frac{\partial u}{\partial x} + c(x)u. \tag{5}$$

If we replace the term $\frac{\partial^2 u}{\partial y^2}$ in Eq. 4 with a general operator D_y of a form similar to that given in Eq. 5, we arrive at another important form of L, namely,

$$Lu = D_x u + D_y u. \tag{6}$$

An example of such an operator is furnished by the axisymmetric Laplacian in spherical coordinates (r,θ) (multiplied by r^2)

$$r^2\nabla^2 u = \frac{\partial}{\partial r} (r^2\frac{\partial u}{\partial r}) + \frac{1}{\sin\theta} \frac{\partial}{\partial \theta} (\sin\theta \frac{\partial u}{\partial \theta}).$$

Note that all forms of L given above represent operators that are essentially the sum of two one-dimensional operators.

The last two forms we consider are truly two-dimensional and shall not, in general, separate out as those given above. The first form is the so-called conservation or divergence form

$$Lu = \nabla \cdot (k(x,y)\nabla u), \tag{7}$$

which in Cartesian coordinates is

$$Lu = \frac{\partial}{\partial x} (k(x,y) \frac{\partial u}{\partial x}) + \frac{\partial}{\partial y} (k(x,y) \frac{\partial u}{\partial y}). \tag{8}$$

Such an operator arises from the study of steady state diffusion, say, of neutrons [Wachspress, 1966]. The last form we note here is the most general, linear second-order elliptic operator

$$Lu = a(x,y) \frac{\partial^2 u}{\partial x^2} + b(x,y) \frac{\partial^2 u}{\partial x \partial y} + c(x,y) \frac{\partial^2 u}{\partial y^2}$$

$$+d(x,y)\,\frac{\partial u}{\partial x}+e(x,y)\,\frac{\partial u}{\partial y}+f(x,y)u\;. \tag{9}$$

The Domain

The second essential quantity in the specification of an elliptic problem is the region in the plane (or space) over which the equation is to be valid. Our survey contains only one routine (D03EAF) that assumes the domain is unbounded. Hence, we shall concentrate entirely on bounded domains.

The simplest domain is the rectangle

$$R=\{(x,y):x_s<x<x_f,\ y_s<y<y_f\}. \tag{10}$$

Note that if polar coordinates are used, R may look like a wedge of a disk, not a rectangle *per se*. The only other distinction we shall make is the general domain, which we shall take to mean non-rectangular.

The Boundary Conditions

To specify the solution of an EBVP completely, we must specify auxiliary conditions along the boundary $\partial\Omega$ of the domain. The conditions often are one or more of the following:

- The solution u is specified, known as a Dirichlet condition.

- The derivative of the solution u in the direction of the outward pointing normal (the "normal derivative" usually written as $\partial u/\partial n$) is specified, known as a Neumann condition.

- The solution is specified to be a periodic function of one of the independent variables, known as a periodicity condition.

- A linear combination of the solution and its normal derivative is specified, the so-called mixed, elastic, or Robin condition. For example, along the line $x=x_s$ of the rectangle R given in (10), we might have the condition

$$a(y)u(x_s,y)+b(y)\frac{\partial u}{\partial x}(x_s,y)=g(y). \tag{11}$$

Of course, other conditions may be specified, e.g., a radiation boundary condition specifying the fourth power of the solution. But for existing software, these seem to be the most important ones.

We next introduce certain terms that are often used in describing elliptic problems:

- A Dirichlet problem is one for which the solution is specified over the entire boundary.

- A Neumann problem is one for which the normal derivative is specified over the entire boundary.

- A separable problem is one for which

  (a) The operator L is the sum of two one-dimensional operators, i.e., it can be written in the form given in Eq. 6;

  (b) The domain is a rectangle in the given coordinate system; and

  (c) The boundary operator B in Eq. 1b on, say, an x boundary is independent of y, and vice versa, as, for example, if in the mixed condition specified in (11), $a(y)$ and $b(y)$ are constant functions.

  Selection of any combination of the first three boundary conditions listed above will give a separable problem. Separability is the critical condition that must be met in order to use one of the so-called "fast direct" solution techniques.

- Self-adjoint problems yield approximations with useful properties. A strict mathematical definition of self-adjoint may be found in Sommerfeld [1949]. For our purposes we note that self-adjoint problems usually have operators that are written in divergence form (see Eq. 7) and for which one of the four boundary conditions listed above is prescribed (the domain is not necessarily rectangular). Approximations to self-adjoint problems usually produce symmetric positive definite systems for which there are specially designed solution algorithms.

There may be singularities in the problem, and there are several types to contend with: coordinate, operator, and coefficient singularities. The first type occurs in coordinate systems containing singular points. At these points certain terms in the operator are undefined (even though the operator itself is well-defined), and special knowledge must be used to define the terms correctly. For instance, in cylindrical coordinates (r,z) the first term of Eq. 3 is undefined at $r=0$. If the solution is not specified along $r=0$, it can be shown that the appropriate condition is

$$\frac{\partial u}{\partial r}(0,z)=0,$$

in which case

$$\frac{1}{r} \frac{\partial}{\partial r} \left(r \frac{\partial u}{\partial r} \right) = 2 \frac{\partial^2 u}{\partial r^2} \; .$$

The second type of singularity is called an operator singularity. It is the result of specifying certain combinations of boundary conditions with an operator in such a way that a non-unique solution results. When this occurs, a singular approximation usually results, and special care must be taken in solving it. Also, non-uniqueness of the solution usually implies some condition on the data of the problem, i.e., on the given functions f and g in Eqs. 1a and 1b. Consider the Neumann problem for Poisson's equation on a domain Ω with homogeneous boundary data, i.e.,

$$\nabla^2 u = f \quad \text{in } \Omega \tag{12a}$$

$$\frac{\partial u}{\partial n} = 0 \quad \text{on } \partial\Omega. \tag{12b}$$

The solution u, if it exists, is not unique. (If u is a solution, then so is $u+c$ where c is any constant.) Furthermore, integrating Eq. 12a over Ω, applying Green's identity and the boundary condition (12b), yields the condition

$$\iint_{\Omega} f d\Omega = 0$$

on f. This condition must necessarily be satisfied if there is to be a solution to problem (12). For this rather simple elliptic problem the derivation of the necessary condition on the data is straightforward. However, for more general problems the necessary condition (or conditions) may be difficult to determine.

The third type of operator singularity occurs when the coefficients of the operator have jump discontinuities. Such problems occur in modelling neutron diffusion wherein one has an operator of the form (7) with u being the neutron density and $k(x,y)$ representing the diffusivity of the material. At a material interface there is a jump discontinuity in $k(x,y)$. In designing an approximation one must know the locations of such discontinuities. Proper treatment of these discontinuities is discussed by Varga [1962, Chapter 6].

The last feature we discuss is the treatment of free boundaries [Cryer, 1977]. Here the boundary (or a portion of it) is determined by the application of some independent condition on the solution. This condition frequently is specified by a nonlinear equation involving u at the boundary. To our knowledge there is no general-purpose software capable of treating free boundaries.

We conclude this section by presenting two examples of elliptic problems that occur in the physical sciences. The field of electrostatics provides the most widely known examples.

Example 1

Suppose a wire is bent into a closed, but arbitrary loop and is flattened between two insulating plates. If the wire is connected to some voltage source, we may ask what the voltage potential is at each point inside the closed loop. Since the loop of wire may be represented as a curve $\partial\Omega$ in the plane, an application of Kirchhoff's current law [Seeley and Poularikas, 1979] yields

$$\nabla\cdot(\sigma\nabla\phi)=0 \quad \text{in} \quad \Omega$$

$$\phi=\phi_s \quad \text{on} \quad \partial\Omega,$$

where ϕ is a scalar potential function from which the voltage may be calculated, σ is the coefficient of electrical conductivity, and ϕ_s is given by the voltage source. (The voltage is determined only up to an additive constant, so voltage at a point is calculated as the difference between the potential at that point and the potential at some reference point.) By various assumptions on Ω and σ we could illustrate each of the operators (2) through (7) with this example.

Example 2

Our second example illustrates the occurrence of Poisson equations in models for fluid flow — an area of computing that has provided a large variety of EBVPs. The Navier-Stokes equations for an incompressible, constant-property fluid [Lamb, 1945] are

$$\nabla\cdot\vec{u}=0 \quad \text{(conservation of mass)}$$

$$\frac{\partial\vec{u}}{\partial t}+(\vec{u}\cdot\nabla)\vec{u}+\frac{1}{\rho}\nabla p=\nu\nabla^2\vec{u} \quad \text{(conservation of momentum)} \qquad (13)$$

where $\vec{u}$ is the velocity vector, ρ is the constant density, p is the pressure, and ν is the kinematic viscosity. By defining the vorticity $\vec{\omega}$ as the curl of the velocity, i.e., $\vec{\omega}=\nabla\times\vec{u}$, and taking the curl of Eq. 13 above, we eliminate the pressure variable to get

$$\frac{\partial\vec{\omega}}{\partial t}+\nabla\times(\vec{u}\cdot\nabla\vec{u})=\nu\nabla^2\vec{\omega}. \qquad (14)$$

In two-dimensional flow

$$\vec{u}=\begin{Bmatrix} u \\ v \\ o \end{Bmatrix} \quad \text{and} \quad \vec{\omega}=\begin{Bmatrix} o \\ o \\ \omega \end{Bmatrix},$$

so Eq. 14 reduces to the scalar equation

$$\frac{\partial\omega}{\partial t}+(\vec{u}\cdot\nabla)\omega=\nu\nabla^2\omega. \qquad (15)$$

We now introduce a stream function ψ by the relationships

$$u = \frac{\partial \psi}{\partial y} \quad \text{and} \quad v = -\frac{\partial \psi}{\partial x} \ .$$

From the definition of the curl, for this special two-dimensional problem, we find the relationship

$$\omega = \frac{\partial v}{\partial x} - \frac{\partial u}{\partial y} = -\nabla^2 \psi. \tag{16}$$

Incorporating the stream function ψ into Eq. 15 gives

$$\frac{\partial \omega}{\partial t} + \frac{\partial \psi}{\partial y} \frac{\partial \omega}{\partial x} - \frac{\partial \psi}{\partial x} \frac{\partial \omega}{\partial y} = \nu \nabla^2 \omega. \tag{17}$$

In a numerical scheme Equation 17 is used to advance the vorticity in time, and Equation 16 — a Poisson equation — is used to calculate the stream function ψ at that time.

ALGORITHMIC ISSUES

In this section we shall discuss briefly several approximation techniques used in software for elliptic problems and the algorithms used to solve the approximating systems. The choice of approximation must be based on several considerations. Is the approximation needed over the entire domain or only at a few points? Is the solution likely to have steep gradients or is it generally smooth? What software exists? (With the growing complexity of algorithms and attendant investment required in programming, this is also a major factor in determining which method to use.)

Approximation Methods

Although many different algorithms have been proposed for approximating elliptic equations [Gladwell and Wait, 1979], reliable software implementations usually are based on either finite difference or finite element approximations. We will describe these two techniques very briefly, giving only enough detail to point out salient features. To keep the presentation focused on the relevant points, we will discuss the application of these two approximations to the so-called "model" problem, i.e., the solution of Poisson's equation

$$\frac{\partial^2 u}{\partial x^2} + \frac{\partial^2 u}{\partial y^2} = f(x,y) \tag{18a}$$

on the unit rectangle

$$R = \{(x,y) : 0 < x < 1, \ 0 < y < 1\} \tag{18b}$$

with the simple Dirichlet condition

$u=0$ (18c)

on the boundaries of R.

To generate a finite difference approximation to problem (18), we select an integer n and define a grid spacing $h=1/(n+1)$ and the grid points (x_i, y_j) by

$$x_i = ih, \quad i=0,1,...,n+1 \tag{19a}$$

$$y_j = jh, \quad j=0,1,...,n+1. \tag{19b}$$

At each interior grid point, we replace the derivatives in Eq. 18a with a second-order central finite difference approximation, resulting in the linear system of equations

$$u_{i-1,j} + u_{i+1,j} + u_{i,j-1} + u_{i,j+1} - 4u_{i,j} = h^2 f_{i,j}, \tag{20a}$$

where i and $j=1,2,...,n$. The boundary condition (Eq. 18c) yields the simple equations

$$u_{i,0} = u_{i,n+1} = u_{0,j} = u_{n+1,j} = 0, \tag{20b}$$

which are incorporated into Eq. 20a. To obtain the approximations $u_{i,j}$ to the solution values $u(x_i, y_j)$, we must solve a very sparse (five non-zero coefficients per equation) system of n^2 equations. The algorithms used to solve the system will be discussed in more detail below.

We have presented the simplest case of a second-order accurate finite difference approximation. Higher-order accuracy may be achieved through the use of deferred corrections [Pereyra, 1966] or the HODIE method [Boisvert, 1981]. Of course, more complicated operators require more work in developing the approximation, and care must always be taken when approximating an operator in the vicinity of a curved boundary. However, the basic simplicity of this method has been most attractive to scientists and, consequently, to software producers (cf. CMMPAK, ELLPACK, FISHPAK, FFT9, HELM3D, ICCG, KLEV, MG00, and SLDGL.)

A finite element approximation to an elliptic problem uses two sets of functions — a set of basis functions $b_i(x,y), i=1,2,...,N$, and a set of trial functions $t_j(x,y), j=1,2,...,N$. The approximation $v(x,y)$ to the solution u is written as a linear combination of the basis elements, i.e.,

$$v = \sum_{i=1}^{N} c_i b_i.$$

With an inner product (f,g) defined by

$$(f,g) = \int_0^1 \int_0^1 f(x,y) g(x,y) \,dx\,dy,$$

the unknown coefficients c_i are determined by the linear system of equations

$$(Lv, t_j) = (f, t_j), \quad j = 1, 2, \ldots, N .$$ (21)

For the model problem we will illustrate the finite element method by choosing the basis functions to be the tensor product of the one-dimensional "hat" or "Chapeau" functions and choosing the trial functions to be the basis functions. To write down the approximation in this case, we change notation slightly and use a double subscript for enumerating the basis and trial functions, so the basis functions are

$$b_{i,j}(x, y) = a_i(x) a_j(y), \quad i \text{ and } j = 1, 2, \ldots, n$$ (22)

where $a_i(x)$ is the piecewise linear polynomial satisfying

$$a_i(x_j) = \begin{cases} 1, & \text{if } x_i = x_j \\ 0, & \text{otherwise} \end{cases}$$

and the grid points x_i and y_j are given by Eqs. 19a and 19b. We will write the approximation as

$$v(x, y) = \sum_{i=1}^{n} \sum_{j=1}^{n} c_{i,j} b_{i,j}(x, y),$$ (23)

where the $c_{i,j}$ will be determined from Eq. 21. Note that the operator L, the Laplacian, cannot be applied to v since it does not have continuous second partial derivatives. To alleviate this difficulty, we apply Green's theorem to the integral on the left side of Eq. 21 to obtain

$$(Lv, t_j) = \iint_R t_j(x, y) \nabla^2 v(x, y) \, dx \, dy$$

$$= -\iint_R \left(\frac{\partial v}{\partial x} \frac{\partial t_j}{\partial x} + \frac{\partial v}{\partial y} \frac{\partial t_j}{\partial y} \right) dx \, dy.$$ (24)

The boundary integral is zero since the trial functions vanish there. Using this equivalence, we now substitute Eq. 23 into Eq. 21 and, using the definition (22) of the basis functions, arrive at the system

$$\sum_{k,l=-1}^{1} c_{i+k, j+l} - 9c_{i,j} = 3h^2 f_{i,j}$$ (25a)

for i and $j = 1, 2, \ldots, n$. Since this basis is interpolatory, i.e.,

$$v(x_i, y_j) = c_{i,j}, \quad i, j = 1, 2, \ldots, n,$$

we obtain from the boundary condition (18c) that

$$c_{0,j} = c_{n+1,j} = c_{i,0} = c_{i,n+1} = 0 \quad i, j = 0, 1, \ldots, n .$$ (25b)

Equations 25a and 25b represent a sparse system of n^2 equations that must be solved for the approximation. We shall discuss the solution techniques later.

Several general features of the finite element method can be seen from this example. The basis elements are chosen to be non-zero only in a small region of the domain, resulting in sparse systems. Having obtained the coefficients $c_{i,j}$, note that the approximation (23) is defined over the entire rectangle and may be evaluated anywhere in the domain. (Contrast this feature with a finite difference approximation which requires interpolation to obtain approximations off the grid.) Although a uniform rectangular grid was used in this example, a much more common practice is to use a triangular grid. The very attractive feature of local grid refinement is quite easily accomplished then, e.g., connecting mid-points of the edges of a triangle yields four smaller basis elements. This feature is especially important for resolving steep gradients. With the use of error estimates, adaptive refinement is possible and has been implemented in some software (B2DE, PLTMG, TWODEPEP).

Another feature is the use of an equivalent variational formulation of the problem. (We produced this formulation in the process of deriving Eq. 24 from Eq. 21.) The variational formulation allows one to use basis elements with less differentiability than that required of the solution, thereby producing more compact approximations that are generally easier to solve. It also facilitates the construction of "weak" solutions, i.e., those possessing less differentiability than required by the differential equation formulation.

One other approximation technique — collocation — has also been used in software. Collocation can be viewed as the special case of a finite element method wherein the trial functions t_j are Dirac delta functions. Specifically, one chooses N points $p_j = (x_j, y_j)$ in the domain Ω and defines the trial functions

$$t_j(p) = \delta(p_j - p), \quad j = 1, 2, \ldots, N, \tag{26}$$

where $p = (x, y)$. Substitution of Eq. 26 into Eq. 21 produces the condition

$$(Lv)(x_j, y_j) = f(x_j, y_j), \quad j = 1, 2, \ldots, N;$$

i.e., the approximation v is required to satisfy the differential equation exactly at the collocation points p_j. ELLPACK contains some collocation software.

Finally, an efficient technique for solving some EBVPs (principally Laplace equations) utilizes approximations to boundary integral representations of the solution. This technique is useful for problems where the domain is the unbounded exterior of some bounded region, and for problems where the solution is desired at only a few points. D03EAF is based on this technique.

Solution Algorithms

As we have seen, the approximation of an elliptic problem produces a (usually) large, sparse linear system which we denote symbolically as

$$Av = b. \tag{27}$$

Thus, a major task of software development is to provide effective routines that will calculate v or some sufficiently good approximation to it. Most algorithms can be classified as direct or iterative methods, although there is at least one hybrid method. Direct methods are characterized by the condition that in the absence of rounding errors they produce the exact answer in a finite number of arithmetic operations, whereas iterative methods produce a sequence of approximations that converge to the solution v only in the limit.

Direct Methods

The use of direct methods has been primarily restricted to separable elliptic problems for which very fast algorithms have been developed. However, with the increasing availability of sparse matrix software more of these algorithms have been used in solving non-separable problems.

For separable equations the variant of Gaussian elimination known as cyclic reduction [Buzbee, Golub and Neilson, 1970; Swarztrauber, 1974b; Sweet, 1977] has been used extensively. This algorithm requires a very rigid structure of the matrix A. By appropriate linear combinations of equations in (27), half the unknowns may be eliminated. The remaining unknowns form a system with the same structure as the original system. Hence, similar combinations will eliminate half of these unknowns. This process is repeated until a single equation in one unknown remains. Once this equation has been solved, a back-substitution process computes all other unknowns. For our model finite difference approximation a cyclic reduction algorithm computes the solution in $O(n^2 \log_2 n)$ operations. Furthermore, most calculations are done in place, so very little extra storage is required. This is the principal solution algorithm in FISHPAK.

Another powerful algorithm is the matrix decomposition technique [Buzbee, Golub and Neilson, 1970; Swarztrauber, 1977] based on the use of the fast Fourier transform (FFT). This technique applies to operators of the form (4), but not to the more general separable operator (6). In this method we expand the vector b in a finite trigonometric series of the appropriate form (determined by the boundary conditions) using the FFT to calculate the coefficients of the expansion. Assuming that the vector v has a similar expansion, we can now calculate the coefficients for v easily. The FFT is then used to construct v from its trigonometric expansion.

The principal work in this method is done by the FFT routine, and the limiting factors in this algorithm are those imposed by the FFT routine. One limitation is that the number of grid points in one dimension must be a highly composite number, i.e., the n of our model problem should be the product of a relatively small number of prime numbers. In this case the operation count for this method on the $n \times n$ grid is $O(n^2\log_2 n)$. Furthermore, little extra storage is required, as most computations are done in place. FFT9 and FISHPAK have implementations of this technique.

The two algorithms based on the FFT and cyclic reduction described above are very efficient methods for the class of problems to which they apply. However, one frequently encounters systems that are slight perturbations from the rigid structure required by these algorithms. In such cases one can apply the capacitance matrix method [Buzbee *et al.*, 1971; Proskurowski, 1976]. In this algorithm one solves an unperturbed system with arbitrary data values for the equations that have replaced the perturbed equations. The solution, a trial solution, is obtained by use of one of the fast, direct methods. One calculates by how much the trial solution fails to satisfy the perturbed equations and uses this information to calculate the appropriate change in the arbitrary data values. One more solution of the unperturbed system then yields the solution to the original perturbed system. This method extends considerably the applicability of the fast direct methods.

A variant of this method has been developed for solving Poisson's equation (12a) on general irregular regions by embedding it in a rectangle. By appealing to classical potential theory, the implicit capacitance matrix method [Proskurowski, 1979] has become an efficient technique for solving these equations. Existing software now can solve Poisson's equation on an irregular region at the cost of about five to ten fast direct solutions on the embedding rectangle and with only a modest amount of additional storage. These techniques have been implemented in CMMPAK and HELM3D.

The use of direct methods for approximations to elliptic problems with the more general operators given by Eqs. 7-9 requires the use of sparse matrix techniques [Duff, 1977]. Basically all these algorithms are some form of Gauss elimination. They differ in the data structure employed to reduce the storage of "zero" elements and the ordering of the elimination of unknowns to control the creation of non-zero entries. A comprehensive discussion of sparse matrix techniques may be found in Chapter 8 of this book.

To illustrate the need for sparse matrix algorithms, consider solving the finite difference approximation (20) or the finite element approximation (25) to our model problem by elimination. In the notation of Eq. 27, the matrix A is of order n^2 and, assuming the unknowns are ordered in the "natural" or "typewriting" order, is a banded matrix with bandwidth n. Elimination of the unknowns in this order will create non-zero entries in all the positions within the band of A. There are $O(n^3)$ such positions, and they require $O(n^4)$ operations to fill in. Note that in addition to the n^2 locations required to specify the data (vector b of Eq. 27), over n times that many non-zero entries are created. However, the nested dissection ordering [George, 1973], which is theoretically optimal, creates $O(n^2\log_2 n)$ non-zero entries using $O(n^3)$ operations. Comparing these two sets of counts, one sees an enormous difference in the auxiliary storage required and in the total number of operations (a measure of computational speed) required to solve for the approximation. The key word here is "theoretically." Much overhead storage and index calculation are not counted in this algorithm, but must certainly be included in an implementation.

Iterative Methods

Most iterative methods [Hageman and Young, 1981; Wachspress, 1966; Varga, 1962] require only the ability to form matrix-vector products of the form Av, thereby requiring only that the original matrix A and a few vectors the length of v be stored. A disadvantage of this class of methods is the need for a good initial approximation to the solution of Eq. 27. However, it is often available, especially in fluid dynamics models (cf. Example 2).

Software implementations of the classical Gauss-Seidel, SOR, and ADI iterative schemes exist. Much work has been done to automate the selection of acceleration parameters for SOR and ADI. ITPACK (whose programs are in ELLPACK) is an example. However, it should be noted here that most implementations assume the user has derived the approximating linear system and do not provide this preprocessing feature.

Several relatively new iterative techniques for which some software exists or is being developed are the preconditioned conjugate gradient [Concus, Golub and O'Leary, 1976] and multi-grid [Brandt, 1977] iterations. When the system (27) is symmetric as for approximations to the operator (8), the conjugate gradient algorithm may be applied. It constructs a finite sequence of approximations with the property that in the absence of round-off error the last iterate coincides with the solution of the linear system. The number of iterations required is the order of the system, i.e., about the number of grid points, which may be quite large. But, the convergence of the iteration can be accelerated considerably by performing at each iteration an operation known as pre-conditioning. There are many pre-conditionings, e.g., a fixed number of SOR iterations, an application of a fast direct method on some auxiliary elliptic system, or − in the case of the software package ICCG − the application of a sparse, incomplete factorization of the original matrix A.

The multi-grid iteration is based on the fact that an iterative solution technique such as Gauss-Seidel reduces high-frequency components of the error faster than low frequency components. To use this fact, the multi-grid iteration defines a set of nested grids with the finest grid corresponding to the original grid. A fixed number of Gauss-Seidel iterations are performed on the finest grid, the approximate solution and data are transferred to the next coarser grid where more Gauss-Seidel iterations are performed, the process is repeated on the next coarser grid, and so on. When transferring from one grid to another, high frequency errors are introduced again, so the transfers are not to ever coarser grids and then to ever finer grids, but are done in a more complicated pattern. This process reduces the error very rapidly and shows much promise for future work. Although this iteration has been described in terms of finite difference grids, there is a direct analogy with nested finite element triangulations. One frequently hears the name multi-level iteration when this technique is applied to finite element approximations. Examples of software employing multi-level iterations are KLEV, B2DE, and PLTMG.

SOFTWARE ENGINEERING ISSUES

The numerical methods that we have just outlined are the heart of software for EBVPs, and the choice of algorithm is the most important decision a software developer must make. No one wants software that is unreliable or inefficient. However, the effective packaging of the numerical method in software is a complex engineering problem that also can affect the success of the project. No one wants software that is awkward to use, is inadequately documented, or requires volumes of tediously generated data as input. In many cases users will choose less efficient software when it is "friendlier."

In this section we discuss some of these software engineering issues from a user's point of view. We describe the basic organizational structure of such software and show how basic design decisions can influence the types of applications and the class of users. One particularly important aspect for users — the form of input and output — is discussed in detail. Finally we address other aspects of the software life cycle that should be of concern to software developers — testing, evaluation, distribution, and maintenance. Following this section we present case studies that describe how these software engineering questions were resolved in the implementation of two specific software packages.

Basic Software Organization

Software for EBVPs comes in a wide variety of sizes and shapes. The forms that we have identified include single modules, systematized collections of modules, interpretive systems, preprocessor systems, and large-scale application systems. We next describe each of these software classes in detail.

Single Modules

Single modules represent the simplest organization. Here a user writes a program that sets up a call to a library subprogram such as B2DE, D03EAF, FFT9, HELM3D, KLEV, or PLTMG which performs all the required computations. Input is by means of the calling sequence and sometimes also by user-written subprograms that define the operator and the domain. Output is usually a vector of solution components (see Fig. 1). One advantage of this organization is that the solution of the elliptic problem can be easily embedded in a larger calculation (e.g., fluid dynamics codes). Other users must, in effect, provide their own user interface. The calling sequences for such programs are often long and complex, especially if they are designed for a fairly large class of problems.

Many additional subprograms are now available that perform only one phase of the computations involved in solving elliptic problems, for example,

- Domain triangulation (such as D03MAF [NAG, 1978]),

- Equation and unknown renumbering (such as REDUCE [Crane *et al.*, 1976]), and

- Special-purpose linear algebraic equation solvers (such as D03EBF [NAG, 1978], GMA and GMAS [Bank, 1978]).

Such programs can be considered as "software parts" in the context of elliptic equations, since they may be pieced together to form a program for solving an elliptic problem. Although such software parts are important tools for the construction of programs, they are usually inaccessible to most users directly because of the increased complexity of their user interfaces (special-purpose linear equation solvers require that the user provide a complete discretization of the problem as input).

Systematized Collections of Modules

The organization of software parts into a systematized collection that can be used (by calls in a user's program) to construct an algorithm to solve a particular problem has proven successful in such well-known packages as EISPACK [Smith *et al.*, 1976; Garbow *et al.*, 1977] and LINPACK [Dongarra *et al.*, 1979]. This organization is illustrated in Fig. 2. Several choices for modules at each phase of the computation may be possible: domain discretization, operator discretization, equation and unknown renumbering, algebraic equation solution, and output.

Such a collection is made coherent by a systematic naming convention for modules and parameters and by having inter-module communication handled automatically. The collection is complete if programs for the solution of a reasonably large class of elliptic problems can be easily constructed with modules from the collection. The internal problem-solving modules in ELLPACK are organized in this way, as is the multi-grid software of Brandt [1977]. Other packages provide capabilities in just one or two phases of the computation. Examples of these are ITPACK [Grimes, Kincaid and Young, 1979], the Yale Sparse Matrix Package [Eisenstat *et al.*, 1977a and 1977b] and SPARSPAK [George and Liu, 1979] for the solution to large sparse systems of linear algebraic equations typically found in the discretization of partial differential equations.

Although collections of software parts admit great flexibility in allowing programs to be tailored for individual applications problems, they require a great deal of sophistication for proper use. An alternative approach is to give users a collection of specialized interfaces to lower-level collections of software parts (see Fig. 3). Although some flexibility is lost, the resulting package is much easier to use, usually requiring a single subprogram call to solve a given problem. The subprograms at the user-interface level might be, for example,

- Different algorithms for the same problem, or
- Solutions of different PDE problems, or
- Specialized interfaces for problems in a specific applications area.

FISHPAK, ICCG, MG00, and SLGDL are examples of collections of this type.

Small-scale Systems

A software system differs from a collection of subprograms in that the user interface comprises a specialized language for the solution of a class of elliptic problems. This implies the existence of an input processor (a "main program") that reads the problem specification and then either solves it immediately or produces code (such as a Fortran main program) that may be executed in a later job step. The former is called an interpretive system while the latter is a preprocessor (or compiler-based) system.

Interpretive systems are the traditional approach: The user supplies "data cards" to a program that solves the problem. The advantage is that no programming in a procedural computer language need be done by the user, and so non-programmers can be easily taught how to use the program. Unfortunately, the input to such programs is often a large number of fixed format data cards (Fortran NAMELIST input is not much better), and the preparation of these is an intellectual activity akin to programming in machine language.

One advantage of interpretive systems is that they are ideal for interactive use, but PDE problems are usually so large that interactive problem-solving is cumbersome on many systems. (Advances in hardware technology and the growing availability of virtual memory operating systems are already changing this restriction.) In fact, storage problems must often be confronted in the construction of such systems, and the static storage allocation of Fortran is almost always the culprit. As a result such systems usually have large memory requirements so that the most difficult problems can be solved. The only alternative is the use of external files, non-portable constructs such as overlays, or very sophisticated memory management procedures. Often such systems provide very versatile postprocessors for manipulating the computed solution (such as computer graphics) but lack the versatility of library subprograms in that the solution is not readily available for use by other programs, except possibly through the local file system.

Many systems of this type have been written for specific applications problems (see Oden, Clough and Yamamoto [1972] or Pilkey, Saczalski and Schaeffer [1974], for example). We make no attempt to describe these; instead we confine our discussion to general-purpose mathematical software.

Preprocessor systems usually have fewer memory-management problems. As shown in Fig. 4, the input processor reads a problem specification and writes a main program (usually Fortran) that sets up appropriate interfaces to solution modules in its library. The result is a program tailored to the particular application, which may be compiled and executed to solve the problem. Since the input processor executes in a separate job step, it can be made much more complex than in interpretive systems without reducing the storage available to solution modules. As a result, the input can be in a much higher-level language, and thus much easier to use. Such very high-level languages come in two forms: problem statement languages and extensions to existing languages (usually Fortran). In a problem statement language, the specification is given in free format using common mathematical terminology. For example, the specification of a Poisson problem might look like

```
EQUATION.    UXX + UYY = 6.*X*(Y-1.)*(1.-X)*(2.-Y)*EXP(X+Y)
DOMAIN.      RECTANGLE  (0.,1.) X (1.,2.)
BOUNDARY CONDITIONS.    U=1.  ON X=0.
                 U=Y   ON X=1.
                 U=1.+X ON Y=2.
                 U=1.  ON Y=1.
GRID.    UNIFORM, 17 LINES IN X, 25 LINES IN Y
OUTPUT. PLOT SOLUTION
```

Both ELLPACK and TWODEPEP have problem statement languages to describe their input, and both produce Fortran programs as output.

Language extension preprocessors read Fortran programs with special declaration statements for elliptic problems and special executable statements for solving them. The preprocessor replaces the special statements by Fortran statements that call programs in its library, while leaving the rest of the Fortran code intact. This approach maintains the flexibility of using a library of subprograms since the solution of the problem may be embedded in a larger calculation easily. A version of ELLPACK with these facilities is planned for release in early 1983.

Large-scale Systems

The remaining class of software is the large production-oriented applications system. Examples of such systems are the structural analysis systems NASTRAN [McCormick, 1972], ASKA [Schrem, 1974], and SAP [Wilson, 1972]. Since these programs are designed for problems in specific applications areas, we do not discuss them in detail (there are many more such codes than general-purpose mathematical software systems). Such systems are usually large (several hundred thousand lines of Fortran is not unknown) and, as illustrated in Fig. 5, very complex. In fact, they take on many functions normally associated with operating systems. Data bases of structural elements are available, as well as data preprocessors to manipulate them. Input language processors read problem specifications and put them in internal format. An executive program controls the solution process, providing dynamic storage allocation, managing external scratch files, calling solution modules, invoking overlays, etc. Postprocessors take output and transform it for human consumption. Such systems are truly the workhorses of many engineering fields, and designers of mathematical software can learn from the successes and failures of these efforts.

Representation of Input and Output

A software user wants a program that takes data and produces the output needed for a particular application (Fig. 6a), thus minimizing effort in solving the problem. The implementation of a numerical method, however, usually requires input in a much different, lower-level form (e.g., a discretized version of the problem). A designer of mathematical software minimizes effort by requiring input to the program to be in exactly this form (see Fig. 6b). The rationale is usually that the resulting software module is then, in principle, applicable to a much larger class of problems; in fact, it probably means that the program will go unused. This was exactly the experience leading to the development of FISH-PAK which is described in the next section. An applications program designer must be concerned with the transformation of data from that available to users in the application area to that required by the numerical method (see Fig. 6c). In practice this job is often quite difficult.

The input common to all elliptic problem solvers consists of a specification of the domain, the operator, and the boundary conditions. In some cases the input is implicit; e.g., the program only solves Laplace's equation, or the domain is always rectangular, or the boundary conditions are always homogeneous. In other cases input must be explicitly supplied, and the multidimensional nature of the input data is the source of many practical difficulties if the production of a "user-friendly" system is desired.

Domain Representations

First, a multidimensional domain must be represented. In many applications the boundary is known explicitly, e.g., a union of rectangles or a cylinder. In this case several representations are possible.

- *Parametric representation of boundary pieces.* Here the boundary (in two dimensions) is the union of k smooth curves defined parametrically as

$$(x_i(s), y_i(s)), \quad 0 \leqslant s \leqslant 1, \ i=1,...,k$$

where

$$x_i(1)=x_{i+1}(0), \qquad i=1,...,k-1$$
$$x_k(1)=x_1(0),$$

with similar relations for the y_i. This is the most general and mathematically elegant representation, requiring the user to provide only a single subroutine, BNDRY(I,S,X,Y), that returns (X,Y) for any input of I and S. In the ELLPACK and TWODEPEP systems the parametric equations are actually entered symbolically. Unfortunately, the flexibility of this system is also its disadvantage. It admits domains that are too complex for any general-purpose software to handle. It is also an awkward representation for domain processing; simply finding the intersection of the boundary with a line (a common operation) is a general zero-finding problem, for example. Finally, a given curve has infinitely many parametric representations, many of which can lead to numerical difficulties in domain processing. In three dimensions these problems are greatly magnified, and a pure parametric representation is rarely used in this case. Instead, simple extensions of two-dimensional objects are allowed, for example, generalized cylinders of the form

$$(x_i(s), y_i(s), z), \ 0 \leqslant s \leqslant 1, a \leqslant z \leqslant b, \quad i=1,...,k$$

or solids of revolution.

- *Selection from a catalog of domain parts.* Many complex domains in two dimensions can be represented by the specification of a sequence of boundary pieces, each selected from a fixed set of allowable shapes. For example, consider the set of domains whose boundaries are made up of lines and arcs of circles. The domain of Fig. 7a could then be represented as follows:

Type	Coordinates
0	(2,3)
1	(4,1)
1	(4,0)
1	(2,0)
2	(0,2), (1.414,1.414)
1	(0,3)
1	(2,3)

As in a catalog we must specify "part numbers" for each piece (0 for an initial point, 1 for a line segment, 2 for a circular arc) and its "size" (the endpoints for each piece and an interior point for an arc — the initial point is the last point of the previous entry). Schemes similar to this are used by B2DE and PLTMG. This scheme also extends to three dimensions using generalized cylinders or solids of revolution. Alternatively, a complex two-dimensional domain may be expressed as the union of simpler two-dimensional parts, such as rectangles and triangles. Such a specification is slightly more complicated since more information is needed to express how large each element is and how it joins up with each of its neighbors. The use of this technique in three dimensions has been pioneered by engineers who have constructed data bases of structural forms such as beams, plates, and solid polyhedra for use in large, general-purpose structural analysis systems [Pilkey, Saczalski and Schaeffer, 1974]. Not all domains of interest can be represented exactly from a domain parts catalog. However, in many applications it is easier to construct an approximate domain in this way than to produce a parametric representation. In addition, by restricting the allowable domains, internal domain processing can be greatly simplified.

• *Discrete domain representations.* A two-dimensional domain may be specified (approximately) by a list of points on its boundary; and if the domain is only known discretely, then this is the most convenient form for the user. In fact, some algorithms require this form explicitly (boundary integral methods such as in D03EAF, for example), although in most cases some continuous boundary curve is derived from this, connecting the boundary points with line segments, for example. Such piecewise polynomial approximations are equivalent to choosing from a (rather restrictive) catalog of boundary parts.

Regardless of the method of domain specification, a user of software for EBVPs must invariably provide some information on how the domain is to be discretized. This is typically a rectangular or triangular grid that transforms the domain into a network of nodes as shown in Figs. 3.7a and 3.7b. The specification of a uniform rectangular grid on a rectangular domain is trivial, requiring only the positions of the horizontal and vertical grid lines to be specified. For non-rectangular domains each grid point must be labelled as interior, exterior, or adjacent to the boundary (with the distances to nearby boundary points supplied). A separate list of points where the grid intersects the boundary must also be provided, and the boundary conditions active at each of these points must be specified. The description of a triangulation is just as complex. A list of node coordinates must be supplied along with a list of the three nodes associated with each triangle. Edges coinciding with boundary segments must also be distinguished and the boundary conditions holding there identified.

The generation of such data is clearly a tedious (and hence error-prone) operation. As a result, many programs have a domain processing phase in which a simple, high-level description of the domain and its discretization is read, and detailed tables specifying the discrete domain are automatically generated. The ELLPACK domain processor, for example, takes a parametric description of the boundary and a list of horizontal and vertical mesh line positions and generates all the information to specify a rectangular node network. The TWODEPEP system starts with a similar parametric representation of the boundary along with a very coarse triangulation and refines the mesh by distributing new triangles uniformly with respect to a user-defined function.

In large finite element systems for structural analysis, the preparation of input by hand to describe a complicated three-dimensional object such as an automobile body can be a staggering undertaking requiring months of effort [Kirioka and Hirata, 1972]. To reduce this effort, a large number of preprocessors, often orchestrated with sophisticated computer graphics, have been written to aid users in defining, checking, and modifying domain discretizations [Napolitano, Monti and Muro, 1974; Everstine and McKee, 1974]. In spite of these advances, many programs for elliptic problems still require users to define their domains implicitly in terms of detailed, low-level discretization tables.

Representation of Operators and Boundary Conditions

We have not yet addressed the question of how the user specifies what the equations to be solved are, that is, the operators L and B of Eq. 1 and the right-hand side functions f and g. In the simplest of cases the operators are fixed by the program (e.g., a program for solving the Dirichlet problem for Laplace's equation, such as D03EAF), so that the explicit specification of these quantities is not required. In more general-purpose programs, the selection from a family of operators may be allowed, with coefficient functions such as $k(x,y)$ of Eq. 7 required as input.

Algorithms employing automatic or adaptive domain discretization (mesh refinement) require such exact knowledge of operator coefficients. Usually, however, the set of points at which the coefficients are evaluated is fixed once the domain is discretized, and some programs require input of these functions as a list of values associated with the nodes of the discretized domain network. This is especially true for the forcing function f which is measured data in many applications or the result of a previous calculation as in fluid flow computations.

Representation of Solutions

The solution of an EBVP is a continuous function satisfying the prescribed equation and boundary conditions, but the numerical solutions can come in several very different forms. As discussed earlier, some solution algorithms calculate the coefficients of a function that solves the problem approximately. This is typically the case in finite element programs such as the finite-element-based PLTMG, whose "answer" is a Fortran-callable function that evaluates the solution on the domain. Other algorithms only produce the values of the solution at the nodes of the discretized domain, such as the finite-difference-based FISH-PAK. The former representation for the solution is clearly more versatile, although neither is usually the ultimate goal: Derivatives of the solution at a set of points may be desired or various integrals involving the solution may need to be computed, each for use in some subsequent computation.

Even when the solution function itself is the desired quantity, a table of the solution comprising thousands of numbers is too overwhelming for human consumption. It is important, then, for software to provide some form of post-processing of the results, and the use of computer graphics is a most effective means. Three-dimensional surface plots, contour plots, and cross-section plots can provide much more insight than pages of numbers, and such facilities are now often an important part of software for elliptic problems (e.g., B2DE, ELLPACK, PLTMG, TWODEPEP).

Testing and Evaluation

After design and implementation it is crucial to determine whether the design goals have been met: Does the program reliably compute the requested quantities? Does it fail gracefully? Is it easily transportable to other machine environments? Is the documentation comprehensible? There is also a class of seemingly inconsequential problems that can prevent the easy integration of software into many computing environments. We next describe some of these.

Ideally, a Fortran program should run correctly on any computer with a Fortran compiler and sufficient main memory. In order to do this, not only must numerical software be written in ANSI Fortran, but it must also be sensitive to differences in machine arithmetic. The simplest method of obtaining such information portably is through the PORT library routines I1MACH, R1MACH, and D1MACH [Fox, Hall and Schryer, 1978].

Further complications arise when programs need to take advantage of special operating system facilities such as extended core storage, dynamic storage allocation and communication with the file system. In this case portability is greatly restricted and special versions of the program may need to be targetted to different machine/operating system environments. Finally, the use of computer graphics, although extremely important, also reduces portability. Programs should use widely available graphics software and should be designed so that they can be run without graphics if necessary.

In addition to portability constraints, some further precautions need also be taken when developing subroutine packages. For example, developers should avoid common names (like "F") for internal routines, especially if the user must supply FORTRAN functions to define various quantities (such as the right-hand side of the Poisson equation). Developers must also be aware that their programs may be merged into local subroutine libraries, and thus must avoid multiple versions of the same subroutine (in different precisions, for example) with the same name. Another common practice that leads to multiply-defined entry points in libraries is the inclusion of slightly-modified versions of generally available software modules as internal routines (linear equation solvers, for example) without changing their names.

Writing good program documentation is much more difficult than most people believe, but without it user acceptance is unlikely. It is impossible to "debug" the documentation without having potential users involved. Both on-line and hard-copy documentation should be provided. A systems implementation manual may also be needed if a program has system-dependent features. All should be readable.

There are many subtle problems that can escape the scrutiny of even the most conscientious program developer. Thus, it is imperative that some independent evaluation of programs be made. We favor a formal pre-distribution of the software to a set of test sites where the program is installed, a standard battery of tests is run, the documentation is read, and some users are allowed to exercise it. Test site representatives would then provide a written certification of the program which would then be included in the program's documentation. This is essentially the NATS model exemplified by the EISPACK and LINPACK projects. In the next section we describe how this process was used in the certification of the FISHPAK package. Unfortunately, this requires a great deal of cooperative effort and is not done often enough by software developers.

Distribution and Maintenance

The emergence of centralized software distribution services (see the section on sources of software at the end of this chapter) has greatly simplified the task of getting software from developers to users. The problem of maintenance of scientific software has not yet been adequately confronted, however. A recent study of 487 business data-processing installations [Boehm, 1980] showed the breakdown of software production effort to be development 43%, maintenance 49%, and other 8%, with the maintenance effort divided almost evenly between fixing errors and installing enhancements for users. Thus, a software project cannot end with distribution. Unfortunately, when software is produced in academia (or in similar research environments), as is much of the software we survey here, long-term formal commitments to software maintenance are rare. We believe that clear lines of communication must be maintained between program users and developers, and the software distribution network employed in the maintenance of the ASKA system is a good example of this [Schiffman, 1974]. An alternative is third party "maintenance centers" which would adopt software developed in research institutions, distribute it, install routine "fixes," and seek advice from authors when substantial new versions are warranted.

CASE STUDIES

In this section we discuss two particular EBVP software packages that have been developed. The first one, FISHPAK, was designed as production software for use in, primarily, fluid dynamics modelling. The second, ELLPACK, was designed as a research tool to facilitate the development and testing of new algorithms for the solution of EBVPs. Both packages illustrate important points about software engineering discussed in the previous section.

Case Study: FISHPAK

FISHPAK,* a systematized collection of FORTRAN subroutines providing finite difference approximations to separable EBVPs, was developed in response to a need for fast and reliable software at the National Center for Atmospheric Research (NCAR). NCAR was chartered to provide, in part, facilities for basic research in the circulation of the earth's atmosphere. By its very nature the research has developed a strong numerical computational component devoted to modelling the motion of fluids. Within this class of numerical models one frequently encounters EBVPs (cf. Example 2). Moreover, these dynamic models are developed as time marching schemes, so one is faced with the problem of solving the same EBVP repeatedly. In fact, in the late 1960's the solution of the EBVP commonly accounted for 70-80% of the computer time in a model.

---

*The name of this package is a play on the French word "poisson" or "fish."

In the majority of the models, the EBVP usually was a standard Poisson or Helmholtz equation defined on a rectangle with simple conditions imposed on its boundaries. Such problems could be solved very efficiently — an order of magnitude reduction in the computation time of the solution was not uncommon — by the recently (in the mid 1960's) published fast direct methods. Yet these algorithms were incorporated in only a very few models, since software implementations of the algorithms had been written for only a few scientists. Because these codes were tailored to the specific problems those scientists were trying to solve, others attempting to use them had the difficult task of modifying a mysterious program to suit their own needs. Quite naturally, the meteorologist or fluid dynamicist chose to code an easily implemented iterative technique such as SOR and experimented to determine the optimum relaxation parameter. As a result, many models were computationally inefficient and some developed spurious instabilities [Torrance, 1979] directly attributable to the solution of the elliptic equation.

In the early 1970's Paul Swarztrauber at NCAR and Roland Sweet, an NCAR consultant, extended the Buneman variant of cyclic reduction as a result of work on applying the cyclic reduction algorithm to Poisson's equation on a disk [Swarztrauber and Sweet, 1973]. These extensions of the Buneman algorithm were coded in FORTRAN and deposited in the NCAR program library with the naive expectation that the codes would be heavily used by NCAR modellers. In fact, since the programs solved only the linear systems arising from finite difference approximations and did not provide the user with the complete solution process, modellers still faced the task of writing a substantial interface program.

In 1974 a project was begun to provide scientists at NCAR with a set of subprograms that accepted as input only the information needed to specify the problem and its finite difference analogue and automatically produced the approximate solution as output. The problem domain was restricted to five commonly-occurring two-dimensional Poisson or Helmholtz equations defined on rectangles in cartesian, cylindrical, and spherical coordinates. The specification of boundary conditions was restricted to combinations of Dirichlet, Neumann, and periodicity. And, no input/output was provided except through the parameter list. These restrictions and work by the authors on the treatment of singularities resulted in a problem domain that was free of special cases.

The design criteria for the software were relatively simple. Based on past experience, the most important criterion was ease of use. This criterion dictated in large measure the parameter list and the attendant documentation of the routines. The second criterion — minimal extra workspace — was a recognition that the solution of a Poisson equation was only part of a larger fluid dynamics model that would be using almost all available core storage. The final criterion was speed. The algorithms that were implemented were not the fastest versions of the direct methods available; but within the framework dictated by the more important second criterion, they were optimal.

In 1975, during the final stages of the development and testing of the package, Bill Buzbee at Los Alamos National Laboratory, recognizing the utility of such a package for national laboratories, proposed an informal testing and certification of the package by staff at the Air Force Weapons Laboratory, Lawrence Livermore National Laboratory, Sandia Laboratories, and Los Alamos [Steuerwalt, 1979]. Their recommendations were incorporated into the package as Version 2 [Swarztrauber and Sweet, 1975]. The package contained about 10,000 lines of source, of which about half were comments. During the period 1977-1980, extensive revisions were made to the package, expanding its capabilities, and Version 3 was issued. Shortly thereafter several significant errors were found and corrected, and a revision was issued in 1981 as Version 3.1, which is the current version. It has doubled in size over the original version.

The informal testing and certification of the first version proved invaluable; the few errors subsequently reported by users were mostly due to particular computing environments. Such was not the case with Version 3, however. No independent testing of this version was arranged, and consequently errors in the package that were undetectable in the NCAR computing environment were later reported to the authors.

The principal purpose of FISHPAK was to provide easy-to-use, fast, and reliable solutions to second-order finite difference approximations of the Helmholtz equation

$$\frac{\partial^2 u}{\partial x^2} + \frac{\partial^2 u}{\partial y^2} + \lambda u = f(x,y)$$

defined on the rectangle $a < x < b, c < y < d$ with some combination of Dirichlet, Neumann, or periodic boundary conditions. The finite difference approximation is defined on one of two grids: centered or staggered. A centered grid may be defined for the interval $[a,b]$ above by specifying an integer M — the number of panels into which the interval is divided. Setting $\Delta x = (b-a)/M$, we define the grid points

$$x_i = a + (i-1)\Delta x, \quad i = 1,2,...,M+1.$$

For a centered grid the extreme grid points coincide with the endpoints of the interval. A staggered grid may be defined for the same interval by again selecting an integer M, setting $\Delta x = (b-a)/M$ and defining the grid points

$$x_i = a + (i-\tfrac{1}{2})\Delta x, \quad i = 1,2,...,M.$$

For a staggered grid the endpoints of the interval are one-half the grid spacing from the extreme grid points. With a similar definition for the y-interval $[c,d]$, the effect of a staggered grid is to place grid points at the centers of the grid rectangles of a centered grid. (Staggered grids lend themselves to the construction of approximations that conserve certain physical quantities and are, therefore, desirable in time evolution models.)

The package consists of a set of drivers, a set of solvers, and an associated package of FFT subroutines. Driver subroutines perform three functions: define the finite difference approximation, incorporate the boundary data, and adjust the right side of the linear system when it is singular. (The treatment of singularities will be discussed in more detail below.) A solver is a subroutine that solves linear systems of equations resulting from finite difference approximations to certain separable elliptic equations. The algorithms are implementations of generalizations of cyclic reduction and of matrix decomposition using the FFT.

There are thirteen drivers. Ten drivers solve a (possibly modified) Helmholtz equation on either a centered or staggered grid in one of five two-dimensional coordinate systems: cartesian, polar, cylindrical, surface spherical, or spherical cross-section. The modified equation results from an attempt to provide some three-dimensional solution capabilities. For instance, a user wanting to solve the three-dimensional Poisson equation in cylindrical coordinates

$$\frac{1}{r}\frac{\partial}{\partial r}(r\frac{\partial u}{\partial r})+\frac{\partial^2 u}{\partial z^2}+\frac{1}{r^2}\frac{\partial^2 u}{\partial \theta^2}=f(r,\theta,z) \tag{28}$$

can use an appropriate FFT (dictated by the boundary conditions on the θ boundaries) to transform out the θ variation in Eq. 28, leaving a set of modified Helmholtz equations of the form

$$\frac{1}{r}\frac{\partial}{\partial r}(r\frac{\partial u_k}{\partial r})+\frac{\partial^2 u_k}{\partial z^2}+\frac{\lambda_k}{r^2}u_k=f_k(r,z).$$

Each of these can be solved by an invocation of the appropriate FISHPAK driver. The solution could then be obtained by an inverse Fourier transform.

One driver solves the three-dimensional Helmholtz equation in cartesian coordinates using a centered grid. Two drivers solve the more general equations with operators given by Eqs. 4 and 6. At the user's option these two drivers also provide fourth order approximations by means of deferred corrections.

There are six solvers — four for real systems and two for complex systems. Five of the six solvers are implementations of generalized cyclic reduction while the sixth one uses the FFT algorithm.

FISHPAK also contains a complete package of FFT subroutines written by Swarztrauber. It provides routines to compute sine, cosine, quarter-wave, real periodic, and fully complex transforms on vectors of arbitrary length. (One must remember, however, that the speed of the FFT is directly related to the prime factorization of the length of the vector. Should that number be prime, there will be nothing fast about the FFT.) This package has been coded in such a way that it is highly vectorizable.

An important feature of the drivers in FISHPAK is the automatic detection and treatment of coordinate and operator singularities. Coordinate singularities require either a special definition of an undefined term in the operator, as was discussed above, or a special approximation to the operator at the singularity. An appropriate combination of input values is used to detect operator singularities. When one is detected, it is possible to determine the correct constant [Swarztrauber, 1974a] that must be subtracted from the right side of the approximation to guarantee that a solution to the finite difference approximation exists. This constant is returned to the user. Furthermore, it can be shown that the procedure followed by the drivers produces a weighted least squares solution to the original unperturbed (and possibly inconsistent) finite difference approximation.

We conclude this section by presenting several examples to illustrate the use of FISHPAK. The examples are arranged in order of increasing user involvement in the construction of the finite difference approximation and, consequently, in the package itself.

Example 3

Using a centered 5-degree grid, approximate the solution of Poisson's equation

$$\frac{1}{\sin\theta} \frac{\partial}{\partial\theta} (\sin\theta \frac{\partial u}{\partial\theta}) + \frac{1}{\sin^2\theta} \frac{\partial^2 u}{\partial\phi^2} = 2 - 6\sin^2\theta\sin^2\phi$$

on the northern hemisphere assuming equatorial symmetry, i.e.,

$$\frac{\partial u}{\partial\theta} (\frac{\pi}{2},\phi) = 0, \quad 0 < \phi < 2\pi.$$

The solution of the problem is

$$u(\theta,\phi) = \sin^2\theta\sin^2\phi + c,$$

where c is any constant and, hence, this is a singular problem.

The finite difference approximation may be obtained by a single call to the FISHPAK routine HWSSSP which solves a modified Helmholtz equation on the surface of a sphere using a centered finite difference grid. Appendix A gives the FORTRAN program for this call. Most of the input parameters are self-explanatory. The approximation is returned in the F array, and the constant by which the forcing function is changed is returned in PERTRB. Detection of an error in the input data is indicated through IERROR.

Example 4

A stream function-vorticity model of tropical rain showers was developed [Takahashi, 1975] which, when written in cylindrical (r,z) coordinates, contains the elliptic equation

$$r \frac{\partial}{\partial r} \left(\frac{1}{r} \frac{\partial u}{\partial r} \right) + \frac{\partial^2 u}{\partial z^2} = rv \tag{29}$$

defined on the rectangle $0 < r < R, 0 < z < Z$ with $u=0$ on the boundaries, where u is the stream function and v is vorticity. This equation may also be solved by one call to a FISHPAK routine — namely, SEPX4, which solves equations with operators given by Eq. 4.

The argument list is similar to the one given in the previous example so it will not be repeated here. The major change is the inclusion of an externally defined subroutine that evaluates the functions a, b, and c given in Eq. 5. For this problem we may differentiate the first term in Eq. 29 to obtain

$a(r)=1$

$b(r)=-1/r$

$c(r)=0.$

The user then merely has to add the subroutine

```
SUBROUTINE COFX(X,AFUN,BFUN,CFUN)
AFUN=1.
BFUN=-1./X
CFUN=0.
RETURN
END
```

to a program of about the same size as the one in the previous example.

The two preceding examples illustrate the use of FISHPAK through calls to the various drivers. The drivers will probably meet the needs of the vast majority of the users. However, there will be special problems that cannot be handled by the drivers. In such cases the user may interface directly to the solver routines. Example 5 illustrates how to construct such an interface.

Example 5

Suppose we want to approximate the solution of Poisson's equation

$$\frac{\partial^2 u}{\partial x^2} + \frac{\partial^2 u}{\partial y^2} = f(x,y) \tag{30}$$

on the rectangle $0<x<1, 0<y<1$ with the boundary conditions

$$u=0 \text{ on } x=0, \ y=0, \text{ and } y=1 \tag{31a}$$

$$\frac{\partial u}{\partial x} = 0. \tag{31b}$$

Such a problem could be handled by the routine HWSCRT, but we impose the additional condition that the right boundary $x=1$ be staggered between two grid lines. (This condition has been contrived for this example, but such minor perturbations occur often.) With this additional condition there is no FISHPAK driver capable of producing the finite difference approximation. We will develop the approximation and solve the system by invoking the solver routine GENBUN which is designed to solve systems of the form

$$a_i v_{i-1,j} + b_i v_{i,j} + c_i v_{i+1,j} + v_{i,j-1} - 2v_{i,j} + v_{i,j+1} = g_{i,j}$$

for $i=1,2,...,m, j=1,2,...,n$. The coefficients a_i, b_i, and c_i are given, and certain conditions are assumed for the boundary values.

For this example we select integers M and N, and define

$$\Delta x = \frac{2}{2M-1} \quad \Delta y = \frac{1}{N}$$

$$x_i = (i-1)\Delta x, \quad i=1,2,...,M+1,$$

$$y_j = (j-1)\Delta y, \quad j=1,2,...,N+1.$$

Note that $x_1=0$, $y_1=0$, and $y_{N+1}=1$ are the boundaries where $u=0$ and that $x_M=1-\Delta x/2$ and $x_{M+1}=1+\Delta x/2$ so the boundary $x=1$ is staggered correctly. Denoting the approximation of $u(x_i,y_i)$ by $u_{i,j}$, we approximate the Laplacian in Eq. 30 by the usual five-point approximation. For $j=2,3,...,N$ and $i=2,3,...,M-1$, the equation may be written

$$s*(u_{i-1,j}-2u_{i,j}+u_{i+1,j})+u_{i,j-1}-2u_{i,j}+u_{i,j+1}=rf_{i,j} \ , \tag{32a}$$

where $s=(\Delta y/\Delta x)^2$ and $r=\Delta y^2$. From Eq. 31a we have

$$u_{1,j}=u_{i,1}=u_{i,N+1}=0. \tag{32b}$$

For $i=M$, we incorporate the central difference approximation to the derivative in Eq. 31b

$$\frac{1}{\Delta x} (u_{M+1,j}-u_{M,j})=0$$

to obtain the equations

$$s^*(u_{M-1,j}-u_{M,j})+u_{M,j-1}-2u_{M,j}+u_{M,j+1}=rf_{M,j}. \tag{32c}$$

The system (32) may be solved using GENBUN by making the identifications

$m=M-1, \quad n=N-1$

$v_{i,j}=u_{i-1,j-1}, \quad i=1,2,...,M-1, j=1,2,...,N-1$

$a_{i+1}=c_i=s, \quad i=1,2,...,M-2$

$a_1=c_{M-1}=0$

$b_i=-2s, \quad i=1,2,...,M-2$

$b_{M-1}=-s.$

Case Study: ELLPACK

In 1975 Garrett Birkhoff proposed a cooperative effort among Harvard University, Purdue University, and The University of Texas at Austin to develop a modest but useful collection of programs for solving elliptic boundary value problems. At that time numerical analysts at Purdue were studying various techniques for approximating elliptic equations. A number of programs had been written as a result of this work, and a systematic evaluation of the methods implemented in this software was under way. Numerical analysts at Texas were primarily interested in techniques for solving linear systems of algebraic equations that result from such approximations, and were already developing a software package (ITPACK) for such problems. Thus, two design considerations immediately emerged. The software should allow researchers interested in one aspect of the solution process to easily use software written by experts in other areas. Also, software components should be easily exchanged and the resulting changes in efficiency and accuracy measured.

In the summer of 1976 James Ortega organized a meeting at ICASE to discuss the feasibility of such a project. It was at this meeting that the goals and framework for ELLPACK were set.* ELLPACK would primarily be a research tool for the development and evaluation of numerical methods for solving elliptic partial differential equations. In addition, it was felt that this software could be useful for the education of students in courses on the numerical solution of PDEs, and the solution to small- to medium-scale applications problems. The development of the system was to proceed in two steps. First a prototype system, called ELLPACK77, for problems on rectangular domains in two and three dimensions would be constructed. It would accept general linear elliptic equations of the form (8) and (9) with mixed linear boundary conditions. Then, if

---

*In attendance were G. Birkhoff, A. Brandt, G. Fix, A. George, G. Golub, J. Ortega, J. Rice, M. Schultz, R. Sweet, R. Varga, O. Widlund, and D. Young.

the basic design was successful, work would begin on ELLPACK78 which would also include capabilities for non-rectangular domains in two dimensions.

Since all software in ELLPACK was to be contributed on a voluntary basis, a modular design was crucial. To accomplish this, ELLPACK breaks the computation into a number of distinct phases as shown in Fig. 8: domain processing, operator discretization, equation/unknown renumbering, algebraic equation solution, and output. The only frozen aspects of the design are the fixed interfaces between modules. Contributors provide "plug-compatible" modules that span one or more phases of the computation. Communication among the software parts is accomplished through fixed interface variables that are (usually) transparent to the user. There are four separate classes of such data:

General control information — This includes a variety of indicators that affect all modules, such as requests for execution traces and module run times.

Problem definition information — This includes FORTRAN functions that return the values of equation coefficients, switches indicating the type of operator and boundary conditions (Poisson, self-adjoint, homogeneous, Dirichlet, etc.), and arrays that describe the rectangular grid.

Module interfaces — These variables contain the output of each module. Domain processors produce tables describing the type of each grid point (inside or outside the domain, near the boundary, etc.) and the locations of all boundary-grid intersections. Discretization modules produce a system of linear equations in a fixed sparse matrix format, along with variables describing properties of the system. Indexing modules produce permutation vectors, and solution modules produce a vector of solution components. (Note that some solution modules may require that the equations be stored in a different format, e.g., band solvers, and some extra overhead is introduced for storage mode conversion).

Workspace — Scratch storage areas are available to each module for arrays whose sizes are problem-dependent.

Since ELLPACK must accommodate a very wide variety of modules, its internal interfaces are quite complex. Thus, a very-high-level problem statement language was designed that would allow users to describe their problems in mathematical notation and invoke modules for execution by simply giving their names. A preprocessor (written in standard FORTRAN) reads the users' input program and produces a FORTRAN main program that allocates the required interface variables and calls modules in the ELLPACK module library to perform the requested computations.

As illustrated in the sample program of Fig. 9, there are two types of statements in the ELLPACK language: declaration statements for defining the problem and executable statements for specifying the sequence of problem-solving modules to be used.* Comment lines start with a star in column one. A brief description of some ELLPACK statements is given below.

Declaration Statements

EQUATION. Defines the PDE operator using mathematical notation. Expressions use FORTRAN syntax; and the strings UXX, UXY, UX, etc. are keywords denoting the derivatives of the unknown function U.

BOUNDARY. Defines the boundary of the domain and the side conditions that apply there. The boundary of a piecewise smooth non-rectangular domain is described parametrically. An additional statement, HOLE, defines holes in domains.

GRID. Defines a rectangular grid for the domain. Both uniform and non-uniform grids are allowed.

SUBPROGRAMS. Gives the user-supplied FORTRAN subprograms. These include any function used in the definition of the equation or boundary conditions.

Executable Statements

DISCRETIZATION. Specifies the module to perform the operator discretization.

INDEXING. Specifies the module to reorder the equations and unknowns of the discretized operator.

SOLUTION. Specifies the module to solve the resulting system of algebraic equations.

OUTPUT. Selects the form of output. The solution and its derivatives may be tabulated or displayed in a contour plot. Various norms may also be computed. If the solution is known analytically, then the error may also be printed or plotted.

With John Rice as project coordinator, the responsibility for defining the standard module interfaces and developing the ELLPACK preprocessor, domain processor, and output modules was centered at Purdue. Contributed modules were integrated into ELLPACK there also, although responsibility for the integrity of individual problem-solving modules rests solely with the module authors. Some milestones in the development of ELLPACK are listed below:

Nov. 1976 ....ELLPACK77 interfaces set
Mar. 1977...................ELLPACK77 first operational

---

*The ELLPACK input language described here differs slightly from the version currently being distributed; it represents a planned revision of the language to be released in early 1983.

```
Mar 1978....................General distribution begun
May 1978 ..................ELLPACK78  interfaces set
Oct 1978....................ELLPACK77  stable;
                           ELLPACK78  first operational
Apr 1979...................General distribution update
Apr 1980...................General distribution update;
                           ELLPACK complete.
```

A preliminary version of ELLPACK77 was operational approximately four months after the initial detailed specifications were set. During the next twelve months the design was evaluated, tests were performed on a variety of computers, and new problem-solving modules were integrated into the system. ELLPACK77 was then released for general distribution. Work on the non-rectangular capabilities was begun several months later.

Coordination of group activities was handled through yearly workshops, intermediate releases of ELLPACK for group members, and the ELLPACK Network. At workshops group members learned of the current system status, discussed improvements, and set the future course of the project. The latest version of ELLPACK was available to group members via the ELLPACK Network, an interactive subsystem of the Purdue University CDC 6500 computing system that was especially friendly to ELLPACK users who knew little about the local operating system.

During the development of ELLPACK77 (Sept. 1976 to July 1978) the Purdue group consisted of three faculty members and two graduate students, each giving part time to the project. The amount of effort required to establish the ELLPACK superstructure during this period was approximately as follows:

```
System design and integration .....4 person-months
ELLPACK preprocessor ..............5 person-months
Output modules.........................4 person-months
```

These estimates do not include the work involved in producing problem-solving modules. This effort is difficult to estimate because of the organization of the ELLPACK group. The issue is further complicated by the fact that many such modules were originally developed independently of ELLPACK. A crude idea of the amount of work required to produce ELLPACK can be obtained from simple counts of FORTRAN statements, and these are given in Table I.

The distribution of ELLPACK is handled by the IMSL software distribution service, although responsibility for the maintenance of ELLPACK remains with the ELLPACK group. Users are notified of new releases through the IMSL newsletter, but are not automatically sent updates. ELLPACK documentation includes a user's guide, a contributor's guide, and a distribution guide which describes the installation procedure.

The problem-solving modules currently available in ELLPACK are described below. Unless otherwise noted, all discretization modules are for rectangular domains in two dimensions.

Discretization Modules

5-POINT STAR (usual finite differences for general domains)
7-POINT 3D (usual finite differences in three dimensions)
HODIE-HELMHOLTZ (high order finite differences for Helmholtz-type operators)
P3C1 COLLOCATION (collocation using bicubic Hermite elements)
P3C1 GALERKIN (Galerkin method using bicubic Hermite elements)
YALE GALERKIN (Galerkin method using tensor product B-spline elements of arbitrary order and smoothness)

Indexing Modules

NATURAL (do not reorder — use same ordering as discretization module)
NESTED DISSECTION (nested dissection ordering for rectangular finite difference grid)
P3C1 COLLORDER (a special ordering for the P3C1 COLLOCATION equations)
RED-BLACK (red-black ordering, if possible; part of ITPACK [Grimes, Kincaid and Young, 1979])
YALE MIN DEG (minimum degree ordering; part of Yale Sparse Matrix Package [Eisenstat et al., 1977a])
YALE RCM (reverse Cuthill-McKee ordering)

Solution Modules

JACOBI CG (Jacobi iteration with conjugate gradient acceleration; part of ITPACK [Grimes, Kincaid and Young, 1979])
JACOBI SI (Jacobi iteration with semi-iterative acceleration; part of ITPACK [Grimes, Kincaid and Young, 1979])
LINPACK BAND (Gauss elimination with partial pivoting for band matrices; part of LINPACK [Dongarra et al., 1979])
LINPACK SPD BAND (Gauss elimination for symmetric positive definite band matrices; part of LINPACK [Dongarra et al., 1979])
REDUCED SYSTEM CG (Jacobi iteration on red-black system with conjugate gradient acceleration; part of ITPACK [Grimes, Kincaid and Young, 1979])

REDUCED SYSTEM SI (Jacobi iteration on red-black system with semi-iterative acceleration; part of ITPACK [Grimes, Kincaid and Young, 1979])

SOR (successive over-relaxation iterative method; part of ITPACK [Grimes, Kincaid and Young, 1979])

SPARSE GE-PIVOTING (sparse Gauss elimination with column pivoting; a version of the ACM algorithm NSPIV [Sherman, 1978])

SYMMETRIC SOR CG (symmetric SOR iteration with conjugate gradient acceleration; part of ITPACK [Grimes, Kincaid and Young, 1979])

SYMMETRIC SOR SI (symmetric SOR iteration with semi-iterative acceleration; part of ITPACK [Grimes, Kincaid and Young, 1979])

YALE ENVELOPE (envelope Gauss elimination)

YALE SPARSE (four variations of sparse Gauss elimination from the Yale Sparse Matrix Package [Eisenstat *et al.*, 1977a and 1977b])

Modules That Both Discretize and Solve

DYAKANOV-CG (symmetric 5-point finite differences for general self-adjoint problems; resulting linear system solved by a preconditioned conjugate gradient method using the generalized marching algorithm to solve the sequence of separable problems [Bank, 1978])

DYAKANOV-CG4 (extends DYAKANOV-CG to fourth order accuracy using Richardson extrapolation [Bank, 1978])

FISHPAK HELMHOLTZ (standard 5-point finite differences for the Helmholtz equation with fast Fourier transform solution to the discrete equations; part of FISHPAK [Swarztrauber and Sweet, 1979])

FFT 9-POINT (2nd, 4th and 6th order compact 9-point finite differences with fast Fourier transform solution [Houstis and Papatheodorou, 1979])

HODIE 27-POINT 3D (high order finite differences for the Poisson equation in 3 dimensions with tensor product solution)

MARCHING ALGORITHM (symmetric 5-point finite differences for separable, self-adjoint equations; generalized marching algorithm for resulting linear system [Bank, 1978])

P2C0 TRIANGLES (6-node triangular quadratic finite elements on general domains with user-controlled mesh refinement and band elimination using out-of-core storage)

This diverse collection of software reflects the varied interests of members of the ELLPACK cooperative group.* The success of ELLPACK demonstrates that such a group of voluntary contributors can be effectively managed under the framework of software parts technology.

---

* See Appendix B for a list of contributors.

The ELLPACK system has now been distributed to more than 100 installations around the world and is being used in a variety of ways. These include (a) an aid in teaching courses on the numerical solution of elliptic problems, (b) a research tool for the evaluation of software for elliptic problems [Boisvert, Houstis, and Rice, 1979], (c) a system for solving "routine" boundary value problems, or for evaluating the appropriateness of various numerical methods on model problems before constructing programs for more complex problems, and (d) a program generator for complex problems. In the last case several problems formally beyond the scope of ELLPACK have been solved by first writing the "closest" ELLPACK program, obtaining the program generated by the Preprocessor, and modifying it to solve the desired problem. This often requires much less effort than programming an entire solution from scratch.

Currently, the ELLPACK input language is being revised, the preprocessor is being rewritten, and the internal interfaces are being refined. The new ELLPACK language will allow users to insert FORTRAN statements among ELLPACK executable statements. In this way user-defined preprocessing and postprocessing will be possible. After a solution is computed, the functions U(X,Y), UX(X,Y), UXX(X,Y), etc. are available to facilitate this. Work on this final version of ELLPACK is expected to be complete by early 1983.

SOURCES OF SOFTWARE FOR PARTIAL DIFFERENTIAL EQUATIONS

In this section we list some of the sources that we have used in gathering information about software for EBVPs. Many applications-oriented programs can also be obtained through these sources. (An excellent survey of sources of engineering-oriented software is given by Shiffman [1974].)

Commercial Libraries

IMSL Inc., Sixth Floor — NBC Bldg., 7500 Bellaire Blvd., Houston, TX 77036
Numerical Algorithms Group Ltd. (NAG), 256 Banbury Rd., Oxford OX2
    7DE, England

Software Clearinghouses

Computer Software Management and Information Center (COSMIC), University
    of Georgia, Athens, GA 30602
IMSL Software Distribution Service, IMSL Inc., Sixth Floor — NBC Bldg., 7500
    Bellaire Blvd., Houston, TX 77036
National Energy Software Center, Argonne National Laboratory, 9700 South Cass
    Ave., Argonne, IL 60439

Journals That Distribute Software

ACM Transactions on Mathematical Software, Association for Computing
    Machinery (ACM Algorithms are distributed through IMSL, Inc.)
Computers and Chemical Engineering, Pergamon Press
Computer Physics Communications, North-Holland

Journals That Publish Descriptions of Programs

Computers and Chemistry, Pergamon Press
Computers and Fluids, Pergamon Press
Computers and Geosciences, Pergamon Press
Computers and Mathematics with Applications, Pergamon Press
Computers and Structures, Pergamon Press
Computing, Springer-Verlag
IEEE Transactions on Magnetics, Institute of Electrical and Electronics Engineers
International Journal of Numerical Methods in Engineering, John Wiley
Journal of Computational Physics, Academic Press
Mathematics and Computers in Simulation (Transactions of IMACS), International
    Association for Mathematics and Computers in Simulation
Mathematics of Computation, American Mathematical Society (AMS)
SIGNUM Newsletter, ACM Special Interest Group in Numerical Mathematics

CATALOG OF SOFTWARE FOR ELLIPTIC PROBLEMS

We next present a catalog of the software that forms the basis of this survey. This represents a substantial update of the software for EBVPs surveyed in
Machura and Sweet [1980], which also contains information on software for
time-dependent problems.

NAME: B2DE
TYPE: FORTRAN subroutine
PROBLEM CLASS: Systems of nonlinear equations of the form

$$-\nabla \cdot (a_m \mathbf{u} \nabla (u_m)) + f_m(\mathbf{u}, \mathbf{u}_x, \mathbf{u}_y) = 0, \quad m = 1, ..., k$$

with nonlinear boundary conditions of the form $g_m(\mathbf{u}) = 0$ or
$(u_m)_n + g_m(\mathbf{u}) = 0$ on bounded two-dimensional domains whose boundaries
are made up of straight lines and circular arcs.
NUMERICAL METHODS: Finite element procedure using continuous
    piecewise-linear triangular finite elements. Coarse user-supplied mesh is
    refined adaptively. Nonlinear finite element equations are solved by a
    damped Newton's method; linearized equations are solved by a multi-level

iterative method.

SPECIAL FEATURES: Includes interactive and batch drivers. Includes portable graphics routines for plotting domain triangulations, contour and flow-line plots, and surface view plots.

PORTABILITY/DISTRIBUTION: Written in RATFOR, leading to portable FORTRAN. Single or double precision. Available from James L. Blue, Scientific Computing Division, National Bureau of Standards, Washington, DC 20234 (no charge).

AUTHOR(S)/VERSION: James L. Blue. Estimated release in late 1982. Data structures and much code adapted from PLTMG [Bank and Sherman, 1981].

NAME: CMMPAK

TYPE: Systematized collection of FORTRAN subroutines

PROBLEM CLASS: Helmholtz equation with Dirichlet or Neumann boundary conditions on general bounded two-dimensional domain.

NUMERICAL METHODS: Three forms of the capacitance matrix method (CMM) can be used to transform the problem to a solution on a rectangle where fast direct methods can be applied. CMMEXP and CMMSIX use the explicit CMM (best for coarse meshes, many right-hand sides), while CMMIMP uses the implicit CMM (fine meshes, few right-hand sides). Both CMMIMP and CMMEXP employ finite differences yielding second-order accuracy for the Dirichlet problem and first-order accuracy for the Neumann problem. CMMSIX attains fourth- or sixth-order accuracy for the Dirichlet problem using deferred corrections.

PORTABILITY/DISTRIBUTION: Portable FORTRAN. Single precision. Available from Wlodimierz Proskurowski, Department of Mathematics, University of Southern California, Los Angeles, CA 90007 (no charge).

REFERENCE: Proskurowski [1978 and 1979]

AUTHOR(S)/VERSION: Wlodimierz Proskurowski, 1978.

NAME: D03EAF

TYPE: FORTRAN subroutine

PROBLEM CLASS: Laplace's equation on an arbitrary two-dimensional domain bounded (internally or externally) by one or more closed contours on which either the value of the solution or its normal derivative is prescribed.

NUMERICAL METHODS: Boundary integral method

PORTABILITY/DISTRIBUTION: FORTRAN targetted for most major computer types. Available as part of the NAG Library (yearly lease) from NAG (USA) Inc., 1250 Grace Ct., Downers Grove, IL 60515.

REFERENCE: NAG [1978]

AUTHOR(S)/VERSION: Numerical Algorithms Group, Mark 7, Dec. 1978.

NAME: ELLPACK
TYPE: FORTRAN preprocessor system
PROBLEM CLASS: General linear equation with mixed boundary conditions on rectangular domains in two and three dimensions; also general bounded two-dimensional domains with holes. Some quasi-linear problems can be solved using fixed-point iteration.
NUMERICAL METHODS: Finite differences, Galerkin, and collocation discretizations. Natural, red-black, minimum degree, reverse Cuthill-McKee and nested dissection orderings. Iterative, general sparse, band and envelope solutions. Fast direct solvers (when applicable).
SPECIAL FEATURES: Mathematically oriented problem-statement language for input, graphics for contour plotting, large variety of numerical methods.
PORTABILITY/DISTRIBUTION: Portable FORTRAN. Single precision. Available through the IMSL Software Distribution Service (small service charge).
REFERENCE: Rice [1977]
AUTHOR(S)/VERSION: Cooperative development effort among Purdue University, University of Texas at Austin, Yale University, and others. Current version dated Feb. 1981. Substantial rewrite of system under way, to be completed by early 1983.

NAME: FISHPAK
TYPE: Systematized collection of FORTRAN subroutines
PROBLEM CLASS: Helmholtz equation in cartesian, polar, and spherical coordinates. Modified Helmholtz equation in cylindrical and spherical coordinates. Both centered and staggered finite difference grids are allowed. Any combination of Dirichlet, Neumann, and periodic boundary conditions. Rectangular domains. In addition, a solver is provided for general separable problems in two dimensions (and the Helmholtz equation in three dimensions (cartesian coordinates).
NUMERICAL METHODS: Second-order accurate finite differences (fourth order optional in the general separable case). Fast equation solution using cyclic reduction and fast Fourier transform methods.
SPECIAL FEATURES: Contains associated package of fast Fourier transform subprograms (FFTPAK) with special drivers for sine, cosine, quarter-wave and periodic transforms.
PORTABILITY/DISTRIBUTION: Portable FORTRAN. Single precision. Available from the NCAR Program Library, National Center for Atmospheric Research, P. O. Box 3000, Boulder, CO 80307 (small service charge).
REFERENCE: Swarztrauber and Sweet [1979]

AUTHOR(S)/VERSION: J. Adams, P. N. Swarztrauber, and R. A. Sweet, Version 3.1, 1981.

NAME: FFT9
TYPE: FORTRAN subroutine
PROBLEM CLASS: Solves

$$au_{xx}+bu_{yy}+cu=f$$

for constant a,b,c on rectangular domains with Dirichlet boundary conditions.
NUMERICAL METHODS: Fourth-order nine-point finite difference discretization (or sixth order in case of Poisson equation) with fast Fourier transform solution.
PORTABILITY/DISTRIBUTION: Portable FORTRAN. Single precision. Available from the Collected Algorithms of the ACM as Algorithm 543 (small service charge).
REFERENCE: Houstis and Papatheodorou [1979]
AUTHOR(S)/VERSION: E. N. Houstis and T. S. Papatheodorou, 1979.

NAME: HELM3D
TYPE: FORTRAN subroutine
PROBLEM CLASS: Helmholtz equation on a general bounded three-dimensional domain with Dirichlet boundary conditions.
NUMERICAL METHODS: Second-order accurate finite differences; capacitance matrix technique is used to reduce the problem to a solution on a cube, where fast direct methods apply.
PORTABILITY/DISTRIBUTION: Portable FORTRAN. Single precision. Available from the Collected Algorithms of the ACM as Algorithm 572 (small service charge).
REFERENCE: O'Leary and Widlund [1981]
AUTHOR(S)/VERSION: D. P. O'Leary and O. Widlund, 1978.

NAME: ICCG
TYPE: Systematized collection of FORTRAN subroutines
PROBLEM CLASS: Linear variable coefficient self-adjoint problems on rectangular domains with mixed or periodic boundary conditions.
NUMERICAL METHODS: Standard finite difference approximation; resulting linear system solved by conjugate gradient iteration with an incomplete Cholesky factorization for preconditioning [Meijerink and Van der Vorst, 1977].
SPECIAL FEATURES: Versions for space or time economization included.

PORTABILITY/DISTRIBUTION: Portable FORTRAN. Single precision. Available from ACCU-Reeks nr. 29, Academic Computer Centre, Budapestlaan 6, de Uithof - Utrecht, the Netherlands.
REFERENCE: Van Kats and Van der Vorst [1978]
AUTHOR(S)/VERSION: J. M. Van Kats and H. A. Van der Vorst, 1978.

NAME: KLEV
TYPE: FORTRAN subroutine
PROBLEM CLASS: Linear variable coefficient equation of the form

$$(Pu_x)_x + (Qu_y)_y + Vu_x + Wu_y + Su = f$$

on rectangular domains with boundary conditions of the form

$$Bu_n + Au = g \ .$$

NUMERICAL METHODS: Standard finite difference approximation with modified upwind differencing for non-symmetric systems. A variety of fixed multi-grid solution techniques are available (including user-defined strategies).
SPECIAL FEATURES: Interactive driver program available.
PORTABILITY/DISTRIBUTION: Portable FORTRAN. Single precision. Available from the Department of Computer Science, Yale University, New Haven, CT 06520 (no charge).
REFERENCE: Douglas [1981]
AUTHOR(S)/VERSION: C. C. Douglas, estimated release in 1982.

NAME: MG00
TYPE: Systematized collection of FORTRAN subroutines
PROBLEM CLASS: Modules are provided for the solution to equations of the form

1) $-(u_{xx} + u_{yy}) = f$,
2) $-(u_{xx} + u_{yy}) + cu = f$,
3) $-au_{xx} - bu_{yy} + cu = f$, and
4) $-\nabla \cdot (p \nabla u) + cu = f$,

where a, b, c, and p are variable coefficients on rectangular domains with either Dirichlet, Neumann or Robbins (mixed) type boundary conditions.
NUMERICAL METHODS: Standard finite differences with solution by multigrid iterative methods (iterative solution, full multigrid truncation error accuracy, accurate solution of the discrete problem to round-off).
PORTABILITY/DISTRIBUTION: Portable FORTRAN. Single and double

precision. Available from GMD, Postfach 1240, D-5205 St. Augustin 1, Federal Republic of Germany (no charge).
REFERENCE: Foerster and Witsch [1981]
AUTHOR(S)/VERSION: H. Foerster and K. Witsch, release 2.0, 1982.

NAME: PLTMG
TYPE: FORTRAN subroutine
PROBLEM CLASS: Solves linear variable coefficient equation of the form

$$-\nabla \cdot a \nabla u + b \cdot \nabla u + cu = f$$

with mixed linear boundary conditions on bounded two-dimensional domains whose boundaries are made up of straight lines and circular arcs.
NUMERICAL METHODS: Finite element procedure using continuous piece-wise linear triangular finite elements. Coarse user-supplied grid is refined adaptively. Discrete equations are solved by a multi-level iterative method.
SPECIAL FEATURES: Includes portable graphics routines for producing plots of the domain triangulation and contour plots of the solution.
PORTABILITY/DISTRIBUTION: Portable FORTRAN. Single precision. Available from Randolph E. Bank, Department of Mathematics, University of California at San Diego, La Jolla, CA 92093 (no charge).
REFERENCE: Bank and Sherman [1981]
AUTHOR(S)/VERSION: R. E. Bank and A. H. Sherman, 1979.

NAME: SLDGL
TYPE: Systematized collection of FORTRAN subroutines
PROBLEM CLASS: General nonlinear elliptic systems on rectangular domains in two and three dimensions. Experimental version of program for general two-dimensional domains also available. SLDGL also solves ordinary and initial boundary value problems as well as parabolic problems in two and three dimensions.
NUMERICAL METHODS: Adaptive mesh and order selection based on family of central difference formulas and Newton iteration. Produces estimates of truncation error.
PORTABILITY/DISTRIBUTION: FORTRAN targetted to several computer types including Burroughs, CDC, IBM, and Univac. Available from W. Schonauer, Rechenzentrum der Universitat Karlsruhe, D-7500 Karlsruhe, West Germany (small service charge for non-profit organizations, fee for commercial usage).
REFERENCE: Schonauer, Raith and Glotz [1981]
AUTHOR(S)/VERSION: W. Schonauer, K. Raith and G. Glotz, 1981.

NAME: TWODEPEP

TYPE: FORTRAN preprocessor system

PROBLEM CLASS: Nonlinear systems of elliptic equations with nonlinear Dirichlet or Neumann boundary conditions on general bounded two-dimensional domains. Also solves eigenvalue problems and parabolic problems.

NUMERICAL METHODS: Finite element procedure using triangular elements. Quadratic, cubic, and quartic isoparametric elements are available. User-controlled refinement of user-supplied initial triangulation. Band elimination with frontal method to organize out-of-core storage. Newton's method for nonlinear problems.

SPECIAL FEATURES: FORTRAN preprocessor for problem statement input language. Portable graphics output package for scalar, vector and stress fields. Also provides three-dimensional perspective plots of solutions as well as printer plot of initial triangulation for error checking.

PORTABILITY/DISTRIBUTION: FORTRAN targeted for most major computers (single and double precision). Available from IMSL, Inc., 7500 Bellaire Blvd., Houston, TX 77436 (yearly lease).

REFERENCE: Sewell [1981]

AUTHOR(S)/VERSION: G. Sewell, Edition 4, 1982.

Table I

Approximate Size of ELLPACK in No. of Card Images

	July 1978 (a)			Feb. 1981 (b)		
	FORTRAN Statements	Comment Cards	Total	FORTRAN Statements	Comment Cards	Total
Preprocessor	3000	3500	6500	5300	2200	7500
Domain processor	--	--	--	1200	700	1900
Output modules	700	1000	1700	4400	3800	8200
Module library	13200	8300	21500	23000	14600	37600
Total	16900	12800	29700	33900	21300	55200

(a) ELLPACK77 version containing 28 problem-solving modules.
(b) ELLPACK78 version containing 32 problem-solving modules.

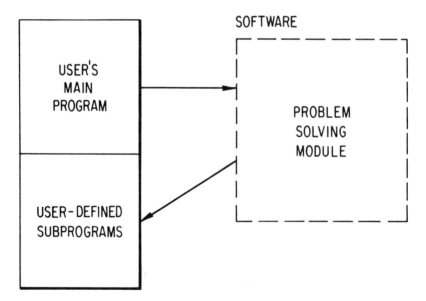

Figure 1. Single module organization

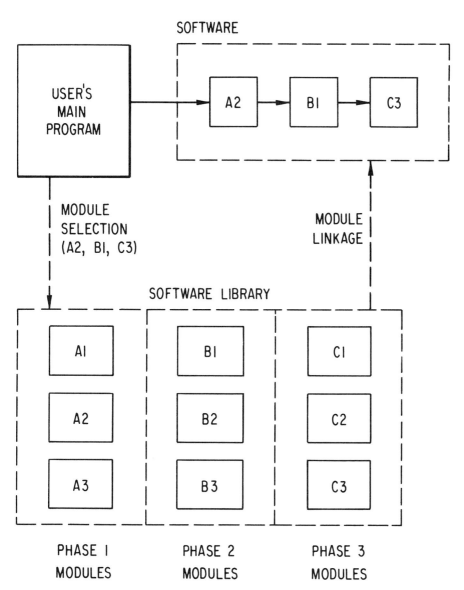

Figure 2. Systematized collection of software parts

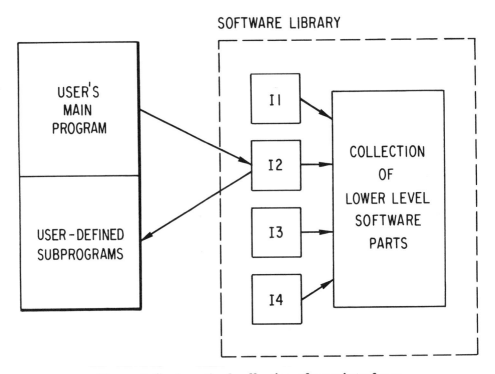

Figure 3. Systematized collection of user interfaces

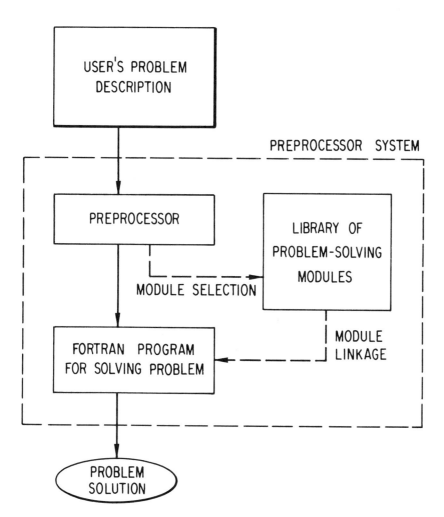

Figure 4. Preprocessor system organization

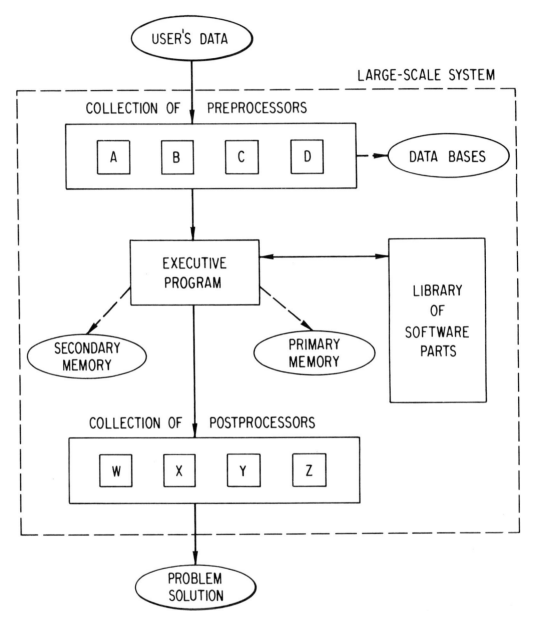

Figure 5. Large-scale system organization

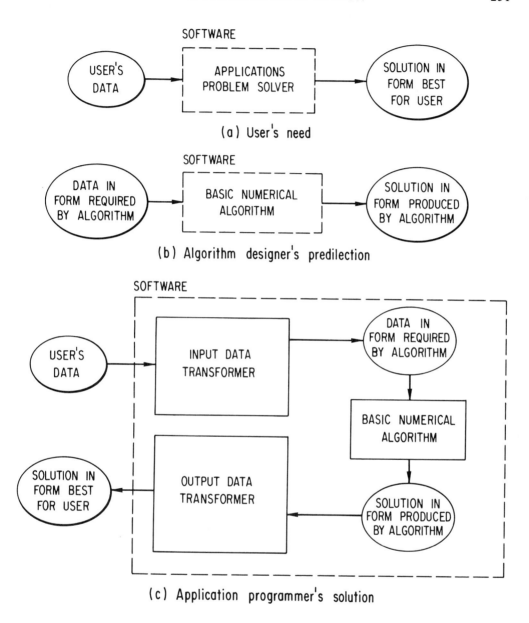

(a) User's need

(b) Algorithm designer's predilection

(c) Application programmer's solution

**Figure 6. Representation of I/O:
An application programmer's dilemma**

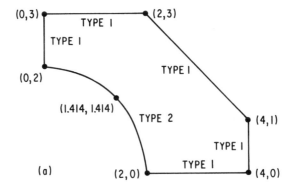

(a)

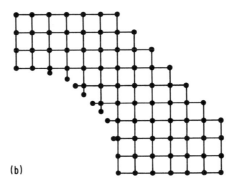

(b)

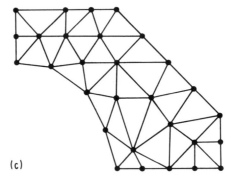

(c)

Figure 7. Typical non-rectangular domain and two discrete versions

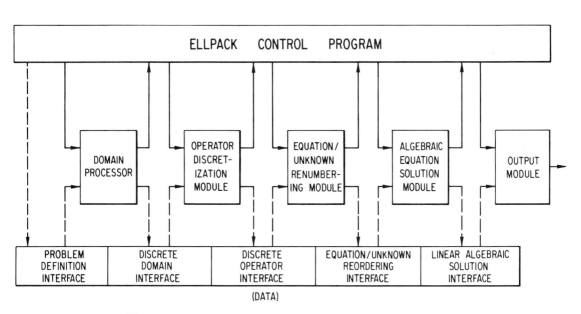

Figure 8. Internal organization of an ELLPACK run

```
*     SAMPLE ELLPACK PROGRAM
*  USING PROPOSED SYNTAX OF 1983 VERSION
*
EQUATION.   UXX + UYY - EXP(X+Y)U = G(X,Y)
*
BOUNDARY.  U  = 0.0  ON  X = 0.0
      U  = 0.0  ON  X = 1.0
      UY = 0.0  ON  Y = 0.0
      UY = 0.0  ON  Y = 1.0
*
GRID.        13 X POINTS
          13 Y POINTS
*
DISCRETIZATION.    5-POINT STAR
INDEXING.        YALE MIN DEG
SOLUTION.        YALE SPARSE
*
OUTPUT.         TABLE(U)
          PLOT(U)
          PLOT(UX)
*
SUBPROGRAMS.
   REAL FUNCTION G(X,Y)
   IF (X .GT. 0.5) THEN
     G = 0.0
   ELSE
     G = 2.0*(X-0.5)
   ENDIF
   RETURN
   END
END.
```

Figure 9. Sample ELLPACK program

REFERENCES

Bank, R. E. [1978]. "Algorithm 527: a FORTRAN implementation of the generalized marching algorithm." *ACM Trans. on Math. Soft.* 4:165-176.

Bank, R. E., and A. H. Sherman [1981]. "An adaptive, multi-level method for elliptic boundary value problems." *Computing* 26:91-106.

Boehm, B. W. [1980]. "Software engineering — as it is." *Software Engineering.* Ed. H. Freeman and P. M. Lewis II. Academic Press, New York, pp. 37-74.

Boisvert, R. F. [1981]. "Families of high order accurate discretizations of some elliptic problems." *SIAM J. Sci. Stat. Comp.* 2:268-284.

Boisvert, R. F., E. N. Houstis, and J. R. Rice [1979]. "A system for performance evaluation of partial differential equations software." *IEEE Trans. on Soft. Eng.* 5:418-425.

Brandt, A. [1977]. "Multi-level adaptive techniques (MLAT) for partial differential equations: ideas and software." *Mathematical Software III.* Ed. J. Rice. Academic Press, New York, pp. 277-318.

Buzbee, B. L., G. H. Golub, and C. W. Neilson [1970]. "On direct methods for solving Poisson's equations." *SIAM J. Numer. Anal.* 7:627-656.

Buzbee, B. L., *et al.* [1971]. "The direct solution of the discrete Poisson equation on irregular regions." *SIAM J. Numer. Anal.* 8:722-736.

Concus, P., G. H. Golub, and D. P. O'Leary [1976]. "A generalized conjugate gradient method for the numerical solution of elliptic partial differential equations." *Sparse Matrix Computations.* Ed. J. R. Bunch and D. J. Rose. Academic Press, New York.

Crane, H. L., *et al.* [1976]. "Algorithm 508: matrix bandwidth and profile reduction." *ACM Trans. on Math. Soft.* 2:375-377.

Cryer, C. W. [1977]. *A Bibliography of Free Boundary Problems.* University of Wisconsin Math. Res. Ctr. Technical Summary Report 1793.

Dongarra, J. J., *et al.* [1979]. *LINPACK Users' Guide.* SIAM, Philadelphia.

Douglas, C. C. [1981]. *A Multi-grid Optimal Order Elliptic Partial Differential Equation Solver.* Yale University Department of Computer Science.

Duff, I. S. [1977]. "A survey of sparse matrix research." *Proc. IEEE* 65:500-535.

Eisenstat, S. C., *et al.* [1977a]. *Yale Sparse Matrix Package: I. The Symmetric Codes.* Yale University Department of Computer Science Research Report No. 112.

Eisenstat, S.C., *et al.* [1977b]. *Yale Sparse Matrix Package: II. The Nonsymmetric Codes.* Yale University Department of Computer Science Research Report No. 114.

Everstine, G. C., and J. M. McKee [1974]. "A survey of pre- and postprocessors for NASTRAN." *Structural Mechanics Computer Programs — Surveys, Assessments and Availability.* Ed. W. Pilkey, K. Saczalski, and H. Schaeffer. University of Virginia Press, Charlottesville, pp. 825-848.

Foerster, H., and K. Witsch [1981]. "On Efficient Multigrid Software for Elliptic Problems on Rectangular Domains." *Math. and Comp. in Simulation (Trans. of IMACS)* 23.

Forsythe, G. E., and W. K. Wasow [1960]. *Finite Difference Methods for Partial Differential Equations.* John Wiley and Sons, New York.

Fox, P. A., A. D. Hall, and N. L. Schryer [1978]. "Algorithm 528: framework for a portable library." *ACM Trans. on Math. Soft.* 4:177-188.

Garbow, B. S., *et al.* [1977]. *Matrix Eigensystem Routines — EISPACK Guide Extension.* Lecture Notes in Computer Science, Vol. 51. Springer-Verlag, New York.

George, J. A. [1973]. "Nested dissection of a regular finite element mesh." *SIAM J. Numer. Anal.* 10:345-363.

George, J. A., and J. W. H. Liu [1979]. "The design of a user interface for a sparse matrix package." *ACM Trans. on Math. Soft.* 5:139-162.

Gladwell, I., and R. Wait, eds. [1979]. *A Survey of Numerical Methods for Partial Differential Equations.* Clarendon Press, Oxford.

Grimes, R. G., D. R. Kincaid, and D. M. Young [1979]. *ITPACK 2.0 User's Guide.* CNA-150, Center for Numerical Analysis, University of Texas at Austin.

Hageman, L. A., and D. M. Young [1981]. *Applied Iterative Methods.* Academic Press, New York.

Houstis, E. N., and T. S. Papatheodorou [1979]. "Algorithm 543: FFT9, fast solution of Helmholtz type partial differential equations." *ACM Trans. on Math. Soft.* 5:431-441.

Kirioka, K., and T. Hirata [1972]. "Computing system for structural analysis of car bodies." *Advances in Computational Methods in Structural Mechanics and Design.* Ed. J. T. Oden, R. W. Clough, and Y. Yamamoto. University of Alabama Press, Huntsville, pp. 529-550.

Lamb, H. [1945]. *Hydrodynamics.* 6th ed. Dover, New York.

Machura, M., and R. A. Sweet [1980]. "A survey of software for partial differential equations." *ACM Trans. on Math. Soft.* 6:461-488.

Meijerink, J. A. and H. A. Van der Vorst [1977]. "An iterative solution method for linear systems of which the coefficient matrix is a symmetric M-matrix." *Math. Comp.* 31:148-162.

McCormick, C. W. [1972]. "The NASTRAN program for structural analysis." *Advances in Computational Methods in Structural Mechanics and Design.* Ed. J. T. Oden, R. W. Clough, and Y. Yamamoto. University of Alabama Press, Huntsville, pp. 551-573.

NAG [1978]. *NAG FORTRAN Library Manual, Mark 7, Volume 1.* Numerical Algorithms Group, Oxford.

Napolitano, L. G., R. Monti, and P. Murio [1974]. "Preprocessors for general purpose finite element programs." *Structural Mechanics Computer Programs — Surveys, Assessments and Availability.* Ed. W. Pilkey, K. Saczalski, and H. Schaeffer. University of Virginia Press, Charlottesville, pp. 807-824.

Oden, J. T., R. W. Clough, and Y. Yamamoto, eds. [1972]. *Advances in Computational Methods for Structural Mechanics and Design.* University of Alabama Press, Huntsville.

O'Leary, D. P., and O. Widlund [1981]. "Solution of the Helmholtz equation for the Dirichlet problem on general bounded three-dimensional regions." *ACM Trans. on Math. Soft.* 7:239-246.

Pereyra, V. [1966]. "On improving an approximate solution of a functional equation by deferred corrections." *Numer. Math.* 8:376-391.

Pilkey, W., K. Saczalski, and H. Schaeffer, eds. [1974]. *Structural Mechanics Computer Programs — Surveys, Assessments and Availability.* University of Virginia Press, Charlottesville.

Proskurowski, W. [1978]. *Four Fortran Programs for Numerically Solving Helmholtz's Equation in an Arbitrary Bounded Planar Region.* Lawrence Berkeley Laboratory Report LBL-7516.

Proskurowski, W. [1979]. "Numerical solution of Helmholtz's equation by implicit capacitance methods." *ACM Trans. on Math. Soft.* 5:36-49.

Proskurowski, W., and O. Widlund [1976]. "On the numerical solution of Helmholtz's equation by the capacitance matrix method." *Math. Comp.* 30:433-468.

Rice, J. R. [1977]. "ELLPACK, a research tool for elliptic partial differential equations software." *Mathematical Software III.* Ed. J. Rice. Academic Press, New York, pp. 319-341.

Schiffman, R. L. [1974]. "Current software dissemination practices and organizations." *Structural Mechanics Computer Programs — Surveys, Assessments and Availability.* Ed. W. Pilkey, K. Saczalski, and H. Schaeffer. University of Virginia Press, Charlottesville, pp. 591-614.

Schrem, E. [1974]. "Development and maintenance of large finite element systems." *Structural Mechanics Computer Programs — Surveys, Assessments and Availability.* Ed. W. Pilkey, K. Saczalski, and H. Schaeffer. University of Virginia Press, Charlottesville, pp. 669-685.

Schonauer, W., K. Raith, and G. Glotz [1981]. "The SLDGL program package for the self-adaptive solution of nonlinear systems of elliptic and parabolic PDE's." *Advances in Computer Methods for Partial Differential Equations - IV.* Ed. R. Vichnevetsky and R. S. Stepleman. IMACS, New Brunswick, New Jersey, pp. 117-125.

Seeley, S., and A. D. Poularikas [1979]. *Electromagnetics — Classical and Modern Theory and Applications.* Marcel Dekker, New York.

Sewell, G. [1981]. *TWODEPEP, A Small General Purpose Finite Element Program.* IMSL Technical Report No. 8102.

Sherman, A. H. [1978]. "Algorithm 533: NSPIV, a FORTRAN subroutine for sparse Gaussian elimination with partial pivoting." *ACM Trans. on Math. Soft.* 4:391-398.

Smith, B. T., *et al.* [1976]. *Matrix Eigensystem Routines — EISPACK Guide.* Lecture Notes in Computer Science, Vol. 6. 2nd ed. Springer-Verlag, New York.

Sommerfeld, A. [1949]. *Partial Differential Equations in Physics.* Academic Press, New York.

Steuerwalt, M. [1979]. "Certification of algorithm 541." *ACM Trans. on Math. Soft.* 5:365-371.

Swarztrauber, P. N. [1974a]. "The direct solution of the discrete Poisson equation on the surface of the sphere." *J. Comput. Phys.* 15:46-54.

Swarztrauber, P. N. [1974b]. "A direct method for the discrete solution of separable elliptic equations." *SIAM J. Numer. Anal.* 11:1136-1150.

Swarztrauber, P. N. [1977]. "The methods of cyclic reduction, Fourier analysis, and the FACR algorithm for the discrete solution of Poisson's equation on a rectangle." *SIAM Rev.* 19:490-501.

Swarztrauber, P. N., and R. A. Sweet [1973]. "The direct solution of the discrete Poisson equation on a disk." *SIAM J. Numer. Anal.* 10:900-907.

Swarztrauber, P. N., and R. A. Sweet [1975]. *Efficient FORTRAN Subprograms for the Solution of Elliptic Partial Differential Equations.* National Center for Atmospheric Research, Boulder, Colo., Tech. Note TN/IA-109 (also, Errata, 1976).

Swarztrauber, P. N., and R. A. Sweet [1979]. "Algorithm 541: efficient Fortran subprograms for the solution of elliptic partial differential equations." *ACM Trans. on Math. Soft.* 5:352-364.

Sweet, R. A. [1977]. "A cyclic reduction algorithm for solving block tridiagonal systems of arbitrary dimension." *SIAM J. Numer. Anal.* 14:706-720.

Takahashi, T. [1975]. "Tropical showers in an axisymmetric cloud model." *J. Atmos. Sci.* 32:1318-1330.

Torrance, K. E. [1979]. "Natural convection in thermally stratified enclosures with localized heating from below." *J. of Fluid Mechanics* 95:477-495.

Van Kats, J. M., and H. A. Van der Vorst [1979]. *Software for the Discretization and Solution of Second Order Self-adjoint Elliptic Partial Differential Equations in Two Dimensions.* TR 10, ACCU, Utrecht.

Varga, R. S. [1962]. *Matrix Iterative Analysis.* Prentice-Hall, Englewood Cliffs, New Jersey.

Wachspress, E. L. [1966]. *Iterative Solution of Elliptic Systems and Applications to the Neutron Diffusion Equations of Reactor Physics.* Prentice-Hall, Englewood Cliffs, New Jersey.

Wilson, E. L. [1972]. "SAP — a general structural analysis program for linear systems." *Advances in Computational Methods in Structural Mechanics and Design.* Ed. J. T. Oden, R. W. Clough, and Y. Yamamoto. University of Alabama Press, Huntsville, pp. 625-640.

Appendix A

SOLUTION OF EXAMPLE 3 USING FISHPAK

```
C                         PROGRAM TO ILLUSTRATE EXAMPLE 3
C
C     WORK IS AN ARRAY USED FOR AUXILIARY STORAGE. IT MUST BE
C     DIMENSIONED AT LEAST 4*(N+1) + (16+INT(LOG2(N+1)))*(M+1).
C     THE ACTUAL NUMBER OF LOCATIONS REQUIRED BY HWSSSP IS RE-
C     TURNED IN LOCATION WORK(1).
      DIMENSION   F(19,73),BNDDRV(73),WORK(710)
C                         THE FIRST (OR ORW) DIMENSION OF ARRAY F
      IDIMF = 19
C     PIMACH IS A FISHPAK UTILITY THAT PROVIDES PI TO MACHINE
      PRECISION PI = PIMACH(DUM)
C                         DEFINE RECTANGLE ON SURFACE OF SPHERE
      THSTRT = 0.
      THSTOP = PI/2.
      PHSTRT = 0.
      PHSTOP = 2.*PI
C          DEFINE NUMBER OF PANELS FOR FINITE DIFFERENCE GRID
      M = 18
      N = 72
      DTHETA = (THSTOP-THSTRT)/M
      DPHI = (PHSTOP-PHSTRT)/N
C                         INDICATE THE TYPE OF BOUNDARY CONDITIONS
      MBDCND = 6
      NBDCND = 0
C                         THE HELMHOLTZ PARAMETER LAMBDA IS ZERO
      ELMBDA = 0.
C                         COMPUTE RIGHT SIDE OF EQUATION
      DO 101 J=1,N+1
C                         STORE EQUATOR DERIVATIVE DATA
      BNDDRV(J) = 0.
      DO 101 I=1,M+1
  101 F(I,J)=2.-6.*(SIN((I-1)*DTHETA)*SIN((J-1)*DPHI))**2
C                         DUMMY REPRESENTS UNUSED BOUNDARY ARRAYS
      CALL HWSSSP (THSTRT,THSTOP,M,MBDCND,DUMMY,BNDDRV,
     1             PHSTRT,PHSTOP,N,NBDCND,DUMMY,DUMMY,
     2             ELMBDA,F,IDIMF,PERTRB,IERROR,WORK)
C     COMPUTE DISCRETIZATION ERROR. SINCE PROBLEM IS SINGULAR,
C     THE SOLUTION IS NORMALIZED TO BE ZERO AT THE POLE.
      ERR = 0.
```

```
      DO 102 J=1,N+1
      DO 102 I=1,M+1
      Z = ABS(F(I,J)-F(1,1)-(SIN((I-1)*DTHETA)*SIN((J-1)*DPHI))**2)
      ERR = AMAX1(ERR,Z)
  102 CONTINUE
      IW = INT(WORK(1))
      WRITE (9,1001) IERROR,ERR,PERTRB,IW
 1001 FORMAT(9H IERROR =,I2/23H DISCRETIZATION ERROR =,E12.5 /
     1  9H PERTRB =,EI2.5/32H REQUIRED LENGTH OF WORK ARRAY
     =  ,I4)
      STOP
C               OUTPUT IS
C
C        IERROR = 0
C        DISCRETIZATION ERROR = .33813-002
C        PERTRB = .63217-003
C        REQUIRED LENGTH OF WORK ARRAY = 600
      END
```

Appendix B

LIST OF CONTRIBUTORS TO ELLPACK

The principal members of the project were as follows:

Randolph Bank	University of California at San Diego
Garrett Birkhoff	Harvard University
Ronald Boisvert	National Bureau of Standards
Stanley Eisenstat	Yale University
William Gordon	Drexel University
Elias Houstis	Northwestern University
David Kincaid	University of Texas at Austin
Robert Lynch	Purdue University
John Rice	Purdue University
Donald Rose	Bell Telephone Laboratories
Martin Schultz	Yale University
Andrew Sherman	Exxon Research
David Young	University of Texas at Austin

Substantial contributions of software were made by several others, including the following:

John Brophy	Granville Sewell
Wayne Dyksen	Paul Swarztrauber
Roger Grimes	Roland Sweet
William Mitchell	Linda Thiel
Cleve Moler	William Ward
John Respess	Alan Weiser

Chapter 10

THE IMSL LIBRARY

Thomas J. Aird
IMSL, Inc.
7500 Bellaire Boulevard
Houston, Texas 77036

INTRODUCTION

This chapter discusses the IMSL Library. The presentation is divided into three sections. The first section covers the IMSL company history, providing information about the general company organization and growth. The second section describes the IMSL Library products. Library contents, documentation, support, availability, and subscription fees are covered. The third section deals with library development techniques. The IMSL Research and Development group organization and the program development environment are described.

IMSL Company History

IMSL was incorporated in Houston, Texas, on October 8, 1970, and began business with the intent of providing high-quality, supported Fortran subroutine libraries in mathematics and statistics. The co-founders were Charles W. Johnson (Chairman of the Board) and Edward L. Battiste (first IMSL President). During 1970 and the early part of 1971, IMSL had seven employees, and they were all involved in producing IMSL's first product: a subroutine library for IBM 360/370 computers.

The process of becoming a financially viable organization took about six years. IMSL's first Library, for IBM computers, was released in July, 1971, and the second Library, for Univac computers, was released later that year. Only seven orders were received during 1971. In 1972, a CDC Library was released and during that year an additional seventy-eight orders were received. At this point in the company history survival was questionable. Additional funding was provided, however, and progress continued. At the end of 1974 there were 236 Library subscribers. In 1975 three new Libraries were released for Honeywell, Xerox, and Digital Equipment 10/20 computers. Also, the marketing division was formed in 1975, and this laid the foundation for some solid growth. In June of 1976 the Burroughs Library was released. At the end of 1976, IMSL had 430 Library subscribers and, for the first time, the company showed a slight profit.

IMSL Library products are summarized in Table I, and the release dates are given. Table II shows the company growth from 1970 through 1980.

During the four-year period from 1977 through 1980, IMSL's growth continued in a very satisfactory way, and it became clear that the subroutine library leasing business had been well established. Five more Libraries were released in this period, and they were for Data General Eclipse, Digital Equipment PDP 11, Hewlett Packard 3000, Digital Equipment VAX, and Prime computers. The number of new subscribers increased steadily each year and, in 1980, 298 new sales brought the total number of subscribers to 1,200. At the beginning of 1981, IMSL had over fifty employees and was organized into four divisions and an administrative group consisting of the president, executive secretary, and a receptionist. Information about the four divisions is summarized in Table III.

Table I

Library Product Release Dates

Computer	Year of First Release
IBM 360/370	1971
Univac 1100	1971
CDC 6000/7000	1972
Honeywell 6000	1975
Xerox Sigma	1975
Digital Equipment 10/20	1975
Burroughs 6700/7700	1976
Data General Eclipse	1977
Digital Equipment PDP 11	1978
Hewlett Packard 3000	1979
Digital Equipment VAX	1979
Prime 300, 400, 500, 50	1980

Table II

IMSL Ten-Year Growth Period

Year	Number of Employees	Number of Library Versions	Number of Library Routines	Number of Library Customers
1970	7	0	0	0
1971	12	2	200	7
1972	12	3	240	85
1973	14	3	280	155
1974	14	3	330	236
1975	21	6	370	312
1976	19	7	370	430
1977	22	8	400	559
1978	24	9	400	713
1979	38	11	460	902
1980	48	12	500	1,200

Table III

IMSL Employees, by Division, January 1981

Division	Date Formed	Number of Employees
Administration	1970	3
Management Information Systems	1980	2
Marketing	1975	12
Operations	1973	16
Research and Development	1970	20

Company Organization and Advisory Board

The organization described above is typical of many young corporations, but IMSL has a special need not shared by many others. That need is to have access to experts in a broad area covering mathematics and statistics.

IMSL has arranged such access by retaining an Advisory Board, and the following members of that Board aid IMSL in the Library area:

R. L. Anderson	Professor of Statistics, Assistant to the Dean for Statistical Services, College of Agriculture, University of Kentucky
W. J. Conover	Professor of Statistics, Department of Information Systems and Quantitative Sciences, Texas Technical University
C. de Boor	Professor of Mathematics and Computer Science, Mathematics Research Center, University of Wisconsin - Madison
W. A. Fuller	Professor of Statistics, Department of Statistics, Iowa State University
T. E. Hull	Professor of Computer Science and Mathematics, Department of Computer Science, University of Toronto
M. W. Johnson, Jr.	Professor of Mechanics and Mathematics, Department of Engineering Mechanics, University of Wisconsin - Madison
W. M. Kahan	Professor of Computer Science and Mathematics, Department of Computer Science, University of California at Berkeley
C. D. Kemp	Professor of Statistics, School of Mathematical Sciences, University of Bradford, England

W. J. Kennedy   Professor of Statistics, Statistical Laboratory, Iowa State University

G. G. Koch   Professor of Biostatistics, Department of Biostatistics, University of North Carolina

P. A. W. Lewis   Professor of Statistics and Operations Research, Naval Postgraduate School, Monterey, California

C. B. Moler   Professor of Mathematics and Computer Science, Department of Computer Science, University of New Mexico

V. Pereyra   Professor of Computer Science, Department of Computer Science, Universidad Central de Venezuela, Venezuela

M. J. D. Powell   Professor of Applied Numerical Analysis, Department of Applied Mathematics and Theoretical Physics, University of Cambridge, England

J. R. Rice   Professor of Computer Science and Mathematics, Department of Computer Sciences, Purdue University

J. R. Thompson   Professor of Statistics, Department of Mathematical Sciences, Rice University

Facts about the IMSL Library

The IMSL Library is a collection of 495 user-callable Fortran subroutines (some are function subprograms) which perform mathematical and statistical calculations. It is the purpose of this section to discuss the IMSL Library. Topics such as Library content, documentation, support, availability, and subscription fees are discussed.

The IMSL Library has evolved in eight editions. Edition 1 was released in 1971. Each new edition contains more subroutines than its predecessor and incorporates improvements to existing subroutines. The version of the Library released in July 1980 is Edition 8. It contains approximately 100,000 lines of Fortran code including comments.

IMSL also provides, as a subscriber option, a Minimal Test Package to aid in verifying that the Library has been properly installed on the subscriber's computer. The Minimal Test Package is a set of Fortran subroutines that execute Library routines using simple example problems which are designed to minimally test each user-callable routine in the IMSL library. The examples used are those in the *IMSL Library Reference Manual* [1980] in the example section of each document. The tests are designed to run individually or as a group and require no argument list. With the exception of a few tests in Chapter U, all minimal tests are input-free and print one line which consists of the IMSL routine name, the routine label, the precision tested, and the message PASSED or FAILED.

Library Contents

Subroutines in the IMSL Library are arranged to fall within chapter subgroups according to their purpose. There are seventeen chapters with the following titles:

CHAPTER	TITLE
A	Analysis of Variance
B	Basic Statistics
C	Categorized Data Analysis
D	Differential Equations; Quadrature; Differentiation
E	Eigensystem Analysis
F	Forecasting; Econometrics; Time Series; Transforms
G	Generation and Testing of Random Numbers
I	Interpolation; Approximation; Smoothing
L	Linear Algebraic Equations
M	Mathematical and Statistical Special Functions
N	Nonparametric Statistics

O	Observation Structure; Multivariate Statistics
R	Regression Analysis
S	Sampling
U	Utility Functions
V	Vector, Matrix Arithmetic
Z	Zeros and Extrema; Linear Programming

The complete list of Library subroutines is given in the *IMSL Library Reference Manual* [1980] and in the *IMSL Library Contents Document* [1980]. The list given here represents a sampling, chosen to give the reader an idea of the types of routines to be found in the Library.

ANALYSIS OF VARIANCE

AGBACP	Analysis of balanced complete experimental design structure data
AGLMOD	General linear model analysis
ACTRST	Contrast estimates and sums of squares
AGXPM	Expected mean squares for balanced complete design models

BASIC STATISTICS

BDCOU2	Tally of observations into a two-way frequency table
BECOR	Estimates of means, standard deviations, and correlation coefficients — out of core version
BECVL	Variances and covariances of linear functions — out of core version
BEIGRP	Estimation of basic statistical parameters using grouped data

BEMDP   Median polish of a two-way table

CATEGORIZED DATA ANALYSIS

CTRBYC   Analysis of a contingency table

CTLLF   Log-linear fit of contingency table

DIFFERENTIAL EQUATIONS; QUADRATURE; DIFFERENTIATION

DVERK   Differential equation solver — Runge Kutta-Verner
                         fifth and sixth order method

DGEAR   Differential equation solver — variable order
                         Adams predictor corrector method or Gear
                         method

DCADRE   Numerical integration of a function using cautious
                         adaptive Romberg extrapolation

DBCEVU   Bicubic spline mixed partial derivative evaluator

EIGENSYSTEM ANALYSIS

EIGRS   Eigenvalues and (optionally) eigenvectors of a real
                         symmetric matrix in symmetric storage mode

EIGRF   Eigenvalues and (optionally) eigenvectors of a real
                         general matrix in full storage mode

EIGCC   Eigenvalues and (optionally) eigenvectors of a
                         complex general matrix

EIGZC   Eigenvalues and (optionally) eigenvectors of the
                         system $A*x =$ lambda $*B*x$ where A and B are
                         complex matrices

FORECASTING; ECONOMETRICS; TIME SERIES; TRANSFORMS

FTAUTO   Mean, variance, autocovariances, autocorrelations
                         and partial autocorrelations for a stationary time
                         series

FTCMP	Nonseasonal ARIMA (Box-Jenkins) stochastic model analysis for a single time series with full parameter iteration and maximum likelihood estimation
FTRDIF	Transformation, differences and seasonal differences of a time series for model identification
FFTCC	Compute the fast Fourier transform of a complex valued sequence
FFTRC	Compute the fast Fourier transform of a real valued sequence
FTKALM	Kalman filtering

GENERATION AND TESTING OF RANDOM NUMBERS

GFIT	Chi-squared goodness of fit test
GGUBS	Basic uniform (0,1) pseudo-random number generator
GGNML	Normal or Gaussian random deviate generator
GGNSM	Multivariate normal random deviate generator with given covariance matrix
GGAMR	One-parameter gamma random deviate generator, and usable as the basis for two-parameter gamma, exponential, chi-squared, chi, beta, t, and F deviate generation
GGBTR	Beta random deviate generator

INTERPOLATION; APPROXIMATION; SMOOTHING

IFLSQ	Least squares approximation with user supplied functions
ICSCCU	Cubic spline interpolation (easy-to-use version)
IQHSCU	One-dimensional quasi-cubic Hermite interpolation

ICSEVU   Evaluation of a cubic spline

LINEAR ALGEBRAIC EQUATIONS

LEQT1F   Linear equation solution — full storage mode —
                      space economizer solution

LEQT1P   Linear equation solution — positive definite matrix —
                      symmetric storage mode — space economizer
                      solution

LEQT2F   Linear equation solution — full storage mode —
                      high accuracy solution

LEQT2P   Linear equation solution — positive definite matrix —
                      symmetric storage mode — high accuracy solution

LLSQF   Solution of a linear least squares problem

LSVDF   Singular value decomposition of a real matrix

MATHEMATICAL AND STATISTICAL SPECIAL FUNCTIONS

MDBETA   Beta probability distribution function

MDCH   Chi-squared probability distribution function

MDFD F probability distribution function

MDNOR   Normal or Gaussian probability distribution function

MDTD   Student's t probability distribution function

MDTN   Noncentral t probability distribution function

Chapter M also contains numerous routines for the inverses of probability distribution functions and routines for Bessel functions, complete elliptic integrals, and Kelvin functions.

NONPARAMETRIC STATISTICS

NAK1   Kruskal-Wallis test for identical populations

NKS1   Kolmogorov-Smirnov one-sample test

NKS2   Kolmogorov-Smirnov two-sample test

NRWRST   Wilcoxons rank-sum test

NDMPLE   Nonparametric probability density function (one
                dimensional) estimation by the penalized likeli-
                hood method

OBSERVATION STRUCTURE; MULTIVARIATE STATISTICS

OCLINK   Perform a single-linkage or complete-linkage
                hierarchical cluster analysis given a similarity
                matrix

ODNORM   Multivariate normal linear discriminant analysis
                among several known groups

OPRINC   Principal components of a multivariate sample
                of observations

REGRESSION ANALYSIS

RLMUL   Multiple linear regression analysis

RLFOR   Fit a univariate curvilinear regression model
                using orthogonal polynomials with optional
                weighting

RLPRDI   Confidence intervals for the true response and for
                the average of a set of future observations on the
                response — in core version

RLRES   Perform a residual analysis for a fitted regression
                model

RLLAV   Perform linear regression using the least absolute
                values criterion

SAMPLING

SSSAND	Simple random sampling with continuous data
SSSBLK	Stratified random sampling with continuous data
SSSCAN	Single-stage cluster sampling with continuous data
SSSEST	Two-stage sampling with continuous data and equisized primary units

UTILITY FUNCTIONS

UERTST	Print a message reflecting an error condition
UERSET	Set message level for IMSL routine UERTST
USPLT	Printer plot of up to ten functions
USHIST	Print a histogram (vertical)
USPC	Print a sample pdf, a theoretical pdf and confidence band information; plot these on option

VECTOR, MATRIX ARITHMETIC

VBLA =SDOT	Compute single precision dot product
VCVTFS	Storage mode conversion of matrices (full to symmetric)
VMULFF	Matrix multiplication (full storage mode)
VSRTA	Sorting of arrays by algebraic value

ZEROS AND EXTREMA; LINEAR PROGRAMMING

ZXSSQ	Minimum of the sum of squares of M functions in N variables using a finite difference Levenberg-Marquardt algorithm
ZXMIN	Minimum of a function of N variables using a quasi-Newton method

ZSCNT	Solve a system of nonlinear equations
ZPOLR	Zeros of a polynomial with real coefficients (Laguerre)
ZX4LP	Solve the linear programming problem via the revised simplex algorithm

Documentation

Documentation for the IMSL Library is contained in a three-volume (approximately 1800 pages) looseleaf manual. The *Library Reference Manual* exists in printed and microfiche form and is organized to have introduction, contents, KWIC index chapters, and the 17 Library chapters.

The Introduction Chapter gives the user general information that pertains to the entire manual. Library subroutine characteristics and documentation conventions are discussed.

The Contents Chapter gives a brief description of each sub- routine, listing such items as purpose, precision/hardware, and required IMSL routines.

The KWIC Chapter — keyword in context — is an aid to help the user locate documentation for routines for a particular application.

Chapter Documentation

Each of the 17 Library chapters begins with an introduction. This contains a quick reference guide for chapter routines, a featured-abilities section, special instructions for using the chapter routines, and a subtleties-to-note section. As an example, the introduction material for Chapter I, "Interpolation, Approximation, and Smoothing," is reproduced in Appendix A.

Subroutine Documentation

Subroutine documentation consists of two parts. The first is a copy of comment lines that appear at the beginning of each subroutine source deck. The format of the comment lines is as follows:

IMSL ROUTINE NAME - routine name.

PURPOSE - a statement of the purpose of the routine.

USAGE - the form of the subprogram CALL with arguments listed.

ARGUMENTS - a description of the arguments in the order of their occurrence in USAGE.

PRECISION/HARDWARE - environment specific information giving the precision of the routine — SINGLE, or DOUBLE and corresponding hardware environment codes.

REQD. IMSL ROUTINES - a list of all IMSL routines called (directly and indirectly) by this routine.

NOTATION - reference to manual introduction and IMSL routine UHELP.

REMARKS (optional) - details pertaining to code usage.

The second part of the document (which does not appear in the source code) includes the following sections:

ALGORITHM - a brief statement of the algorithm and references to detailed information.

PROGRAMMING NOTES (optional) - programming details not covered elsewhere.

ACCURACY (optional) - a statement about the accuracy of the routine.

EXAMPLE - an example showing subroutine input, required dimension and type statements and output.

The first part of each subroutine document (the comment lines) is part of the distribution materials sent to each subscriber. Many IMSL subscribers use this computer-readable material to build an on-line IMSL documentation system.

As an example, the documentation for subroutine ZPOLR, a subroutine to find the zeros of a polynomial, is reproduced in Appendix B.

Support

The IMSL Library is supported by both standard and optional materials and services. These materials and services are summarized here.

A. STANDARD MATERIALS PROVIDED WITH A LIBRARY SUBSCRIPTION
  1. The IMSL Library in Fortran on magnetic tape.
  2. Library accessing, generation, and maintenance utility programs on magnetic tape, to supplement selected operating system capabilities.
  3. One fully maintained *Library Reference Manual* in printed loose-leaf form.
  4. One fully maintained *Library Reference Manual* in microfiche form at a reduction ratio of 42 to 1.
  5. The IMSL *User News*, distributed quarterly.
  6. One fully maintained *General Information Manual* in printed loose-leaf form.

B. STANDARD SERVICES PROVIDED WITH A LIBRARY SUBSCRIPTION
  1. The automatic distribution of new Library editions containing new and/or improved routines (new tape and modification pages for the printed and microfiche forms of the *Library Reference Manual*).
  2. Maintenance of the routines and the Reference Manuals between editions, if necessary.
  3. Telephone, telex or mail consultation on Library usage.
  4. A mechanism for requesting new routines for future Library editions.

C. OPTIONAL LIBRARY MATERIALS AND SERVICES
  1. In selected cases, load module library tapes corresponding to A.1. above, with a complete new tape sent with each new edition or maintenance transmittal.
  2. The IMSL Library Minimal Tests in Fortran on magnetic tape, including:
     a. The automatic distribution of maintenance and new editions for the Minimal Tests, and
     b. Telephone, telex or mail consultation on Minimal Tests usage.

The consultation service provided with a Library subscription allows any user who is a member of a subscribing organization to contact IMSL by letter, phone, or telex concerning

A. Subroutine usage,
B. IMSL Library related programming problems,
C. Choice of the correct IMSL subroutine for a particular application, or
D. Future Library development plans.

The consultation service has proven to be extremely valuable to IMSL subscribers. It gains quick answers to questions that otherwise could involve long delays and bother to the user.

Availability

The IMSL Library is intended to operate on most large-scale scientific computers and on some scientific minicomputers. The following environments are fully supported by IMSL:

Computer Hardware	**Compiler [Reference, Appendix C]**
IBM 360/370	Fortran IV (G1,H) [6] WATFIV [10]
Xerox Sigma	Extended Fortran [9]
Data General Eclipse	Fortran 5 [11]
Digital Equipment PDP 11	Fortran IV-PLUS [14]
Digital Equipment VAX	Fortran IV-PLUS [15]
Hewlett-Packard 3000	Fortran 3000 [13]
Prime 300, 400, 500, 50	Fortran [16]
Univac 1100	Fortran V [8]
Honeywell 6000	Fortran Y [7]
Digital Equipment 10/20	Fortran 10 [5]
Burroughs 6700/7700	Fortran [2]
CDC 6000/7000	Fortran Extended [3]

Compatible hardware systems are also supported.

IMSL supports the Library for these environments as follows:

(a)   Any problem reported to IMSL receives top priority by IMSL person-
      nel. Corrections are made and transmitted to all subscribers.

(b)   IMSL testing procedures are intended to eliminate problems of
      incorrect operation.

(c)   The IMSL Library is modified, if necessary, as computer hardware
      and software change.

The IMSL Library operates correctly in many computer-compiler environments not listed above. IMSL handles problems arising in these environments
on an individual basis, and no commitment to full support is given.

Subscription Fees and Policies

The IMSL Library subscription fee is based on the type of computer and
type of organization. Large-scale computing systems are classified, by IMSL, as
Type I, and smaller scale systems as Type II. Referring to the list of supported
environments given above, the Data General Eclipse, Digital Equipment PDP
11, and Hewlett Packard 3000 are classified as Type II. All others are Type I.

Subscription fees and library reference manual fees for these two types are
given below in Tables IV and V, respectively.

In order to become an IMSL subscriber, an official of the organization
must sign a contract agreeing to abide by the Library subscription policies.

The IMSL Library subscription policies are designed to protect the
proprietary nature of the product while not necessarily restricting the members of
the subscribing organization from making full use of the Library.

Table IV

IMSL Library Subscription Fees, May 1981

Time Period	Type I		Type II	
	Non-University	University	Non-University	University
1 Year	$2,000	$1,200	$1,500	$900
2 Years	$4,000	$2,400	$3,000	$1,800
3 Years	$6,000	$3,600	$4,500	$2,700

Table V

IMSL Library Reference Manual Fees, May 1981

Reference Manual	Non-University	University
Printed	$40	$25
Microfiche	$12	$ 8

Library subscription policies are explained fully on one page of the IMSL Library Products Order Form, IMSL F-S0008, May 1981. In summary, a subscriber agrees (a) to use the Library on only one computational system for members of the subscribing organization, (b) to return Library Product materials to IMSL if the subscription is discontinued, (c) that title to IMSL Library products remains with IMSL, (d) to notify IMSL if application programs which use IMSL Library routines are distributed outside of the subscribing organization, and (e) that IMSL Library product materials will not intentionally be made available to persons or organizations not authorized for access under the subscription.

LIBRARY DEVELOPMENT

This section discusses how IMSL develops its Library products. The Research and Development Division organization and the computer systems used are covered.

Before going into the details of the current state, a brief historical sketch of the IMSL development process is given.

During the period 1970 to 1973, all programs developed by IMSL were prepared in punched card format. There were no software tools to aid program preparation. Jobs were run at a local service bureau with the aid of a pickup/delivery service. In 1973, a Data 100 communications terminal with a card reader and line printer was acquired to support program development. Jobs were still prepared in punched card format, but turnaround time was significantly reduced by the use of the Data 100 terminal. This procedure continued until 1976 when IMSL installed a Data General Eclipse computer for in-house computing. Program preparation media changed from punched cards to video display terminals connected to the Data General computer. At the time of the conversion from punched cards to terminal entry, IMSL had about 1,000,000 punched cards in storage cabinets.

The use of the Data General computer brought with it the automation of many of the tasks associated with program development and testing. The need for automation in the area seems absolutely mandatory for any large-scale software development activity such as the one in which IMSL is involved.

Table VI provides some rough idea of the cost of developing and maintaining the IMSL Library. These figures are derived from the total IMSL Research and Development expense. All costs associated with Library development and maintenance are included. Related costs such as marketing, customer relations, printing, and tape copying are not included. For each year from 1971 to 1980, total expenses are accumulated (e.g., the 1975 total expense is the sum of all such expenses from 1970 through 1975). This number is divided by the number of routines to produce the cost per routine. It is divided by the number of lines (approximately 200 lines per routine) in the Library to produce the cost per line. The costs are broken down into the cost for the basis version and the cost for each additional version.

At present, the average cost for the basis version of an IMSL Library subroutine is $22 per line, and the cost for each additional version is $1 per line. For the twelve supported environments, the cost is about $33 per line ($= $22 + 1×11).

The next two sections discuss the organization of personnel and the current IMSL program development system.

Table VI

IMSL Library Development and Maintenance Costs

Year	No. of Versions	No. of Routines	Cost per Routine	Cost per Line	Cost per Line Basis	Cost per Line Additional
1971	2	200	$1636	$8.18	$6.13	$2.05
1972	3	240	2656	13.28	7.96	2.66
1973	3	280	3243	16.22	9.73	3.24
1974	3	330	3395	16.98	10.19	3.40
1975	6	370	3847	19.23	10.58	1.73
1976	7	370	4584	22.92	13.29	1.60
1977	8	400	5068	25.34	16.47	1.27
1978	9	400	6183	30.91	23.49	.93
1979	11	460	6129	30.64	21.45	.92
1980	12	500	6573	32.86	22.02	.99

The Research and Development Organization

The Research and Development Division of IMSL has evolved over the past ten years to meet the needs of a rapidly growing product set and market-place. The division is organized into six groups which are summarized here.

Research and Design Group

This group is responsible for the design level activities in the software development process. Members of the group hold Ph.D. degrees. They represent IMSL at technical meetings and interact with members of the IMSL Advisory Board to keep up with the current state-of-the-art in numerical computations. Library subroutines are designed and initial documentation is prepared by members of this group. Consultation calls from subscribers are often handled by a member of the Research and Design group.

Development Group

This group is responsible for programming and testing of all library subroutines. Members of this group hold B.S. and M.S. degrees. They work with members of the Research and Design group in designer/programmer teams. Each IMSL subroutine is supported by a designer/programmer team.

Consultation calls from subscribers are handled by the appropriate member of the assigned team.

Internal Data Processing Group

This group is responsible for the operation of IMSL in-house computing equipment. Currently the equipment consists of a Data General Eclipse computer and a Data 100 communications terminal. This group is also responsible for the preparation of tape copies of products that are sent to IMSL subscribers.

External Data Processing Group

This group is responsible for activities related to data processing equipment not owned by IMSL. Part of the activities involve support for the Development group. Test sites must be found, and RJE procedures must be established for each of the IMSL supported environments listed above. The other activities of this group involve packaging products. Activities such as preparation of master tapes, writing of systems documents, and consultation with customers on Library installation procedures are included in this category.

Product Systems Group

This group is responsible for development of non-numeric programs that are used to support products or product development. Development system tools are prepared by this group.

Special Products Group

This group was formed to allow IMSL to develop new products. This group is responsible for development of Library subroutines for microcomputers.

The Research and Development organization is effective in dealing with the diverse work activities that are part of Library development. It allows each group to concentrate on specific areas without excessive breadth while not being greatly affected by activities outside of the group's control. It does require cooperation among groups, especially among those with high interaction needs.

The Development System

The IMSL Development group is responsible for managing over 12,000 Fortran program units (495 Library subroutines + 495 Minimal Test routines) × 12 computing systems) + exhaustive tests. Each new Library subroutine and its associated minimal test routine must be run on each of the supported computer systems. Whenever an existing Library subroutine is changed, its test programs must be run on these computers. Without the aid of the development system, this would be an enormously difficult task.

The IMSL development system is discussed in Aird and Kainer [1978]. A brief sketch is given below, and extensions of the earlier system are discussed.

The IMSL software development system is implemented on a Data General Eclipse C330 computer with 512K bytes of main memory and two 192M byte disk storage units using the AOS operating system [Data General, 1977]. A Data 100 communications computer provides remote job entry to the external computers where IMSL runs test programs. The development system is designed around the Data General AOS Command Language Interpreter (CLI), the AOS file system, and an IMSL-designed file naming convention.

The Data General AOS CLI provides the means of joining together the file system, the IMSL file naming conventions, and a set of software tools (utility programs) into a unified program development environment.

Prior to explaining how the CLI-based development system works, a brief explanation is given of the IMSL Fortran Converter. This is a program which performs automatic conversion of Fortran programs and subprograms from one computer-compiler environment to another. The conversion is done automatically via built-in features of the Converter and specifically via Converter instructions inserted in the program by the programmer. The code for all versions of a program or subprogram is contained in one file called a basis deck. The programmer works with the basis deck only. Distribution decks are generated by the development system as needed. This general approach to handling portability problems is explained in [Aird, 1977] and [Aird, Battiste, and Gregory, 1977].

The file naming convention used by IMSL is designed to clearly distinguish the different types of files used in the development of the Library.

The naming convention consists of the following:

name = \<IMSL routine name. 6 characters or less\>
prog = \<IMSL test program or associated subprogram name. 6 characters or less\>
Computer = \<IBM|XEROX|DGC|DEC11|HP3000|VAX|PRIME|UNIVAC| HIS|DEC10|BGH|CDC|H32|H36|H48|H60|ALL\>
ALL →portability across the entire computer set.
H32 →portability across H32 computer set
(etc. for H36, H48, and H60). The H32 computer set consists of IBM, XEROX, DGC, DEC11,

HP3000, VAX, and PRIME. The H36 set consists
of UNIVAC, HIS and DEC10. The H48 set consists of BGH. The H60 set consists of CDC.
The number following H is the number of bits
in a single-precision floating number.

File Name	Description
name.BD	Basis deck for IMSL routine "name"
name.computer	Distribution deck source file
name.DGC.OB	DGC object files
name.ET.PR name.MT.PR	DGC program files
RJE.name.ET.computer RJE.name.MT.computer	RJE jobfile as produced by the system
name.ET.DGC.LISTFILE name.MT.DGC.LISTFILE	Output listfile as produced by the system for DGC
name.ET.prog.BD name.ET.prog.computer	Exhaustive test program or subprogram
name.ET.LL.BD	Required programs and routines list for the exhaustive test
name.ET.DD.computer	Data for exhaustive test
name.MT.prog.BD name.MT.prog.computer	Minimal test program or subprogram
name.MT.LL.BD	Required programs or routines list for the minimal test
name.MT.DD.computer	Data for the minimal test

This naming convention gives a unique name to each file which enables
identification of the file type by the programmer and the system.

The development system uses the Data General convention of logically splitting disk space into independent sections called directories. Each programmer has a working directory where all modifications to IMSL programs take place. Modification is not permitted elsewhere. The main directory in the development system is called DEV. This directory contains all of the programs used by the system as well as the subdirectories used.

The LIB directory contains one subdirectory for basis decks, one directory for each of the supported computer systems, and three other directories called MAINT, EDITION, and TESTS. Each of the directories for the supported computer systems contains the source distribution files for the Library subroutines and minimal test routines appropriate for that computer system. The DGC directory also contains the compiled object files of the DGC product set. The EDITION directory contains all of the files for new and modified Library subroutines and minimal test routines which have been prepared for the next edition. The MAINT directory contains files for the versions of Library subroutines and minimal test routines which have been modified for the next maintenance. The TESTS directory contains the source and DGC object files for all versions of the exhaustive test programs, their associated subroutines, and data. The directory structure is shown in Figure 1.

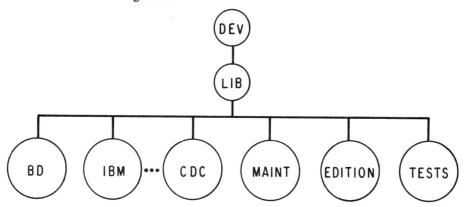

Figure 1. IMSL development system directory structure

The development programmers work at video display terminals and use the development system by executing CLI commands. A list of some of the development system commands is given at the end of this section. The RUN.LIB command is basic to the development system. It is used to make test runs of Library subroutines and automates a very tedious task that must be performed many times each day.

The RUN.LIB command is issued in the following way:

RUN.LIB/C=computer name.MT (or name.ET)

where *computer* is one of the distribution environments and *name* is the name of the IMSL routine to be tested. The development system begins the execution of the command by locating the appropriate list (prepared by the programmer) of the required routines. The system then determines which (if any) basis decks located in the programmer's working directory must be converted by examining the time last modified. If the time last modified of a basis deck is later than the time last modified of its corresponding distribution file, conversion must take place. Next, for all computers other than Data General, a job is built consisting of the routines required for running and the appropriate job control language. For every computer, there is a default site for running. Unless specified otherwise by the programmer via a /COMPILER= type switch, the default job control language is used. The JCL is contained in files named

computer.compiler.cc

These files contain the commands necessary for compilation, binding, and execution of a Fortran job at the job site. After the job is built, it is submitted (unless otherwise specified) to the default site queue for that particular computer or to a specific queue via the /QUEUE=site switch. From here, an operator copies the job to magnetic tape and places the tape on the Data 100 for input to the remote site. For Data General, after conversion, the required source files are compiled, if necessary. This determination is made in the same manner as the determination for conversion. This is followed by binding and execution of the program file. The programmer is informed upon completion of the job (execution for Data General, queue submission for other computers). The system monitors itself during the execution of a RUN.LIB command and, if at any time during the execution, an error occurs, the system will terminate execution of the command and send the user an appropriate error message.

The system will perform conversions, compiles, binds, and execution only in the working directory (that is, the directory from which the programmer gave the command). If a basis deck or its corresponding distribution file is not found in the working directory, the system will try to locate the file by searching the TEST directory, the MAINT directory, the EDITION directory, and finally the appropriate distribution directory of LIB. This enables the programmer to run a test without necessarily having all of the required routines in the working directory. The programmer can alter which directories are searched by specifying the /SL switch and setting up his own list of directories. In addition, the programmer can specify the /CURRENT switch which negates the search of MAINT and EDITION. This is particularly useful if the programmer is trying to duplicate a problem which occurred in the current release of the Library because it uses the version of the Library which is out in the field at that time.

Both the Fortran converter and the development system are written in modular form to ensure easy readability and debugging of the code. The system itself was designed in such a way as to easily facilitate the addition of a new computer into the product set. It also allows the programmer unlimited flexibility in varying specific runs. The command generally requires 30 to 60 seconds of CPU time and takes between 15 and 20 minutes of elapsed time to execute to completion. Therefore, the programmer usually issues the command from one of four batch streams, freeing the terminal for other uses.

The production and support of a multi-environment subroutine library is a complicated task, requiring many tedious details. The IMSL development system makes life much easier for the programmers undertaking this task. Through the issuance of a simple, one-line command, the programmer can cause the execution of a job, without concern for annoying, but necessary, details such as obtaining the right version of a required code. This enables the programmer to concentrate on the task at hand: supporting and enhancing a state-of-the-art subroutine library.

Other commands (software tools) available through the IMSL software development system are listed below:

Command	Purpose
BRNANL	Submit a code to be run through the branch analysis program
COMPARE	Compare two codes and list differences
EDIT	Edit a file using one of Data General's text editors, SPEED or LINEDIT
HELP	Provide on-line documentation for all system commands
MAGCARD	Send computer readable documentation to an IBM mag card typewriter
PFORT	Submit a code to be run through the PFORT verifier — runs on an IBM computer
POLISH	Reformat source code according to IMSL conventions

REVIEW   Review a code and list deviations from
              IMSL conventions

RJE   Submit a job for execution at one of
              the RJE sites

RJE.S   Status of RJE activity

RUN.CVT   Execute the Fortran converter to
              produce a specific distribution deck

SEQUENCE   Sequence a deck

SPECS   Insert explicit declarations for all
              variables

SPLIT   Split a file into separate program units

STRIP   Remove sequence numbers and trailing
              blanks from a deck

TRACER   Interactive debugging tool — Information
              Processing Techniques Corporation

REFERENCES

Aird, T. J. [1977]. "The IMSL Fortran converter: an approach to solving portability problems." *Portability of Numerical Software.* Ed. Wayne Cowell. Springer-Verlag, New York, pp. 368-388.

Aird, T. J., E. L. Battiste, and W. C. Gregory [1977]. "Portability of mathematical software coded in Fortran." *CM Trans. on Math. Soft.* 3:113-127.

Aird, T. J., and D. G. Kainer [1978]. "Conference on the programming environment for development of numerical software." *Proceedings.* JPL, Pasadena, Cal.

Data General [1977]. *Advanced Operating System (AOS), Command Line Interpreter, User's Manual,* Pub. No. 093-000122.

IMSL Library Contents Document [1980]. Edition 8.

IMSL Library Reference Manual [1980]. IMSL LIB-0008.

Appendix A

SAMPLE CHAPTER DOCUMENTATION

"INTERPOLATION, APPROXIMATION, AND SMOOTHING"

This chapter contains several subroutines for interpolation, approximation, and smoothing of tabulated data, in one and two independent variables. Cubic spline functions are used for the one dimensional case and bicubic splines are used for two-dimensional data. A routine to produce a rational function which approximates a given function in the Chebyshev sense is also included. The following discussion summarizes and clarifies the set of abilities that are included in Chapter I.

Quick Reference Guide to Chapter Abilities

This chapter contains the following subroutines:

Interpolation - One Independent Variable

    ICSCCU - Interpolation by cubic splines; easy-to-use version
    ICSICU   - Interpolation by cubic splines; user-supplied end conditions
    ICSPLN - Interpolation by cubic splines; periodic end conditions
    IQHSCU - Interpolation by quasi-Hermite piecewise polynomials

Smoothing - One Independent Variable

    ICSSCU - Smoothing by cubic splines
    ICSSCV - Smoothing by cubic splines; easy-to-use version
    ICSMOU - Smoothing by cubic splines; error detection and elimination
    ICSFKU - Least squares approximation by cubic splines; fixed knots
    ICSVKU - Lease squares approximation by cubic splines; variable knots
    IFLSQ   - Least squares approximation with user-supplied basis functions

Interpolation - Two Independent Variables

    IBCIEU - Interpolation by bicubic splines
    IBCICU - Interpolation by bicubic splines - coefficient calculator

IQHSCV - Interpolation to irregularly distributed data points

Other Routines

ICSEVU - Evaluation of cubic splines output by ICS-routines
IBCEVU - Evaluation of bicubic spline output by IBCICU
IRATCU - Rational Chebyshev approximation of a continuous function

Featured Abilities

IMSL features the cubic spline interpolation, approximation, and smoothing routines.

ICSSCV is based on work by Grace Wahba, Paul Merz and F. Utreras Diaz. It produces a smoothing spline to approximate data with experimental or random errors, using statistical considerations to decide how much smoothing to do.

The subroutine ICSMOU is based on a code by Victor Guerra developed under the direction of R. A. Tapia to deal with contaminated data. It allows the user to specify conditions for detecting and correcting points that are in error.

The routine ICSVKU is designed to compute a least squares cubic spline approximation to a set of points. The cubic spline approximation has a fixed number of knots, but the knot locations are determined by ICSVKU in order to minimize the least squares error.

ICSFKU computes a least squares cubic spline approximation with a given fixed set of knots. The routine is designed to facilitate experimentation with various knot sets.

These least squares routines are based on the SPLINE algorithm by Carl de Boor and John R. Rice.

Each of the cubic spline routines (also, IQHSCU) in this chapter produces coefficients in a standard format. This allows them to be used interchangeably. The cubic spline evaluation subroutine ICSEVU can accept input from any of these routines. The following Chapter D subroutines may be used to further process cubic splines and bicubic splines:

DCSEVU - Evaluation of first and second derivatives of a cubic spline
DCSQDU - Integration of a cubic spline
DBCEVU - Evaluation of mixed partial derivatives of a bicubic spline
DBCQDU - Integration of a bicubic spline

Name Conventions for This Chapter

All names in this chapter start with the letter I. For the cubic spline routines the second and third letters are C and S, respectively. For the bicubic spline routines the second and third letters are B and C, respectively.

Special Instructions on Usage

The following general guidelines are provided to aid the user in the selection of the appropriate subroutine. It is assumed that tabulated data $(X(I), F(I))$, $I=1,...,NX$ exist. Three general problem area settings are as follows:

P1. Least squares smoothing

The values $F(I)$ are assumed to contain small random errors and a function S is to be selected (from a certain set of functions) in order to minimize

$$\sum_{I=1}^{NX} (F(I)-S(X(I)))^2 * WT(I)$$

for some weight set $WT(I)$.

The associated IMSL subroutines are ICSSCV, IFLSQ, ICSFKU, ICSVKU and ICSSCU.

P2. Interpolation

The values $F(I)$ are assumed to be accurate and, in some sense, they define a smooth curve. A function S is desired for which $S(X(I))=F(I)$ and which can be evaluated at other points that lie between $X(1)$ and $X(NX)$. The associated IMSL subroutines are ICSCCU, ICSPLN, ICSICU and IQHSCU.

P3. Smoothing by error detection

Most values $F(I)$ are assumed to be accurate but some (perhaps 10%) contain errors. It is desired to detect the points in error and compute new values that give a smooth curve. Also, this smoothing technique may be used as a preprocessor stage for least squares techniques or interpolation techniques. The associated IMSL subroutine is ICSMOU.

Subtleties to Note

Let $X(1) = XK(1) < XK(2) < ... < XK(NXK) = X(NX)$ define a partition of the interval $[X(1), X(NX)]$. A cubic spline function is any function S defined on this interval that is continuous, has continuous first and second derivatives, and is equal to a third degree polynomial on each of the subintervals $(XK(I), XK(I+1))$, $I = 1,...,NXK-1$. The points $XK(I)$ are called knots.

A cubic spline function can be represented in several different ways, and the following representation is used by IMSL subroutines:

$$S(P) = C(I,3)*(P-XK(I))^3 + C(I,2)*(P-XK(I))^2 + C(I,1)*(P-XK(I)) + Y(I)$$

where $XK(I) \leqslant P < XK(I+1)$. In order to evaluate the cubic spline at a point P, the appropriate subinterval $[XK(I), XK(I+1))$ containing P is located and then the cubic polynomial coefficients for that interval are used to compute the value.

Each of the IMSL subroutines, ICSCCU, ICSPLN, ICSSCV, ICSICU, ICSFKU, ICSVKU, ICSSCU produces a cubic spline function with this representation. Their output consists of a vector Y of length NX and an NX-1 by 3 matrix C which uniquely define a cubic spline function S.

The IMSL subroutine ICSEVU evaluates a cubic spline, with this representation, using the coefficients Y, C and the knots XK. It follows from the definition that $S(XK(I)) = Y(I)$. In some cases $Y(I) = F(I)$ and $XK(I) = X(I)$.

For the purposes of producing a curve that looks like one drawn manually, it is sometimes necessary to relax the continuity condition on the second derivative of S. Subroutine IQHSCU produces such a function and it is called a quasi-Hermite piecewise polynomial. It is similar to a cubic spline with regard to representation and evaluation, but, the second derivative is not continuous. These functions have useful application in the interpolation area. IQHSCV produces an analogous approximation to two-dimensional data.

Cubic splines and quasi-Hermite piecewise polynomial functions are computationally stable, easy to produce and evaluate, and are quite flexible for approximation, interpolation, and smoothing.

Some of the routines in Chapter I require that input vectors be in ascending order. (That is, for a vector X of length NX, $X(I)$ must be less than $X(I+1)$ for $I = 1,2,...,NX-1$.) The user can reorder a vector by using IMSL sorting routines. For example, if the user has a vector X of length NX containing abscissae and a vector Y of length NX containing the corresponding ordinate values, the following Fortran statements will reorder the X vector keeping the desired relationship between X and Y:

for single precision:

    CALL VSRTR (X,NX,IR)
    CALL VSRTU (Y,1,1,NX,0,IR,WK)

for double precision:

    CALL VSRTRD (X,NX,IR)
    CALL VSRTUD (Y,1,1,NX,0,IR,WK)

(Note: IR is a vector of length NX initialized as follows: $IR(I)=I$ for $I=1,2,...,NX$. WK is a work vector of length NX. For additional information, see appropriate Chapter V documentation.)

For further information concerning the properties of the cubic splines, one should consult one of the following references.

See references:

1. Ahlberg, Nilson, and Walsh, *The Theory of Splines and Their Applications,* Academic Press, New York, 1967.

2. de Boor, C., *A Practical Guide to Splines,* Springer-Verlag, New York, 1978.

Appendix B

SAMPLE SUBROUTINE DOCUMENTATION

IMSL ROUTINE NAME	- ZPOLR
PURPOSE	- ZEROS OF A POLYNOMIAL WITH REAL COEFFICIENTS (LAGUERRE)
USAGE	- CALL ZPOLR (A,NDEG,Z,IER)
ARGUMENTS A	- REAL VECTOR OF LENGTH NDEG+1 CONTAINING COEFFICIENTS IN ORDER OF DECREASING POWERS OF THE VARIABLE.
NDEG	- INTEGER DEGREE OF THE POLYNOMIAL. (INPUT) NDEG MUST BE GREATER THAN 0 AND LESS THAN 101.
Z	- COMPLEX VECTOR OF LENGTH NDEG CONTAINING THE COMPUTED ZEROS OF THE POLYNOMIAL. (OUTPUT)
IER	- ERROR PARAMETER. (OUTPUT) TERMINAL ERROR

IER = 129, INDICATES THAT THE DEGREE OF THE POLYNOMIAL IS GREATER THAN 100 OR LESS THAN 1.

IER = 130, INDICATES THAT THE LEADING COEFFICIENT IS ZERO. THIS RESULTS IN AT LEAST ONE ZERO, Z(NDEG), BEING SET TO POSITIVE MACHINE INFINITY.

IER = 131, INDICATES THAT ZPOLR FOUND FEWER THAN NDEG ZEROS. IF ONLY M ZEROS ARE FOUND Z(J), J=M+1,...,NDEG ARE SET TO POSITIVE MACHINE INFINITY. ZPOLR WILL TERMINATE WITH THIS ERROR IF IT CANNOT FIND ANY ONE ZERO WITHIN 200*NDEG ITERATIONS OR IF IT DETERMINES, WITHIN THOSE 200*NDEG ITERATIONS, THAT IT CANNOT FIND THE ZERO IN QUESTION.

PRECISION/HARDWARE - SINGLE AND DOUBLE/H32
                     - SINGLE/H36,H48,H60

REQD. IMSL ROUTINES   - SINGLE/UERSET,UERTST,UGETIO,ZQADC,
                        ZQADR
                      - DOUBLE/UERSET,UERTST,UGETIO,VXADD,
                        VXMUL,VXSTO,ZQADC,ZQADR

NOTATION   - INFORMATION ON SPECIAL NOTATION AND
                        CONVENTIONS IS AVAILABLE IN THE
                        MANUAL INTRODUCTION OR THROUGH
                        IMSL ROUTINE UHELP

Algorithm

ZPOLR computes the NDEG zeros of the polynomial

$$P(Z) = A_1 * Z^{NDEG} + A_2 * Z^{NDEG-1} + ... + A_{NDEG} * Z + A_{NDEG+1}$$

where the coefficients, A_i, $i = 1,2,...,NDEG+1$, are real.

The zeros are stored in the complex array Z with complex conjugate pairs stored contiguously.

ZPOLR uses Laguerre's method. The routine is a modification of B. T. Smith's routine ZERPOL.

ZPOLR iterates toward a zero using Laguerre's method, which is cubically convergent for isolated zeros and linearly convergent for multiple zeros. The maximum length of the step between successive iterates is restricted so that a new iterate lies inside a certain region, about the previous iterate, proved to contain a zero of the polynomial. An iterate is accepted as a zero when the polynomial value at that iterate is smaller than a computed bound for the rounding error in the polynomial value at that iterate. The original polynomial is deflated after each real zero or pair of complex zeros is found, and subsequent zeros are found using the deflated polynomial.

See reference:

Smith, B. T., "ZERPOL, a zero finding algorithm for polynomials using Laguerre's method," Department of Computer Science, University of Toronto, May 1967.

Example

We want to find the zeros of the polynomial

$$X^3 - 3X^2 + 4X - 2.$$

We can proceed as follows:

Input:

```
INTEGER   NDEG,IER
REAL      A(4)
COMPLEX  Z(3)
NDEG =  3
A    = (1.0,-3.0,4.0,-2.0)
CALL ZPOLR (A,NDEG,Z,IER)
```

Output:

```
IER  =  0
Z    = (1.0+i,1.0-i,1.0)  (roots)
```

Appendix C

FORTRAN REFERENCE MANUALS

The Fortran reference manuals related to supported environments (Section 2.4) are listed here.

[1]   *American National Standard FORTRAN — ANSI X3.9-1966,* American National Standards Institute, New York, NY 1966.

[2]   *Burroughs B6700/B7700 FORTRAN Reference Manual,* 5000458, Burroughs Corporation, 1974.

[3]   *FORTRAN Extended Version 4 Reference Manual,* 60497800, Control Data Corporation, July, 1979.

[4]   *FORTRAN Reference Manual,* 60174900, Control Data Corporation, 1972.

[5]   *DEC System 10 FORTRAN-10 Language Manual,* DEC-10-LFORA-B-D, Third Edition, Digital Equipment Corporation, October, 1974.

[6]   *IBM System/360 and System/370 FORTRAN IV Language,* GC28-6515-10, Eleventh Edition, International Business Machines Corporation, 1974.

[7]   *Series 6000 FORTRAN Compiler, BJ67, Rev. 0,* Honeywell Information Systems, Inc., 1972.

[8]   *Univac 1100 Series FORTRAN V Programmer Reference,* UP-4060 Rev. 2, Sperry Rand Corporation, 1974.

[9]   *Xerox Extended FORTRAN IV,* 90 09 56F, Xerox Corporation, April, 1975.

[10] *WATFIV User's Guide, Version 1, Level 5,* University of Waterloo, 1977.

[11] *FORTRAN 5 Reference Manual, 093-000085-04, Revision 4,* Data General Corporation, October 1978.

[12] *PDP-11 FORTRAN Language Reference Manual,* DEC-11-LFLRA-C-D, Digital Equipment Corporation, June 1977.

[13] *HP3000 Series II Computer System FORTRAN Reference Manual,* Part No. 30000-90040, Hewlett-Packard, February 1977.

[14] *PDP-11 FORTRAN IV-PLUS User's Guide,* DEC-11-LFPUA-B-D, Digital Equipment Corporation, December 1975.

[15] *VAX-11 FORTRAN IV-PLUS Language Reference Manual,* Digital Equipment Corporation, August 1978.

[16] *The FORTRAN Programmer's Guide,* FDR 3057-101A, by Anthony Lewis, Prime Computer, Inc., 1979.

Chapter 11

THE SLATEC COMMON MATHEMATICAL LIBRARY

Bill L. Buzbee
Computing Division
Los Alamos Scientific Laboratory
Los Alamos, New Mexico 87545

INTRODUCTION

In 1974 the Computing Departments of Sandia National Laboratory, Albuquerque (SNLA), New Mexico, Los Alamos National Laboratory, Los Alamos, New Mexico, and Air Force Weapons Laboratory (AFWL), Albuquerque, New Mexico, organized themselves into a loose-knit confederation named SLATEC (*S*andia, *L*os Alamos, *A*ir Force Weapons Laboratory *T*echnical *E*xchange *C*ommittee). The purpose of the organization is to foster the exchange of technical information among the three computing departments. In 1980 the committee was expanded to include the computing centers of Sandia National Laboratory, Livermore (SNLL), California, and Lawrence Livermore National Laboratory (LLNL), Livermore, California. In 1981, the National Bureau of Standards (NBS) became a participant on the math library subcommittee.

Because the objective of SLATEC is to exchange technical information, subcommittees have been established under its auspices. One of these is the SLATEC Common Math Library Subcommittee. Initially, the Common Math Library Subcommittee met annually for an informal exchange of technical information, comparing development activities, library contents, objectives, etc. In 1975 the subcommittee considered the development of a common math library for use by all member sites, but rejected the idea because each participating site had an active math library project and the objectives of these projects were significantly different. For example, Sandia supported a library of about 100 thoroughly tested and thoroughly documented subroutines, while Los Alamos supported about 400 routines with relatively brief documentation.

Motivations

In 1977 the proposal for a common math library among the SLATEC sites was reexamined and adopted for the following reasons:

1. **Portability.** Over the past three decades, Department of Energy (DOE) laboratories have received the first of almost every new super-computer. Software development by the vendor for these machines typically lags behind hardware development by one or two years. Further, the vendor is likely to concentrate software expertise on systems and languages. By developing a common math library that is portable, of broad capability, and supported by DOE personnel, we are able to equip these machines with a high-quality math library at essentially the same time a workable Fortran compiler is available.

2. **Model Sharing.** There is significant exchange of scientific software among personnel of these laboratories. A common math library simplifies and facilitates that exchange.

3. **Resource sharing.** By 1977, no single laboratory had sufficient breadth of expertise to develop and maintain a large library of high-quality math software. However, collectively, the three laboratories have relatively broad expertise.

4. **Availability of "free" software.** By 1977 there was a large amount of free, high-quality software available, e.g., LINPACK, FISHPAK. Further, each member site was faced with integrating that software into its facility.

5. **Showcase.** The DOE funds research, development, and software implementation of new numerical algorithms. This library provides a convenient repository for that work.

6. **Proprietary restrictions.** Contracts accompanying commercial software restrict the usage of that software to the purchaser's organization.

Item 1 is a primary reason for SLATEC sites to provide a complete math library. We define as commercial libraries those offerings which require a profit from sales in order to survive. Software is expensive, costing approximately $50 a line debugged and documented in today's market. To amortize this expense, commercial libraries must target their products at a volume market. Supercomputers do not constitute a volume market. For example, the CRAY-1 computer has been operational for over four years, yet less than twenty CRAY's are installed. The vendor of one of the most popular commercial math subroutine libraries has yet to transport its library to this machine.

After we began the development of a common math library, other benefits of it were recognized. These will be discussed in a subsequent section.

ORGANIZATION AND ADMINISTRATION

Informally, the SLATEC Common Math Library Subcommittee consists of all personnel in these laboratories engaged in the development and support of math software. Formally, the subcommittee consists of one voting member from each laboratory. These people normally have administrative responsibilities over math software in their respective organizations. The subcommittee

1. Specifies the content of the library,
2. Assigns responsibility for specific library activities, and
3. Coordinates and reviews library development activities.

In all of these items the subcommittee leans heavily on the recommendations of technical staff. Typically, our mode of operation has been to assign responsibility for an area (e.g., solution of nonlinear systems) to one of the member laboratories, requesting personnel of that laboratory to make recommendations to the subcommittee. Formulation of those recommendations may require a few weeks or several months. If the recommendation includes development of new software, the subcommittee asks that specifications of that software be distributed to member sites for their comment. Final specifications of new software are agreed to in the subcommittee, and development is then done by the responsible site. As we will see in the final section, there are some pitfalls in this area.

Goals

In the spring of 1977 the subcommittee set the following goals for the common math library:

1. Portability.
2. Good numerical technology.
3. Good programming style.
4. Good documentation.
5. Robustness.
6. Careful testing.

However, the availability of free high-quality software subsequently influenced the subcommittee's position on goals 3, 4, and 6; i.e., while continuing to seek high quality, we did not insist on uniformity of style, documentation, and testing throughout the library.

In November of 1977 the subcommittee committed to having the common math library available at each member site by 1981. Perhaps no one is more surprised than the subcommittee that the schedule was met.

LIBRARY DEVELOPMENT

In pursuit of the aforementioned goals, the subcommittee decided to proceed sequentially through the following tasks:

1. Develop standards for handling of errors and for localizing machine-dependent information,
2. Develop programming standards,
3. Develop documentation standards,
4. Develop testing standards,
5. Assign responsibility to individual laboratories for various areas of the library, and
6. Develop format for library abstracts.

This task list is thus an outline for much of the remainder of this paper. In addition we will discuss some unexpected benefits of the library, some "lessons learned," and our future plans.

Portability

The subcommittee adopted the following approach to portability:

1. Each subroutine must pass PFORT [Hall and Ryder, 1973];
2. All machine-dependent information must be concentrated in subprograms, and, when a library routine needs to know such things as the number of bits in the mantissa, calls are made to the subroutines to get that information;
3. All error messages must be processed by a common error handler;
4. All I/O must be localized in subroutines; and
5. All names must be less than seven characters.

The subcommittee adopted the Bell Laboratories machine constants package in total, and it is in use today.

The subcommittee considered use of the SNLA error handler and rejected it because it had insufficient generality. The subcommittee also considered use of the Bell Laboratories error handler [Fox *et al.*, 1976]. We found it inadequate because *m* consecutive executions are required to resolve *m* different errors. Instead, R. E. Jones [Jones and Kahaner, 1982] developed an error-handling package that provides

1. Global enable/disable of error message printing by a single call.
2. Summary reporting of errors.
3. Warning messages.
4. Printing of first *m* occurrences of an error, etc.

Initially, no provision was made for inclusion of numerical values in the error messages; i.e., the routines calling the error handler passed only character information. However, software such as ordinary differential equation solvers need to report changes in step size, etc., so input of numerical values to the error handler was added. Also, the usage of interactive systems resulted in a need to route error messages to several output devices.

Today the evolution of the error handler appears to be complete; however, its current implementation is expensive in memory — so much so that Los Alamos personnel have implemented it directly into the system library, thereby preserving the SLATEC interface while conserving memory. Until all sites have substantial experience with this package, its future is uncertain. As we will note, the subject of error handlers should be approached with caution.

Standards

Following its task list, the subcommittee spent about six months developing an ambitious set of standards for programming, documentation, etc. These are given in Appendix A. As the development was nearing completion, we realized that by using "free software," we would not be able to comply uniformly with our standards. We decided that we would seldom modify software obtained from other sites, but would provide additional features via drivers to it. Thus, new editions of free software can be integrated into the library with nominal effort. Consequently, our "standards" became "recommendations" overnight.

Code Selection

The subcommittee approached code selection by dividing the library into subsections along the lines of the SHARE classification system. The actual selection of the codes to go into a library is a major undertaking. The subcommittee sought individuals at the various labs who were experts in a specified field to make a comprehensive survey of the software that was already available and to recommend what characteristics the codes should contain. These studies were nontrivial and in most cases took several man-months to complete. In this section, we will briefly discuss the selection of codes chosen in each subset.

Dynamic Storage Allocation

The subcommittee considered use of a dynamic storage allocation facility along the lines found in [Fox *et al.*, 1976]. The facility can significantly reduce the number of subroutine arguments throughout the library. We decided against it because memory is frequently a critical resource and thus its allocation is best left in the hands of the applications programmer.

Special Functions

After some deliberation, the subcommittee adopted the FNLIB collection of special functions developed by Wayne Fullerton [1977]. FNLIB is easily ported and offers variable accuracy as a function of mantissa length and exponent range. This collection was augmented by functions from FUNPACK [Cody, 1975], and exponential and Bessel functions from AMOSLIB [Amos and Daniel, 1977]. Initially, we included all elementary functions in this set. However, this led to problems on CDC 6000 and 7000 series equipment because elementary functions used with the FTN compiler must be callable by value or by reference. In the latter case elementary functions must recognize arguments in either large core memory or small core memory. These options were impossible to accommodate in a portable fashion, and, combined with efficiency considerations, caused the subcommittee to adopt a policy that elementary functions (ANSI 1977 intrinsic functions) will always be vendor-supplied.

Polynomial Zeros

Based on experiments comparing accuracy and efficiency, the subcommittee selected polynomial solvers utilizing companion matrix techniques for polynomials of low degree. These routines in turn use EISPACK software. Because of the associated memory requirement when the polynomial degree is high, the subcommittee also adopted routines based on n-dimensional Newton techniques [Kahaner, 1979].

Quadrature

Los Alamos and Sandia Albuquerque have long been centers of expertise in quadrature, and the routines selected are embellishments of routines jointly developed by those laboratories. Recently, there has been growing interest in QUADPACK [de Doncker, 1978]. The original version of these codes did not meet the programming standards set forth by the subcommittee. However, Elisa de Doncker has modified the codes, and they have been accepted for the library.

Interpolation

The growing use of splines prompted the inclusion of an edited version of the de Boor B-spline package [de Boor, 1977] supplemented by selected routines from de Boor [1978]. D. Amos, Sandia Albuquerque, modified the package to conform to library standards. The major effect of this modification was to eliminate restrictions on order and make the package portable. Quadrature routines [Amos, 1979] were also added to accommodate a variety of problems in approximation theory and applied mathematics.

Approximation, Interpolation, and Data Fitting

R. E. Chang, SNLL, was requested to review the current codes available in the area of curve fitting. The study included the determination of the types of approximations that should be incorporated in the library, design of a testing methodology that consisted of the determination of what characteristics should be tested, and their range, performance testing, and recommendation of codes for the library. Details of the study may be found in Chang [1981].

Linear Equations

Anticipating that LINPACK [Dongarra *et al.*, 1979] would become a world standard, the subcommittee adopted it. However, because it does not provide argument checking and error messages and because its storage convention for banded matrices is complicated, drivers to LINPACK were developed by SLATEC personnel.

Ordinary Differential Equations

L. F. Shampine and H. A. Watts of SNLA designed a software interface, DEPAC [Shampine and Watts, 1980], for a package of codes to satisfy the requirements of the subcommittee. They modified three existing codes to provide the capabilities specified in DEPAC. Two of these codes were originally written by Shampine, Watts, and Gordon at SNLA. The remaining one was a code written by A. C. Hindmarsh [1980], LLNL, which uses C. W. Gear's method for the stiff case.

Eigenanalysis

EISPACK [Smith *et al.*, 1976; Garbow *et al.*, 1977] was adopted as the core software for this area. David Kahaner, National Bureau of Standards, coordinated development of additional drivers to enable the user to exploit use of the complex data type of Fortran and to compute either eigenvalues or eigenvectors with a single subroutine call.

Vector Operations

The Basic Linear Algebra Subroutine (BLAS) package [Lawson *et al.*, 1977] was adopted because the subroutines were already in use by at least two of the member sites and because they are used by LINPACK.

Linear Least Squares

After extensive testing of existing software for linear least squares, a package of routines was developed by Tom Manteuffel [1980a and 1980b], formerly at Sandia National Laboratories and now at Los Alamos. This package was subsequently accepted for inclusion in the SLATEC math library.

Fast Fourier Transforms

The subcommittee adopted the Fast Fourier Transform package of Paul Swarztrauber [1978], National Center for Atmospheric Research.

Nonlinear Equations and Linear Least Squares

The subcommittee requested K. L. Hiebert, SNLA, to survey this area. Eight codes that solve systems of nonlinear equations and twelve codes that solve nonlinear least squares problems were collected and studied in depth. All the codes are user-oriented Fortran subroutines that exist in either mathematical software libraries or packages, e.g., IMSL, NAG, NPL, and MINPACK. Extensive testing [Hiebert, 1980 and 1981] was done to determine the robustness and reliability of the codes under various conditions. Recommendations were made to the subcommittee based on the results of the testing and the requirements of the SLATEC Library. For example, the codes from the NPL Library could not be adopted because they are proprietary. Modifications were made to the adopted codes to meet subcommittee specifications.

Fast Poisson Solvers

The subcommittee adopted the FISHPAK collection of Paul Swarztrauber and Roland Sweet [Swarztrauber and Sweet, 1975], National Center for Atmospheric Research. This package was certified by DOE personnel [Steuerwalt and Sweet, 1975].

DOCUMENTATION

When the subcommittee began specifying the details of the documentation in 1979, it was unanimously agreed that we wanted documentation that was easy to maintain and easy to distribute, i.e., machine-readable documentation. A cursory check of the SLATEC source showed that in almost every case the prologue of subroutines constituted a sufficient usage document. Thus the subcommittee decided to add unique comment cards (see Appendix B) to all SLATEC source and to write a program that would key on these cards to extract the prologue from each subroutine, thus producing a writeup for it. In those rare cases where the prologue was inadequate, the subcommittee decided to expand it. The documentation program is distributed with the SLATEC source. Tom Jefferson and Dona Crawford, SNLL, designed the documentation program; Karen Haskell and Walt Vandevender, SNLA, implemented it.

QUICK CHECKS

The subcommittee agreed that the SLATEC Common Math Library should include diagnostic capabilities. We decided to develop a set of quick-check drivers, each of which would test one or more SLATEC routines. For each library routine the typical driver creates some of the diagnosable conditions and verifies that the proper error code and message are produced. In addition, a small but nontrivial test of the accuracy of the routine is made. This set of quick-check routines provides the capability to make a simple and basic test of the library's integrity whenever it is modified or installed in a new environment. Achieving portability of diagnostic routines has proven nontrivial and, as we will note, is an area for future work by SLATEC.

DISCLAIMER

Because all software in the SLATEC Common Math Library is available to anyone on request and because subcommittee members hope that the library will enjoy usage by universities and laboratories engaged in scientific computation, a disclaimer for the software was developed and approved by the legal departments of all participating laboratories. Because of the length of the disclaimer and the number of routines in the library, we do not put the full disclaimer in every routine. Perhaps the best use of the disclaimer was suggested by Andy White, namely, that the error handler be modified to print the disclaimer on each occurrence of an error. (This suggestion is not to be taken seriously.)

CLASSIFICATION SYSTEM

After initially considering the AMS system [Luke, Wimp, and Fair, 1972] and the modified SHARE [1973] system for classifying the contents of the library, the subcommittee decided to use the AMS system. However, the AMS system proved to have inadequate depth, and subsequently we adopted the classification system of Bolstad [1975].

ABSTRACTS

As of January 1981, the SLATEC Common Math Library has a rudimentary abstract in the form of a table of contents containing a short description of each routine with the routines sorted by chapters. As we will note, this is an area for future work and becomes increasingly important with the growth of the library. The subcommittee would like an abstracting system that is interactive.

DISTRIBUTION OF LIBRARY SOURCE

The subcommittee accepted a recommendation that its interchange tape format be unlabeled nine track tape, 1600 bpi, ASCII records blocked with twenty-seven 80-column card images per block. Also the National Energy Software Center (NESC) at Argonne National Laboratory has agreed to serve as distribution point to nonSLATEC sites. Further, NESC will distribute all of the library including those packages which they already distribute separately, such as LINPACK and EISPACK.

MAINTENANCE

SLATEC plans to issue one edition of the Common Math Library each year. When a bug is found, it will be reported to SLATEC sites and to NESC with local fixes for it. A SLATEC laboratory will be asked to assume responsibility for providing a revision or replacement in the next edition.

One of our biggest maintenance problems is dealing with a large volume of source code. As we will note, this is an area for future work by SLATEC.

UNEXPECTED BENEFITS FROM SLATEC

Now that five (1977-1981) years of development work have passed and the library is becoming operational, we recognize some unexpected benefits from this effort. For example, SLATEC's decision to use machine-readable documentation proved fortuitous to Los Alamos when it decided to institute online retrieval of documentation for its software. Also, all of our laboratories have acquired VAX computers. The portability of the SLATEC collection allowed us to equip those machines with software of good quality — in some cases identical to other math libraries that users of these machines were accustomed to. The CRAY-1 computer was the first computer on which the SLATEC error handler was installed. This installation was done at Los Alamos, where the CRAY-1 is used to run numerical simulations that last for several hours. Occasionally these numerical simulations terminate unexpectedly, because of numerical instabilities, etc. When this happens, owners of these simulations must regain control of the computer to close files, print diagnostic information, etc. In some cases they wish to restart the calculation from an earlier dump. Personnel associated with the simulations quickly realized that the error handler could be modified to provide the capability of regaining control of the computer in these situations.

FUTURE PLANS

We have already noted several areas where additional work is needed on the SLATEC library. Among them are an interactive abstracting facility, portable quick checks, and facilities for maintaining a large volume of software. We must add software in such areas as optimization, boundary value problems, partial differential equations, and integral equations. Some of these are areas of research in math software. Other areas for future work include usage statistics and unique naming conventions. Statistics are needed to identify those routines that are used frequently so that they can be made efficient. At the same time, since usage statistics can quickly become voluminous, we need facilities for selective collection of usage statistics on our library.

Because of its size, we already have the probability of naming conflicts within the SLATEC Common Math Library. When it is integrated with graphics libraries, utilities, etc., probability of conflict becomes fairly large. Thus, the subcommittee will continue to consider unique naming conventions for all SLATEC codes. We are reluctant to do so because of the impact on our users. It means changing code that we did not originate and causing incompatibility with in-depth documentation provided by our "software suppliers." Thus, the subject may become a point of discussion with our "suppliers."

LESSONS LEARNED

The SLATEC Common Math Library is a fairly ambitious effort, especially when one considers the amount of source code involved and that it is a collaborative effort. Nevertheless, some of us would be willing to do it again. Therefore, for our benefit and for the benefit of others, we have recorded some of the lessons learned.

1. Ego investment. Success in an effort such as the SLATEC Common Math Library requires persistent, motivated people who desire to get the best code available as opposed to always using code that they developed. We believe that SLATEC has been exemplary in this regard, and we point to the contents of the library as evidence of that.

2. Which version should we use? There are numerous versions of the BLAS, and during the development of the library every member site had at least one version in operation. Thus, selection of a specific version of similar packages is nontrivial and will inevitably cause problems for some sites.

3. Frequent meetings. For most of us, development of the SLATEC library has been an ancillary activity, and we have tended to work on it only immediately before the next meeting. Thus, frequent meetings ensure progress as well as keeping minds fresh.

4. Design, then implement. In those few instances where SLATEC has designed software such as the error handler or the LINPACK drivers, we have been plagued by a tendency to revise a design after implementation has begun. This is deadening to the developers. We believe that the best course is to take as much time as is needed to do the design, providing opportunity for wide comment. Once the design is complete, freeze it and implement.

5. Schedules. Setting a date and commitment to perform certain things by that date has been useful. Of course the people who must do the work must also agree to the schedule.

6. Flexibility. Had we insisted on rigid standards for programming and documentation, we would have precluded use of "free software."

7. Good records. It is essential to keep a written record of all decisions with regard to the library. It has been the subcommittee's custom that the host site for each meeting keep the minutes thereof. A better procedure is to designate a secretary and perhaps rotate that assignment on a systematic basis.

8. Error handler. The numerous difficulties that we have had with the error handler suggest that this may be the wrong approach to that problem. The current package is a patchwork of features evolved over a period of time and has relatively large memory requirements.

9. Experts. The development of a library the size of SLATEC should not be undertaken unless a substantial number of experts are available for code selection and development and for consultation about the codes. Almost 40 people have contributed to this effort.

ACKNOWLEDGMENTS

The SLATEC Common Math Library was made possible by many talented, persistent people working together. We are pleased to acknowledge them here.

R. Allen, AFWL and The University of New Mexico
D. Amos, SNLA
R. Basinger, LLNL
W. R. Boland, AFWL and Clemson University
R. Chang, SNLL
J. Chow, Los Alamos
J. Coffey, Los Alamos
D. Crawford, SNLL
R. Davenport, Los Alamos
K. Fong, LLNL
W. Fullerton, Los Alamos and Bell Laboratories
A. Funk, AFWL
R. Hanson, SNLA
K. Haskell, SNLA
A. Hayes, Los Alamos
K. Hiebert, SNLA
A. Hindmarsh, LLNL
M. Havens, AFWL
R. Huddleston, SNLL
N. Jacobsen, Los Alamos
T. Jefferson, SNLL
R. Jones, SNLA
T. Jordan, Los Alamos
D. Kahaner, NBS
T. Manteuffel, Los Alamos
A. Marusak, Los Alamos
C. Moler, University of New Mexico
D. Montano, Los Alamos
J. Norton, Los Alamos
M. Scott, SNLA
L. Shampine, SNLA
T. Suychiro, LLNL
W. Vandevender, SNLA
E. Voorhees, Los Alamos
L. Walton, SNLA
H. Watts, SNLA
A. White, Los Alamos

REFERENCES

Amos, D. E. [1979]. *Quadrature Subroutines for Splines and B-Splines.* Sandia Laboratories Report SAND79-1825.

Amos, D. E., and S. L. Daniel [1977]. *AMOSLIB, A Special Function Library.* Sandia Laboratories Report 77-1390.

Bolstad, J. [1975]. "A proposed classification scheme for computer program libraries." *SIGNUM Newsletter* 10:32-39.

Chang, R. E. [1981]. *An Evaluation and Comparison of Curve Fitting Software.* Sandia Laboratories Report SAND80-8727.

Cody, W. J. [1975]. "The FUNPACK package of special function subroutines." *ACM Trans. on Math. Soft.* 1:13-25.

de Boor, C. [1977]. "Package for calculating with B-splines." *SIAM J. Numer. Anal.* 14:441-472.

de Boor, C. [1978]. "A practical guide to splines." *Applied Math. Sci. 27,* Springer-Verlag, New York.

de Doncker, E. [1978]. "An adaptive extrapolation algorithm for automatic integration." *SIGNUM Newsletter* 13:12-18.

Dongarra, J. J., *et al.* [1979]. *LINPACK Users' Guide,* SIAM Publishing Co., Philadelphia.

Fox, P. A., *et al.* [1976]. *Basic Utilities for Portable Fortran Libraries.* Bell Laboratories Computing Science Technical Report 37.

Fullerton, W. F. [1977]. "Portable special function routines." *Portability of Numerical Software.* Ed. W. Cowell. Springer-Verlag, New York.

Garbow, B. S., *et al.* [1977]. "Matrix eigensystem routines — EISPACK guide extension." *Lecture Notes in Computer Science,* Vol. 51. Springer-Verlag, New York.

Hall, A. D., and B. G. Ryder [1973]. *The PFORT Verifier.* Bell Laboratories Computing Science Technical Report 12.

Hiebert, K. L. [1980]. *An Estimation of Mathematical Software Which Solves Systems of Nonlinear Equations.* Sandia Laboratories Report 80-0181. (Submitted to *ACM Trans. on Math. Soft.)*

Hiebert, K. L. [1981]. "An evaluation of mathematical software which solves nonlinear least squares." *ACM Trans. on Math. Soft.* 7:1-16. (See also Sandia Laboratories Report 79-0483.)

Hindmarsh, A. C. [1980]. "LSODE and LSODEI, two new initial value ordinary differential equation solvers." *ACM SIGNUM Newsletter* 15:10-11.

Jones, R. E., and D. Kahaner [1982]. *XERROR, The SLATEC Error-Handling Package.* Sandia National Laboratories Report SAND82-0800.

Kahaner, D. [1979]. *Zeros of a Polynomial with Optional Error Bounds.* Los Alamos Program Library Writeup C217.

Lawson, C. L., *et al.* [1979]. *"Basic linear algebra subprograms for Fortran usage."* *ACM Trans. on Math. Soft.* 5:308-323.

Luke, Y., J. Wimp, and W. Fair [1972]. "Subject classification system for the journal *Mathematics of Computation."* *Cumulative Index to Mathematics of Computation.* American Mathematical Society.

Manteuffel, T. A. [1980a]. *An Interval Analysis Approach to Rank Determination in Linear Least Squares Problems.* Sandia Laboratories Report SAND80-0655.

Manteuffel, T. A. [1980b]. *The Weighted Linear Least Squares Problem: An Interval Analysis Approach.* Sandia Laboratories Report SAND80-1260.

Shampine, L. F., and H. A. Watts [1980]. *DEPAC — Design of a User Oriented Package of ODE Solvers.* Sandia Laboratories Report 79-2374.

SHARE [1973]. *SHARE Reference Manuals.* SHARE Program Library Agency.

Smith, B. T., *et al.* [1976]. "Matrix eigensystem routines — EISPACK guide." *Lecture Notes in Computer Science,* Vol. 6, 2nd ed. Springer-Verlag, Berlin.

Steuerwalt, M., and R. A. Sweet [1979]. "Certification report on efficient Fortran subprograms for the solution of elliptic partial differential equations." *ACM Trans. on Math. Soft.* 5:765-771.

Swarztrauber, P. [1978]. *A Fast Fourier Transform of Periodic and Other Symmetric Sequences.* Los Alamos Program Library Writeup F504.

Swarztrauber, P., and R. Sweet [1979]. *Efficient Fortran Subprograms for the Solution of Elliptic Partial Differential Equations.* NCAR-TX/IA 109.

Appendix A

SLATEC PROGRAMMING STANDARDS

Names:

The prefix characters C and D, C for complex and D for double precision, shall be reserved. The subroutine name shall consist of a prefix and five characters; S can be used as a prefix for single precision routines. Names should be esoteric to eliminate possible conflict.

Ordering of subroutine arguments:

1) Input variables (functions usually first).

2) Variables which are both input and output.

3) Output variables.

4) Work arrays.

Array declaration parameters should immediately follow the associated array name.

Prologue format:

1) Abstract. Includes informative description of the routine, *biographical* data, and references.

2) Argument description. Arguments will be discussed in *exactly* the order they are input to the routine, under headings corresponding to the above 4 categories.

3) Array declaration. This could take the form of the actual Fortran DIMENSION statement, or a more descriptive pseudo-Fortran statement (comment) for more complicated array dimensioning information.

4) An end of prologue marker, for example

        C***END OF PROLOGUE

The prologue should contain enough information that it could be used as the main source of documentation.

Appendix B

PROLOGUE COMMENT CARDS FOR DOCUMENTATION PROGRAM

```
            SUBROUTINE BLAT(A,B,C,D,
     1                      E,F,G)
C***BEGIN PROLOGUE  BLAT
C***REVISION               (date of latest subprogram update)
C***CATEGORY NO.           (one or more classification codes,
                            separated by commas with the primary
                            code appearing first.)
C***KEYWORD(S)             (one or more keywords, separated by commas)
C***DATE WRITTEN           (date original program was written)
C***AUTHOR                 (author(s); e.g., Jones R. (SNLA))
C***PURPOSE                (1-3 lines describing what the program does)
C***DESCRIPTION            (program abstract including method used,
                            description of arguments according to
                            calling sequence, dimension information,
                            consultants, etc.)
C***REFERENCE(S)           (primary book, report, journal article, etc.)
C***ROUTINES CALLED (names of routines called; or (NONE) )
C***END PROLOGUE
```

Any number of desired comment cards may come between the C*** cards. Not all of the C*** cards may be present for some routines.

If the routine is subsidiary to another routine, its prologue would be as follows:

```
            FUNCTION BLAM(H,I,J)
C***BEGIN PROLOGUE BLAM
C***REFER TO   BLAT
C***ROUTINES CALLED  (NONE)
C***END PROLOGUE
```

Chapter 12

THE BOEING MATHEMATICAL SOFTWARE LIBRARY

A. H. Erisman
K. W. Neves
I. R. Philips
Boeing Computer Services Company
Tukwila, Washington 98188

INTRODUCTION

The Boeing Mathematical Software Library is a complete offering of mathematical software available to all customers of the Boeing Computer Services (BCS) international network of IBM, CDC, and CRAY computers. The library, or parts of it, is also available on many internal Boeing computers from large-scale mainframes to microcomputers. In this paper we will present a number of different views of the library to attempt to achieve a complete picture of it. We will view the library

- As a product, describing what it is.
- As a user tool, describing the user community and its effect on shaping the product.
- As a project, describing some of the processes in building and maintaining the product.
- Historically, describing how the library has grown over the years and looking briefly into future challenges.

We have taken the position that the primary purpose of a mathematical software library is to provide productivity benefits in applications. Hence, while the primary builders of the library have interests in numerical analysis and mathematical software, the end result is a success only if it can be, and is, used as a tool by those in other fields. The goal is to have a library that provides productive and efficient building blocks for applications software and is not simply a research tool for numerical analysts.

THE BOEING LIBRARY AS A PRODUCT

The Boeing Mathematical Software Library is a collection of libraries, packages, documentation, and support services organized to respond to a broad range of user requirements. At the core of this library is BCSLIB, a collection of subprograms grouped together by function:

- Elementary and Special Functions.
- Interpolation, Approximation, and Transforms.
- Polynomials and Nonlinear Equations.
- Quadrature and Differential Equations.
- Linear Algebra.
- Statistics and Probability.
- Optimization and Mathematical Programming.
- Time Series Analysis.
- Utility Software.

BCSLIB is designed to satisfy the needs of most users for mathematical, statistical, and utility software most of the time. Since this is the entry point for users of the Boeing Mathematical Software Library, usage of routines is designed to be simple. Thus routines in BCSLIB generally have short, simple calling sequences. In addition, choices of "what routine to use" are minimized by limiting the size and scope of this library. User documentation is contained in a single volume and is also available on-line. More detail on the software and documentation features is deferred to later in this section.

Supporting this software are user services: consultation, training, and promotion. BCS maintains a staff of experts to provide a consultation service for users with mathematical or statistical problems. Help is provided concerning available routines, appropriate mathematical/statistical techniques for a particular problem, and software usage difficulties being encountered by a user. Special classes on the use and availability of mathematical software in Boeing are offered through the Boeing training program. In addition, an introduction to the library is taught in most training classes at Boeing. A quarterly newsletter, specializing in mathematical, statistical, graphical, and utilities software, is published as a part of the library project and has become important in the communication between library developers and users. The newsletter provides the following:

- Quick announcement mechanism of new capabilities, courses, and seminars.
- Tutorial articles.
- Mechanism for user feedback (e.g., surveys).

The newsletter, available without charge, has wide distribution at Boeing, among network users, and among those interested in mathematical software.

To reach the goal of having mathematical software used in sophisticated scientific/engineering programs, more specialized software is required. To add this capability to BCSLIB would make the library more difficult for the less sophisticated user. Hence we have adopted a "second level" of software for our library, referred to as Level II.

This, too, is grouped by functions, but is documented separately. Coverage is more specific in Level II, being geared to special user needs rather than generic "problems that can be solved." Some duplication of capability is found here, since the more sophisticated user may select particular trade-offs of one algorithm (e.g., more speed at the expense of more storage). Thus the "core" library, BCSLIB, contains basic computational modules with easy-to-use drivers and default input or output options to simplify calling sequences. Level II software includes more specialized subroutines as well as more flexibility of access to internal computation in the basic computational modules.

We will illustrate the concept of two levels of software with the section on the solution of linear equations. In BCSLIB, this section includes routines to solve equations where the coefficient matrix can have the following properties:

- General (real and complex).
- Positive definite symmetric.
- Positive definite symmetric banded.
- Real symmetric and complex Hermitian indefinite.

Options of factor only, solve only, or factor and solve with one right-hand side are available to the user.

The "second level" of software in this area incudes all of the dense matrix software from LINPACK [Dongarra et al., 1979], SPARSPAK [George and Liu, 1979], general sparse matrix routines [Duff and Reid, 1979], some iterative and special matrix routines, MATLAB [Moler, 1979], and out-of-core equation solvers for many different matrix types. Iterative refinement and solving with many right-hand sides simultaneously are some other options available. This software is referenced in the BCSLIB manual in the capabilities outline, but it is not documented there. A separate document for this linear algebra software is in preparation; at present, loose-leaf usage abstracts for each routine are made available through the consultation service.

The software is not independent between the core and Level II. In fact, some of the routines in BCSLIB are based on LINPACK software with simplified drivers, modified to conform to our standards.

It is also seen from this illustration that the choice between what goes in the core library and what goes in Level II is not easy. We were tempted to eliminate the positive definite symmetric case from the core. Comparisons by Barwell [1974] indicate that the indefinite solver is within a few percent as efficient as the Cholesky algorithm. Why put in two choices when a user may not know if the matrix is positive definite and the benefits of knowing are small? In the end we left it in the core because it is a "traditional" capability which some users expect.

On the other side of this issue, sparse matrix problems are more widely recognized now than ever before, and the gains from using sparse matrix software are enormous. Should such software, with a simple interface, be provided in the core? We decided not to do so at this time. The decisions for both cases may be reversed at a later date.

These second-level specialty modules are referred to as "petals" in reference to the diagram in Figure 1, which illustrates this multilevel software concept. Petals incude all software in the particular specialty area and thus overlap BCSLIB.

Level II contains program packages as well as subroutines. This is especially true in the statistics area. BCSLIB contains many of the basic computational modules in statistics, but Level II is not nearly as exhaustive in coverage. Rather, references are made to major statistical packages offered on our network computers, such as SPSS, BMDP, and SAS. We have found that in the statistics area, many of our users prefer to access packages where links to data-base packages and graphics packages provide a convenient means for doing statistical analysis of large data sets and displaying the result. Data-base and graphics package support is a part of the BCS network offering and is coordinated with the mathematical library project.

Standards for mathematical software in BCSLIB have been established in the areas of usage abstract format, naming conventions, standard error handling procedure, working storage, and software portability. These standards are described in the appendix. The standards also apply to "second level" routines developed or modified by our staff, but are not universal here because of the inclusion of externally generated software such as LINPACK.

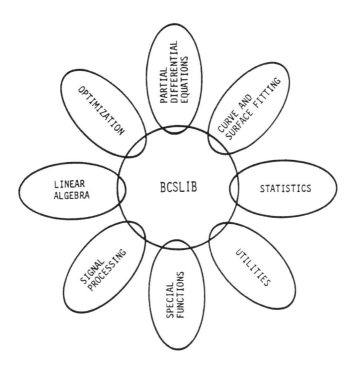

Figure 1. Diagram for the Boeing library structure

We comment here briefly on portability because of its connection with efficiency. To the extent possible, all code is written in transportable Fortran. Since a user of our library may access the software on a variety of computing environments at Boeing, it is important that the names and calling sequences of comparable routines remain standard in all library versions. Historically, this concept of "usage portability" was not enforced by library developers. In recent years it has become a major goal and is an important aspect of the library development standards. In order to achieve maximum efficiency, specialized code (e.g., assembly language) is sometimes developed for a particular mainframe version of a given library routine. In all cases an equivalent transportable Fortran version is maintained to ensure portability to new mainframes, and without exception "usage portability" is maintained. The user can then move a portable program from one machine to another without changing the calling sequence to the library routine, yet that routine may be different internally in a way that achieves efficiency but is otherwise transparent. With the transportable Fortran version also available, the program can also be readily moved to a non-Boeing computer, though some efficiency would be lost.

BCSLIB contains three function subprograms — HSMCON - REAL, HDMCON - DOUBLE PRECISION, and JHMCON - INTEGER — that provide to the user certain mathematical and machine constants at maximal precision. These constants are an extension of the International Federation for Information Processing standard for machine parameters [Ford, 1978]. The effect of their use by both library software and application program is threefold:

- Increase portability.
- Obtain maximum possible machine accuracy.
- Reduce development and maintenance costs.

Some examples of the available constants are machine overflow threshold and relative machine precision.

All of BCSLIB and Level II software is available on major mainframe computers on the BCS network, including IBM, CDC, and CRAY hardware. Appropriate subsets of these libraries are available on computers used for special applications at Boeing, including Harris, Data General, XEROX, SEL, VAX, PDP, Floating Point System Array Processor, and TERAK microcomputers.

THE BOEING LIBRARY AS A USER TOOL

The dominant reason for the two-level structure of the Boeing Mathematical Software Library described in the previous section is the diversity of the user community. The library is accessed by many users of the BCS computer network. Users are, for example, aerodynamics research scientists, system software designers, structural engineers, and business forecasters from Boeing. Non-Boeing users are, for example, geologists, petroleum engineers, land use planners, and bankers.

User needs from this group range from simple descriptive statistics to non-linear regression analysis, from the solution of a few simultaneous linear equations to large out-of-core complex valued systems of equations, from linear interpolation to two-dimensional biquintic spline approximation.

For some users the mathematical software is simply a black box to solve a problem; what happens inside the routine may be irrelevant. For other users, internal computations are important.

Other factors that dictate the requirements for a wide range of software are the trade-offs among ease of use, efficiency, generality, storage, and portability. Any single piece of software must be a compromise of these considerations.

With several different pieces of software, more user needs can be met. There are, for example, computer applications within Boeing that tax all of the resources of the computer. For the designers of such codes to be able to use library subroutines, the routines must be able to respond to their high efficiency demands. Since it is not possible or reasonable to have all potential problem variations of a particular routine in the library, we have adopted a compromise approach. We recommend using a library routine to simplify code development, since the library routine is a tested building block. If more speed or less storage is required in the particular application, then the library routine can be tailored for this application. Finally, if the resulting tailored subroutine has potentially wide application, it can be packaged as a part of the Level II library for use by others. This process is discussed further in the next section.

Mathematical library subroutines are normally used as building blocks in computer codes. These codes are implementations of mathematical models arising from many problem areas, some of which have been mentioned above. To be useful, the mathematical model must generate valid computational results that can be interpreted in terms of the "real life" process or system being modeled. Therefore, the computational process embedded in the model should be efficient in its use of computing resources, should have predictable accuracy, and, when appropriate, should be capable of error control. If it is not efficient, the software may not be used, or the user will not obtain the desired results because his program takes too long to execute or is prohibitively costly to run to completion. Not every physical phenomenon or engineering system can be effectively modeled by a computer simulation. However, where the computer model is used, it is crucial that the software be as reliable and efficient as possible. At Boeing, for example, the preliminary design of air foils is often subjected to a computer simulation before an expensive wind tunnel mock-up is created. Thus, the computer has become an integral part of the design and manufacturing process. Hence, software efficiency, accuracy, and reliability can directly affect the final cost of the product. In other instances, computer models actually replace expensive tests. For example, computer modeling of crashing vehicles has proven to be an accurate, reliable, and cost effective alternative to tests with instrumented mannequins. Another unusual example is the discrete simulation (by computer) of airplane evacuation schemes. The cost of one physical test of a 747 evacuation, employing hundreds of people, an airplane, and a test administration, exceeded the cost of developing a reliable program to eliminate (not all but many) such physical tests. The program is only as effective as the statistical discrete simulation algorithms employed, which all rely on the robustness of a random number generator.

When the computational process is not of predictable accuracy, the modeling results are difficult to relate to the engineering model (which has its own set of limitations). Thus the usefulness and reliability of the mathematical software

are critical to computer simulation.

In addition to these properties, users also expect the computational process in their models to be extendible to support future model and algorithm development. The extensions of an algorithm should also be easily (automatically) implementable in all models in which the algorithm is being used. In this way, most users are likely to benefit from advances in numerical analysis even after their models have been implemented in a computer program.

It is important to remember that many builders and users of mathematical models are experts in the area in which their models arise (e.g., engineering), and not in numerical analysis. By using the right kind of mathematical software building blocks, the user will bring into the model high quality and up-to-date expertise in numerical analysis. This expertise, which is built into the software, will ensure that the solution process has the required properties: efficiency in terms of computing resources, predictable accuracy and, when appropriate, error control capability. This approach will give the modeler flexibility and freedom from the limitations of an inadequate numerical algorithm.

A set of robust mathematical software — documented, maintained, and up-to-date — can satisfy all of the above user requirements and more. Users who incorporate such software as building blocks in the computer implementation of their mathematical models will save time and money. These savings provide the economic justification and basis for the software library project. The savings come about because duplication of effort can be avoided by not recreating building blocks and by simplifying overall model debugging and validation. As a result, model building time is decreased and the overall reliability, efficiency, and effectiveness of the model are increased.

Concomitant with the above philosophy, the mathematical libraries at Boeing are maintained by one organization. Users are discouraged from obtaining the source modules and instead attach a compiled "object" library at program execution time. Thus the user is relieved of version control and validation for all library software. Any improvements in a given subroutine are included in new versions of the library, and all users benefit. Substantive algorithmic changes are, when appropriate, added to the library as new subroutines. Likewise, algorithms that are no longer "state-of-the-art" are gradually phased out.

Mathematical software is an expensive tool to develop; hence many libraries are proprietary and contain some restrictions on their use. Such restrictions are important in protecting software and maintaining version control. Yet too many restrictions can render the library useless as a tool. For example, if library software cannot be used outside of the host computer, then it cannot be used to develop software for delivery elsewhere. Because of this user requirement, Boeing has established a mechanism for release of the proprietary software when there is a genuine need.

Boeing users are required to process a release-of-data form with appropriate approvals and release restrictions. Generally the software is approved for release with the constraint that it may be used within the environment of the host program only. Non-Boeing users of the BCS network may want to develop software on the BCS network and then use that software on their own in-house computers. The software release process also works for this case, although there is generally some charge for the software.

We conclude this section with a brief discussion of usage statistics. The figures given are lower bounds because of the way the data are gathered; discussion of how the monitoring is done is deferred to the next section.

In 1980, BCSLIB was used at least 3.5 million times, 2.3 million on the more scientifically oriented CDC-based computers; accesses have shown a steady growth over the years of monitoring since 1976. Level II software was formally introduced to the BCS network in 1978; accesses exceeded 10,000 in 1980.

The most heavily accessed routine in the library continues to be TBLU1, a subroutine that does one-dimensional interpolation. The subprograms for machine constants, two-dimensional interpolation, sorting, and the solution of general linear simultaneous equations are next most heavily accessed. A polynomial least squares subroutine is also in the top ten most-accessed list.

From these statistics we observe that most library usage is in computationally simple areas. Individual subroutines from BCSLIB and Level II, which are computationally more complex, have fewer accesses. This fact raises the question of whether or not a library should include the lower use, more computationally complex (hence more expensive to produce) subroutines.

The answer is clearly yes, since number of accesses alone is not an adequate measure of the usefulness of a library routine. A linear interpolation routine could be programmed, perhaps somewhat less robustly and efficiently but still reasonably, by many. A subroutine for computing eigenvalues and eigenvectors for the large, sparse generalized eigenvalue problem, on the other hand, is a very difficult task which could be written by only a few people. Hence access by a small number of users will allow them to compute something that could not be done without library software. This discussion is pictured in Figure 2.

Obviously there remains a danger of developing very complex mathematical software to solve problems that no one wants to solve. On the other hand, there may be no demand for (or use of) a particular capability because of a lack of awareness of its existence. Thus, usage alone cannot dictate the library scope or coverage. The approach that we have taken to this problem of "coverage" is discussed in the next section.

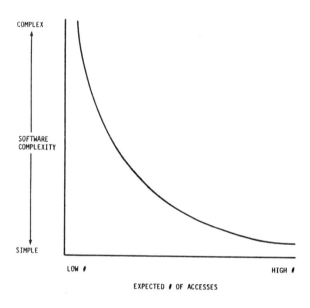

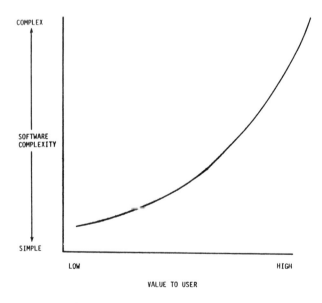

Figure 2. Number of users and value of use vs. complexity of math software

THE BOEING LIBRARY AS A PROJECT

The Boeing Mathematical Software Library project is one of many projects performed by the Mathematics and Modeling staff of Boeing Computer Services. This staff consists of approximately 100 professionals in numerical analysis, applied mathematics, statistics, and operations research performing applied mathematical analysis and research in support of all elements of The Boeing Company and contract research work for external agencies. The library project, under the direction of a project leader from the Numerical Analysis group, draws on the resources of a large part of this staff. In 1980, for example, 40 individuals made some contribution to this project, though there were never more than 6 contributors at any time.

This mode of operation has the advantage that analysts supporting the library project are in touch with user needs through other projects. Because of this the library project serves as a means of technology transfer. A specialized software capability such as a complex valued out-of-core equation solver or a two-dimensional biquintic spline approximation may be developed for a particular project. Through the library, this capability can then be transferred to other users by packaging the software and testing it according to the library standards. This ensures the quality and usefulness of such software for future uses.

Another way in which software is added to the library is through evaluation and subsequent adoption of externally developed software. Sources include packages produced by Argonne (e.g., EISPACK, LINPACK), the Harwell Library, Sandia software, and *TOMS* algorithms. The evaluation process is generally motivated by a problem-solving need. Either a new capability is available for an area that has high use in our library (such as LINPACK), or software is available in a new area of customer need (as was the case with SPARSPAK). Because subroutines are generally added to Level II based on identified need, coverage in different areas (e.g., linear algebra or approximation) is not necessarily uniform within Level II.

We have discussed our means of communicating with users and assessing their needs, working jointly on projects. This can also lead to seminars in a particular area. Recently, members of the numerical analysis staff presented a seminar on sparse matrix computation (solvable problems, available software, and research directions) to a research staff in the Boeing Commercial Airplane Company. Other more general seminars are offered through the training division of Boeing Computer Services.

The consultation service provides another means of assessing user needs.

Users with particular problems can often be directed to problem formulations that can use existing software, sometimes in the "proof of concept" mode described in the previous section. Occasionally, a real hole in the coverage of the library is identified, and then a task is defined to fill this hole. Priorities — depending on budget, the expertise of available staff, and the importance or applicability of the need — are established by the library project manager.

Another aspect of the library as a project is the establishment of standards (discussed earlier and outlined in the appendix) and the use of software tools. Two of the tools, COMPARE [Harwell, 1981] and the PFORT Verifier [Ryder and Hall, 1973], not only are useful in developing mathematical software for the library but receive a great deal of use as a part of the utilities portion of the library. COMPARE, developed by Harwell, compares two files, identifying differences. It is smarter than many straightforward implementations. As a missing line in one file is identified, the remaining lines are matched. This has become a surprisingly popular utility with the users. The PFORT Verifier, from Bell Laboratories, identifies deviations from portable Fortran 66 (ANSI X3.9-1966) and is also widely used beyond the support of the mathematical software library.

For high-use routines, software tailoring is done to attain efficiency for a particular machine while maintaining portable usage. This has become especially important for the CRAY-1 where even the vector compiler (CFT) cannot be expected to take full advantage of the hardware capability from Fortran. This is discussed in more detail by Dodson, Lewis and Poole [1981].

Efficient portability of usage is essential for user productivity. Transportability of software is essential for productivity of the library development staff. Thus the proper balance between these is very important.

Measuring the accesses to the Boeing Mathematical Software Library has been based to date on sampling the loader. Thus the usage figures cited in the previous section indicate only projected subroutine loads based on a sample period. A routine that has become a part of a user load module is not counted, and a routine accessed once but used many times within a program is counted only once. Thus the access figures given in the previous section represent a gross underestimate of the actual use of this software. We are currently considering other implementations from which we could get more useful statistics with negligible penalty to the user.

Finally, we mention the problem of purging old routines from the library. Even when a new and better capability replaces an outdated routine, the outdated version cannot simply be dropped. Some users do not have the resources or interest to change their programs to call a new routine. Others have "certified" programs which must be able to recompute precisely the same results on certain test problems. A new routine, even if it gives more accurate results more efficiently, cannot be used without "recertification."

We have taken the approach of dropping a routine from the user document first, so that new users will be directed to the improved capability. The old software, however, remains available on the system. When usage reaches a low level, we announce that this routine will be dropped. Users requiring it can then have the subroutine, for which they become responsible.

Deleting software as improved capability is added is a very real problem because of the significant results in research and development in mathematical software in recent years. It is an even more difficult problem for the Boeing Mathematical Software Library because of its historical development, a subject we consider in the next section.

THE BOEING LIBRARY: HISTORY AND FUTURE

The present Boeing Mathematical Software Library is based on two distinct libraries: BSLIB (Boeing Scientific Library) and LMR (Library of Mathematical Routines). Both started in the mid 1960's, developed by separate mathematical staffs on different computers in different Boeing companies. BSLIB was initially Univac-based, then converted to IBM hardware under the direction of Ken Wiegand, who managed the Numerical Analysis group in the Boeing Aerospace Company. LMR [Newbery, 1971] was developed for CDC hardware under the direction of A.C.R. Newbery in the Boeing Airplane Company.

When BCS was founded in 1971, the mathematics staffs were combined under S.L.S. Jacoby in BCS, and we began to merge the two libraries. The two were based on quite different philosophies. LMR contained over 700 routines and was documented in nine volumes. BSLIB had only about 150 routines documented in a single volume aimed at the "common use" areas. First steps in the merger involved comparison of some of the high-use routines and the transfer of capability between the libraries. BCSLIB became the name of the IBM library, EKSLIB the name of the CDC library.

By 1974, when the library was under the project leadership of Esko Cate, the multilevel structure was developed. Cate, Erisman, Lu, and Southall [1974] defined the structure of the library, and maintaining this structure has continued to be the goal of all work in the project. Libraries on the two machines were about 40% compatible at that time, and the problem of deleting routines was being addressed in earnest.

Kenneth Neves took over the project leadership of the library in 1975, followed by Ivor Philips in 1979.

The first minicomputer version of BCSLIB was installed on VAX and PDP machines within Boeing in 1978. Subsets of BCSLIB were selected based on usage of these computers. Another subset was adapted for the TERAK microcomputer in 1980. With the issuance of new user documents for the IBM- and CDC-based libraries in 1981, both now named BCSLIB, compatibility between them will be at about 90%. When the CRAY was added to the network with the CDC computers in 1981, the CDC version of BCSLIB was converted to the CRAY, leaving common libraries in these areas.

The objective of the library project is to offer a transportable Fortran-based core library, common to all large-scale machines, with appropriate subsets of this on all Boeing computers including micros, minis, and array processors. This will allow portability between all machines. To the extent reasonable, tailored versions of the software will be provided to achieve efficiency on the different machine types, especially for the CRAY and array processors.

In the second level software, further development of the "petal" documents is required. It is anticipated that the second-level software coverage will be more dynamic than the core library and therefore more frequently updated and revised.

A further goal is the development of better usage statistics. Better data on usage would be a helpful guide in deciding what routines to tailor and where expanded capability may be beneficial.

The CRAY and the Floating Point System Array Processor, from our experience, have had two interesting impacts on mathematical software. Because their hardware architecture is significantly different from previous hardware, they seem to have upset the dream of providing adequate mathematical software across machine types with one portable Fortran version. Some tailoring is required for reasonable efficiency. Because of the more complex architecture, however, more interest has been expressed in mathematical software for these computers than we ever have observed for more traditional hardware. Users seem more reluctant to "program it themselves" on these high performance machines and are looking for building block software. It appears to us, then, that the new hardware types simultaneously create a challenge and an opportunity for mathematical software libraries (see Neves [1981]).

The multilevel library structure plays a central role in our approach to meeting the users' efficiency needs in a vector library, as well as providing a mechanism for a "phased" conversion of the core library and other petals. The early phase of the vector library effort was simply conversion of existing libraries from CDC CYBER computers; no attempt was made to optimize code for the CRAY. The purpose was to maximize available capabilities at the earliest possible time and to aid users in their own conversion of programs that already use library software. This initial conversion was performed with minor effort. The process involved converting the library first to FTN5, the CDC ANSI '77 compiler (meeting this CYBER need), and then converting to the CRAY compiler, CFT. The core library, which is quite portable, was converted first. A CDC product, F45, was used to aid in conversion to FTN5. Another tool, FMTFIX (a BCS-developed tool), was helpful in automatic conversion of Hollerith expressions in FORMAT statements to FTN5/CFT acceptable form. Machine dependencies such as machine epsilon, maximum floating point number, and other environmental constants alluded to earlier were replaced by the appropriate CRAY-1 equivalents. Similar conversion activities related to other specialty libraries are under way or completed.

This conversion phase met the users' requirements for capability and compatibility. The second phase, which is ongoing, is aimed directly at efficiency. Three techniques are being employed simultaneously:

1. Evaluating and implementing generally available CRAY vector software.
2. Tailoring existing software to take advantage of CRAY hardware.
3. Offering a library of key computational modules, CRAYPACK, coded in CRAY Assembly Language (CAL).

The three techniques build on each other. For example, much of the generally available software (e.g., BLAS) is used in tailoring existing software; and, quite often, in tailoring a specific software module (e.g., sparse equation solvers), the inner-loops written become key computational modules (e.g., CAL-coded loops) made available in CRAYPACK.

The trend in computer hardware and supporting languages seems destined to put even heavier demands on mathematical software libraries and their developers. The next decade could very well bring a revolution in "nontraditional" computing. For example, library support to a multiple-instruction/multiple-data (MIMD) stream computer could add a whole new dimension to the battle between portability and efficiency. Currently, the move to distributed computing (intelligent terminals) is already being felt in the Boeing environment. This trend could truly test the multilevel library structure and place even stronger demands on it. Finally, through our participation in the X3J3 ANSI Fortran committee, we have observed new trends in the fundamental language for scientific computation. The structure for Fortran being contemplated has a core language with specialty modules (not unlike our library structure). Such a structure in Fortran can pose several difficult issues for library developers to resolve. For example, should libraries be developed entirely within the core language to avoid the use of "optimal" (and, hence, non-portable) vector extensions of Fortran? If the answer is yes, a second question occurs. Will there be a loss of efficiency on vector hardware that supports such "optimal" extensions of Fortran by using only "scalar" Fortran? The answer to this question will depend on both time and manufacturer.

SUMMARY

In this paper we have discussed several views of the Boeing Mathematical Software Library. We have concentrated, in this discussion, on the challenge of meeting the needs of a diverse community of users and described how this has led to a multilevel library structure. A large part of the library activity has not been discussed in detail — in particular, testing, version control, user feedback mechanisms, and usage monitoring. This omission was due to space constraints, not to our view of their relative importance.

A large part of the value of the library is the total service provided, not simply the individual subroutines. Ideally this library can be a significant means of technology transfer and productivity improvement for users.

The library is not "almost finished." New hardware technology and software language trends add to both the challenge and the need for effective mathematical software.

ACKNOWLEDGMENTS

The authors would like to thank Dr. S.L.S. Jacoby and Dr. James L. Phillips for careful reading of the manuscript and many helpful suggestions.

REFERENCES

Barwell, V. [1974]. Private communication.

Cate, E. G., A. M. Erisman, P. Lu, and R. M. Southall [1974]. "A user-oriented multi-level main library." *Mathematical Software II, ACM-SIAM Conference.* Purdue University.

Dodson, D. S., J. G. Lewis, and W. G. Poole [1981]. "Tailoring mathematical software for the CRAY-1." Submitted to the JPL/ACM-SIGNUM Conference — The Computing Environment for Mathematical Software, July 1981.

Dongarra, J., J. Bunch, C. Moler, and G. Stewart [1979]. *LINPACK Users' Guide.* SIAM, Philadelphia.

Duff, I., and J. Reid [1979]. "Some design features of a sparse matrix code." *ACM Trans. on Math. Soft.* 5:18-35.

Ford, Brian [1978]. "Parameterization of the environment for transportable numerical software." *ACM Trans. on Math. Soft.* 4:100-103.

George, A., and J. Liu [1979]. "The design of a user interface for a sparse matrix package." *ACM Trans. on Math. Soft.* 5:139-162.

Harwell [1981]. The Harwell Library.

Moler, C. B. [1979]. *MATLAB — An Interactive Matrix Laboratory.* University of New Mexico.

Neves, K. W. [1981]. "The vector computer challenge." To appear in *EPRI Parallel Processing Workshop Proceedings.* Dallas, Texas.

Newbery, A.C.R. [1971]. "The Boeing library and handbook of mathematical routines." *Mathematical Software.* Ed. John R. Rice. Academic Press, New York, pp. 153-169.

Ryder, B. G., and A. D. Hall [1973]. *The PFORT Verifier.* Bell Laboratories Computing Science Technical Report 12 (Rev. April 1979).

Appendix

BCSLIB STANDARDS

The BCSLIB standards (extracted from the BCSLIB User's Manual) for documentation, naming conventions, standard error procedure, simplified working storage, and software portability are reproduced here.

DOCUMENTATION

A user manual is provided for each computer system on which BCSLIB is available. For ease of reference the software provided has been categorized into sections. Thus, there is a linear algebra section, a statistics section, etc. Each section of the manual begins with an introduction to the software in that section. The section introductions describe the software in that section in general terms and contain some guidelines on appropriate usage. Each introduction is followed by a detailed index, giving the names, title (descriptive of the purpose), and location of specific documentation for each subprogram in that section.

Each subprogram is documented in the same way:

SUBPROGRAM NAME and TITLE   Each subprogram starts on a new page with its name and a one-line title. The title is descriptive of the purpose of the subprogram.

PURPOSE   A brief one-paragraph description of the purpose of the subprogram is given, sufficient to specify exactly what operations it performs.

METHOD   A concise description of the method employed is given. It generally ranges between one paragraph and three paragraphs.

USAGE   A brief illustration is given of the appropriate DIMENSION, COMMON, and CALL statements needed.

INPUT   A detailed description is given of each input argument in the calling sequence.

OUTPUT   A detailed description is given of each output argument in the calling sequence.

USER-SUPPLIED SUBPROGRAMS   Some BCSLIB subprograms require the user to supply an external subprogram. In these cases a detailed description of the input to and required output from the user-supplied subprogram is given. For example, a BCSLIB numerical quadrature subprogram requires the user to supply a subprogram that, at any specific point, evaluates the function to be integrated.

Some subprograms are supplied in several different precisions. On long word-length computers, usually only SINGLE PRECISION versions of subprograms are provided; while on shorter word-length computers, DOUBLE PRECISION versions are also available. In this latter case, there will be extra sections under the heading of OTHER PRECISIONS. The differences in usage for other precisions will be pointed out, and the USAGE section will be repeated (but no other).

In addition to the standard sections, where appropriate, REMARKS, WARNING, and USAGE EXAMPLE sections may be added:

REMARKS   Clarifying comments are given where some ambiguity as to input, output or method may exist.

WARNING   A warning is given when there is a danger of misusing or misinterpreting a subprogram.

USAGE EXAMPLE   A specific application of a subprogram is given when usage is complicated or difficult.

With few exceptions, all subprograms and parameters are named in accordance with FORTRAN 66 (ANSI X3.9-1966). When exceptions occur, they are clearly indicated.

BCSLIB NAMING CONVENTIONS

A naming convention has been established for subprogram names. Specifically, all subprograms whose input and output are primarily REAL have names that start with HS; those that are primarily DOUBLE PRECISION start with HD. Similar conventions exist for other types and precisions. The HS,HD,... prefixes in subprogram names are chosen to help minimize the possibility of name conflicts with operating system and user subprograms.

The FORTRAN coding in the usage abstracts has the following characteristics:

- FORTRAN 66 (ANSI X3.9-1966) is used. IBM extensions are used only when necessary (e.g., COMPLEX*16).

- The default naming convention for FORTRAN variables is assumed. This is the equivalent of the FORTRAN statement

$$\text{IMPLICIT REAL(A-H,O-Z), INTEGER(I-N).}$$

Any exceptions to this convention are noted when they occur.

STANDARD ERROR PROCEDURE

The BCSLIB subprograms test for and report detected errors to the user. Examples of errors tested for and reported are as follows:

- Usage errors, such as argument out of range.

- Computational errors, such as a singular matrix.

- Possible computational problems, such as loss of significance.

Usage errors and computational problems detected by subprograms are processed in a standardized three-step manner. This multiple reporting of errors is intended to ensure that the user is clearly aware of input errors or computational problems.

1. **Error Message Printed**

   When a library subprogram detects an error or warning condition, it calls HGERR - Standard Error Handler. HGERR prints a standard message which includes the subprogram name and the error code number (see Step 2 below). The output unit on which the messages are printed and the maximum number of printed messages are set to default values, which may be changed by a user.

2. **Success/Error Code Returned**

The subprogram call list contains an output argument IER — Success/Error Code — which defines the success or failure of the call. The convention used for IER is as follows:

IER$=0$ The call was successful; results were computed.

IER>0 Results were computed, but may not be useful.

IER<0 An error was detected, or the computation failed; no results are returned.

Each individual subprogram abstract defines the specific error code numbers and related conditions for IER $\neq 0$.

3. **Fatal Error Action**

A fatal error is indicated by IER<0. When a fatal error is detected, the primary output is set equal to a "clobber" constant. If the primary output is an array, only the first element is set equal to a "clobber" constant. The intent of "clobbering" the output is to present the user with an answer that is very distinctive, so that, if it is inadvertently used in a subsequent calculation, or is printed, it will become obvious immediately that an error has occurred. The "clobber" constants usually used are the largest machine-representable numbers of the appropriate type.

SIMPLIFIED WORKING STORAGE

A simplified working storage concept has been implemented in a number of easy-to-use driver subprograms. These drivers often call several lower-level subprograms, each of which may require one or more working storage arrays. The working storage part of the call to any such driver requires definition of only one working storage array. It will be named either WORK or HOLD.

WORK   A WORK array is used solely for scratch storage; that is, its contents do not have to be preserved after return from the driver using it. Its related length NWORK is always specified in the call to the driver.

HOLD   The contents of a HOLD array are not of interest to a user, but its contents must be preserved between successive calls to the same driver or to a related driver. Its related length is NHOLD.

The driver tests NWORK or NHOLD to determine if enough total working storage has been provided and acts as follows:

YES   The driver partitions the work array WORK or HOLD as needed and calls the lower level subprogram or subprograms.

NO   The driver calls HGERR - Standard Error Handler to print out an error message that defines the name of the driver subprogram and the amount of additional working storage required. The driver also sets the success/error code IER < 0 and returns to the calling program.

Note that if a user program calls more than one driver requiring working storage of the type WORK, the same WORK array may be used by each driver. In this case, the working storage array WORK should be dimensioned to the maximum required by the drivers sharing WORK, and NWORK should be set equal to that maximum.

SOFTWARE PORTABILITY

All significant programs eventually face the problem of conversion to a different system (on either the same or a different computer) from the one on which they were initially created. Either their usefulness outlives the system on which they were developed, or it is necessary to make them function on more than one system. The key to minimizing conversion effort is to consider problems associated with software portability as early as possible in the program design and development stages.

PORTABILITY CODING GUIDELINES

A set of guidelines for developing portable software (i.e., software that can, without modification, be compiled and executed on several different computer systems) is defined below. Use of these guidelines will help to minimize any future conversion effort when software is moved from one system to another. Portable software can be developed at the same cost as nonportable software and with no loss of efficiency at execution time. These guidelines are part of the standards that were used to develop the software in BCSLIB.

- **Use only portable FORTRAN (or COBOL).** The portability of FORTRAN code can be determined by using the industry standard verifier PFORT.

- **Use BCSLIB subprograms.** BCSLIB is available on all major BCS TSO/CTS systems including MAINSTREAM-TSO and MAINSTREAM-CTS. It is also available on MAINSTREAM-EKS. Many of its subprograms are designed to be easily portable. Names and calling sequences of almost all mathematical and statistical and many utility subprograms are unchanged from system to system. That is, it may sometimes be necessary to use double precision versions of some subprograms when moving from a long word-length computer to a shorter one, or vice versa. There is also a release procedure for source (or object) code for BCSLIB subprograms. Thus it is possible to develop portable programs on any major BCS computing system.

- **Isolate nonportable code.** Place any required nonportable code into individual subprograms.

- **Plan precision changes.** This situation is inherent in the hardware precision of various computer types. For example, it is common practice to change the precision from REAL to DOUBLE PRECISION when converting a program from the CDC Cyber series to the IBM 370 series, and vice versa. Such conversions can be performed more easily if the possible need for precision change is considered when the original coding is done.

- **Use BCSLIB for mathematical and machine constants.** Mathematical and machine constants, such as π, or the machine floating point overflow bounds, are provided in a portable manner. BCSLIB subprograms themselves use these mathematical and machine constants to improve their maintainability and portability.

PORTABLE SUBPROGRAMS IN BCSLIB

BCSLIB contains standardized portable library subprograms, which can be identified by the naming conventions. These subprograms were generally developed using the portability coding guidelines. However, in order to simplify the interface between library routines and user programs, exceptions were made to FORTRAN 66 (ANSI X3.9-1966) and the more restrictive PFORT requirements. First, as many as four alphanumeric characters were permitted to be stored in a single-precision floating-point word. Secondly, it was assumed that an integer will never require more space than a single-precision floating-point number, and that double precision and complex numbers will never require more space than two single-precision floating-point numbers. This permits the input of one single-precision array for work space into BCSLIB programs rather than separate work arrays for integer, single-precision, double-precision, and complex variables. It is felt that this is very unlikely to cause problems on any current or future installation of library software.

Each portable subprogram in BCSLIB falls into one of two categories of portability:

- **Portable code - portable usage.** Most of the subprograms in BCSLIB fall into this category. That is, the code can be installed on many different computer types (and has been at Boeing) with coding changes.

  It must be realized that, while there is no change in usage, there may be slight changes in output. This is due to arithmetic and other differences between computer systems. These differences are usually of little importance to a user since they typically affect only the least significant bits of returned results.

  In a different category are random number generators. The output from such subprograms normally will vary from one computer type to another. This makes correct conversion of programs that employ random number generators difficult to verify. The random number generators HSRPUN, HSRPNR, and HSRPEX provide the same output on all computer types that support at least 16-bit INTEGER and 21-bit REAL arithmetic. The output from HSRPUN (uniform random number generator) will be identical, and the outputs from HSRPNR (normal random number generator) and HSRPEX (exponential random number generator) will agree to about five decimal digits on different computer types. These generators are about 30 times slower than the standard random number generators and are intended primarily as a conversion aid.

- **Portable usage only.** A few specialized subprograms in BCSLIB provide basic capabilities whose nature is inherently nonportable, i.e., machine dependent. These subprograms are coded specifically for each computer type. The usage is the same, but internally they vary from system to system. An example is the provision of a mathematical constant such as π, which requires a decision on how many digits to specify. The appropriate number of digits varies from one computer to another and from single to double precision. The subprograms HSMCON, HDMCON, and JHMCON provide mathematical and machine constants to maximum accuracy in single-precision, double-precision, and integer form, respectively. Thus, executing the FORTRAN statement PI = HSMCON(12) will provide the most accurate single-precision value of π for a given computer type, and will continue to do so on any computer system on which BCSLIB is available.

A second example is the provision of date, time-of-day, or elapsed cpu-time subprograms. These often have different forms on different systems. Hence HGDATE, HGTIME, and HGSEC are provided to aid program portability between computers (on which BCSLIB is available).

A third example is the provision of several random-number generators that produce the same output sequences independent of the computer system they are on.

Chapter 13

THE PORT MATHEMATICAL SUBROUTINE LIBRARY

Phyllis Fox
Bell Laboratories
Murray Hill, New Jersey 07974

MOTIVATION AND HISTORY OF DEVELOPMENT

Bell Laboratories is a multi-computer environment with several large mainframe computers, and an increasing proliferation of midicomputers and mini-computers. Programmers typically start developing a program on a small local computer such as a DEC VAX, and then, particularly for large numerical calculations, move to the Honeywell 6000, or an IBM machine, or the CRAY-1. Because all these small- and large-scale machines have differing arithmetic structures and various flavors of Fortran, there is a danger that users could waste a great deal of time tailoring code to each new computer/compiler setting.

Fortunately, by the early 1970's, as more computers began to appear in Bell Labs, it was clear that the development of a portable numerical library over which numerical programs could be built would have tremendous advantages. Such a library, written in Fortran and designed to run on any computer with a compiler capable of accepting ANS Fortran, would enable users to move freely between machines. Even more significant, the model set by the library demonstrating the possibility of writing portable code would encourage users to adhere to good practice.

The PORT Mathematical Subroutine Library developed from this push for portability. In the sections below, we discuss first the approaches taken to make the library portable; then, to give the reader some idea of the scope of the library, we describe the effect of its setting in a research laboratory on the contents of the library.

Portability

Suppose we define portability as that quality of a computer program which affects how easily it can be moved from one computer to another. Then we find that portability is affected by the following incompatibilities between computers:

Language and compiler differences

Differences between operating systems

Variations in arithmetic:

  static number representation

  dynamic computation

Language and Compiler Differences

In order to avoid Fortran constructs that are not accepted by all compilers, PORT programs are written in that particular subset of 1966 ANS Fortran specified by the PFORT verifier. This language-checker was developed by B. G. Ryder [1974] at Bell Labs and has been released to the world at large where it sees steady and appreciative use. Programs that pass the PFORT verifier will compile almost everywhere.

Differences Between Operating Systems

The two aspects of operating systems most likely to affect Fortran programmers wishing to write portable programs are the problems of specifying file names, and the installation and loading of subroutine libraries.

For users whose programs involve input-output, the PORT library provides a mechanism, described in the next section, for designating the numbers of the READ and WRITE and error-message files. More complicated file usage is not considered to fall in PORT's purview.

So far as libraries and loaders are concerned, it is assumed that recipients are able to compile PORT programs onto a loadable library. To accommodate users whose loaders require a sequential ordering of programs, with called programs coming after their calling programs, the tape is sent out with programs given in this order.

Numerical Considerations

To enable programs to run on computers differing in floating-point number representation, a way must be found to specify the largest number, the smallest number, etc. for each computer. The process has two stages. First, there is the matter of choosing correct values; secondly, there must be a way to specify the values within a program.

Because PORT deals primarily with numerical computation, the values it uses for floating-point quantities, such as the largest and smallest numbers, and the number of significant digits, are not determined from static hardware machine representation, but rather are based on a PORT model that takes into account the dynamic behavior of computation. W. S. Brown [1981] has put the model on a firm theoretical foundation, and N. L. Schryer has developed an extensive probing test program which can be run on a computer to check the validity of a set of prescribed values. In Appendix A, Schryer describes the structure and use of his program.

In PORT, machine-dependent values are specified for the following real (single-precision) and double-precision floating-point quantities: the smallest positive magnitude and the largest, the smallest relative spacing between two consecutive floating-point numbers and the largest spacing (essentially the dynamic roundoff value), and the log to the base 10 of the base b. Machine-dependent values specified for integer quantities include the following: the number of bits per integer storage unit (word); the number of characters per word; the logical numbers for the READ, WRITE and error message units; the largest and smallest integer values; the base b of the floating-point arithmetic; the effective number of base-b digits in the mantissa; and the effective largest and smallest exponents.

A programmer or a program wishing to gain access to these values can call on three Fortran functions provided in PORT. These are named R1MACH for reals, D1MACH for double-precision values, and I1MACH for integer values. (Notice the PORT convention of using a digit as the second character of a function name to help prevent name conflicts.) The argument of the Fortran function specifies which value is called for. For example, since an argument of 2 for R1MACH specifies the largest real floating-point value, a programmer can call for the value using the statement

HUGE = R1MACH(2) .

A listing of the program for R1MACH is included as Appendix B. The error handling in the program, represented by the call to SETERR near the end of the program, is discussed in the following section, and the procedures for activating the constants for a given target computer are discussed under installation procedures in the third section.

The partial redundancy the reader may have noticed in the integer specification of floating-point values in I1MACH and the values provided by R1MACH is used during installation by a program that checks for consistency of the two specifications (although, of course, such consistency does not guarantee correctness: Schryer's test program is the appropriate tool to use).

All the programs in the PORT library which make use of machine-dependent values use the three Fortran functions to find them.

The approaches to portability discussed here have proved very successful. In fact a recent book containing programs for digital signal processing published by the IEEE [1979] specifies as a standard for the published programs, to ensure portability, the PFORT language subset of ANS Fortran and the use of the PORT functions for obtaining any machine-dependent values used in the programs.

Is There a Penalty for Portability?

The developers of the PORT library are asked from time to time to discuss the penalties that must ensue from the portable approach. The questions generally revolve around the restriction of the language, the use of calls to obtain machine-dependent constants, and the special care needed to develop portable numerical algorithms.

So far as the use of the PFORT subset of ANS Fortran is concerned, we have found that the benefits to be gained from using special local Fortran compiler options which deviate from the standard are trivial or nonexistent. If one is restricted to Fortran anyway, the standard variety is good enough.

The PORT method of obtaining machine-dependent values via calls to Fortran functions may be more expensive than the use of constants, but of course if the cost becomes noticeable, a first-time switch or other mechanism can be used to cut down the overhead. In general the cost in running time is minute.

The algorithms used in PORT are seldom influenced by the stricture of portability. However, one of the exceptions which is worth mentioning is the portable random number generator written by A. M. Gross and described on page S-2 of the IEEE book [Programs, 1979] mentioned above. This program guarantees to provide the same exact sequence of random numbers on any computer with at least 16 bits. The algorithm is based on Marsalgia's "super-duper" generator which combines a congruential generator and a Tausworth shift type generator. Of course, the generator can be written more efficiently by taking into account the characteristics of a specific computer; but even as it stands, the program is a remarkable *tour de force* and is tremendously useful in moving programs between machines.

Aside from such special cases, we and, more importantly, our users have detected few inefficiencies in the PORT library stemming from its portability.

PORT's Development in a Research Setting

The PORT library has been under development since 1974. The first edition was a small library of 151 subprograms, many of them lower-level routines without user documentation. However, at this release the portability and all the basic structure of the library (which will be further discussed below) was established and implemented. The second, and current, edition of PORT was released in the summer of 1977. This version contains 550 subprograms, of which 125 are documented in user reference sheets. Again, the other subprograms are lower-level modules. There are 41,000 lines of Fortran in the library source. The first printing of the user reference manual for PORT 2 was 700 copies; an additional 800 manuals have been printed since. There are 475 pages of documentation in the manual.

PORT has the advantage, not available to many commercial libraries, of being developed in the context of its user community. Program counseling at the various computer centers in Bell Labs provides a continuous flow of suggestions for easing the path of a user. But even more, the actual contents of PORT, its programs, are due to its setting. A chemist or physicist at Bell Labs will develop a need for a particular type of numerical computation and will confer with the research numerical analysts. This interaction may result in an extant program being modified, if necessary, to solve the problem or, in some cases, a new algorithm and program being designed and written. The programs in the PORT library have been developed in applications of this sort, and have been put into the library only after considerable experience has been gained with them.

A Case History of a Program

To cite a particular case history of this sort, consider the PORT program for numerical quadrature, QUAD, written by J. L. Blue [1977]. QUAD is an automatic quadrature routine with additional embellishments acquired during its use in a research setting. It does automatic recognition of endpoint singularities, which arise frequently, and the program even makes an automatic change of variable to handle them. This aspect was refined in use on actual problems, and so was the ability of QUAD to recognize the presence of noise in a function being integrated and to limit the attempted accuracy accordingly. After a period of test on actual problems, QUAD was put into PORT.

Among the other programs in PORT that have been developed the same way are differential equations programs and spline routines designed and written by N. L. Schryer, rational approximation routines written by D. D. Warner, a package of nonlinear equation solvers again from Blue, and, in a coming edition, linear algebra programs and optimization routines designed by L. C. Kaufman.

There are a few programs in the PORT library that have been adapted from outside sources, but only, of course, with the concurrence of their authors.

Scope

PORT does not claim to cover the entire front of numerical computation; gaps exist and generally can be attributed to one of three causes: the need is well covered by other libraries developed in Bell Labs or outside (e.g., statistical libraries), or the request and the pressures for something haven't appeared, or — perhaps — it is something we haven't yet had time to develop.

THE PORT FRAMEWORK

The foundation of the PORT library consists of the machine constants package discussed above, a package for automatic error handling, and a package to effect dynamic storage allocation, in a Fortran sense, by using a stack in a named COMMON region. A. D. Hall and N. L. Schryer have contributed heavily to the major design decisions and the implementation of these packages, and the programs are available as Algorithm 528 of the Collected Algorithms of the Association for Computing Machinery [Fox, Hall, and Schryer, 1980b]. A general overview of the PORT library is given in Fox, Hall, and Schryer [1980a]; the main points of the error handling and storage allocation are reviewed below.

Error Handling

In PORT, there are two types of error that can occur in a program: *fatal* and *recoverable*. A user can elect to recover from the second type of error, but, unless the option is taken, both types are considered fatal. In any case an error message is printed, giving the name of the subroutine in which the error occurred and the number of the error as given in the user reference sheets described in the following section. For fatal, or effectively fatal errors the run is terminated and a call is made to a dump routine. The dump routine itself is a local option: At Bell Labs, Murray Hill, a symbolic dump [Honeywell, 1976] is provided that lists the names of the variables and their values when the dump was called, including values in the list of active subprograms.

For the user who wishes to recover from an error and to gain control over the error-handling process, a "recovery mode" is provided. At any point in a run the user can enter the recovery mode and, while in this mode, can do any of the following: 1) determine whether a recoverable error has occurred, and, if so, obtain the error number; 2) print any current error message; 3) turn off the error state; and 4) leave the recovery mode. The user is fully responsible, while in recovery mode, for querying for errors and dealing with them. The occurrence of two unrecovered errors in a row causes a fatal error.

What Kind of Errors Are Fatal?

A design decision was made in PORT to treat as fatal any errors representing unrecoverable situations or user blunders such as setting an input parameter to an impossible value. Recoverable errors are reserved for situations which cannot be determined in advance, such as failure of convergence within an expected number of function evaluations, or the occurrence of a singular matrix.

Dynamic Storage Allocation Using a Stack

A dynamic storage allocator has been integrated into the basic PORT library structure. We consider this method for providing scratch space greatly superior to other methods; the historical approach of compiling workspace directly into individual subprograms is clearly inefficient, and the other general method of passing names of scratch arrays puts a considerable naming and dimensioning burden on the user. We have found that use of dynamic storage allocation in PORT leads to more clearly structured programs, cleaner calling sequences, improved memory utilization, and better error detection. The allocator is implemented as a package of simple portable Fortran subprograms which manipulate a dynamic storage stack.

In general, the casual PORT user need not be concerned about the operation or even the existence of the dynamic storage stack; the fact that the PORT subprograms are using the stack is invisible. However, for a user wishing to take advantage of the stack mechanism, a subroutine is available for obtaining space of a given type (logical, integer, real, double-precision or complex) and length on the stack — a pointer to the start of the allocated space is returned. There is a matching subroutine which can be called to deallocate the space by returning it to the stack, and it is also possible to query the mechanism to find out how much stack space is currently available. Another useful option, which has proved invaluable for debugging purposes, is the ability to list stack statistics: the number of outstanding allocations (there should be none at the end of a run), the current active length of the stack, the maximum active length achieved, and the maximum active length permitted.

The inclusion of the stack mechanism in PORT presents the possibility of several inventive applications which we have begun to explore. The current version of the stack has back-pointers marking off allocated regions, and when pointers are overwritten, a fatal error is signaled. For the next edition of PORT, in order to provide more detailed post-mortem information, D. D. Warner has written a stackdump which is called before the symbolic dump; it is very helpful in debugging. We realize that it would be possible also to include directly on the stack the names of active subprograms and even the CPU clock time at which they were called, leading to the development of a portable trace and timing mechanism. Finally, for computers with true operating-system dynamic memory allocation, it should be possible to coordinate the PORT stack with the local allocation scheme to minimize stack overflow problems.

Language Considerations

There are two non-ANS standard Fortran assumptions made in PORT. The first assumption is that there is no runtime subscript range checking. The second assumption is that variables initialized in DATA statements retain their most recently assigned values.

The assumption on subscript ranges allows dummy arrays in subroutines to be given a last subscript dimension of 1 under the assumption that larger values can actually be used. The extension of the assumption to cover arrays in COMMON means that although subprograms in PORT have been compiled with a default stack size of 500 double-precision words, the stack can be initialized in the main program to a larger size. Use of the "*" option of Fortran 77 will resolve this difficulty.

The second assumption, that a variable which is initialized by a DATA statement in a subprogram and then changed within the subprogram keeps the latest value from one invocation of the subprogram to the next, is used in PORT's error-handling and stack allocation packages. The values in the DATA statements are used to record various active statuses, such as the presence of an unrecovered error. It would have been possible, as the reader may have noticed, to store such values in the stack, rather than in DATA statements, but this was avoided in order to keep the error-handling and stack-management packages totally independent — a design decision made as a precaution against future design incompatibilities.

Modularity

PORT is structured like an onion. The programs most visible, on the outer layer of the library, are the simplest. The calls to these top-level routines need few parameters and are documented in brief reference sheets. The top-level routines, in turn, may set default values and call lower level routines containing more parameters. Some of the lower level routines are also documented and available to the more sophisticated user, who may wish, for example, to influence the details of the step-size monitoring in differential equation solution. Blue's description of the implementation of QUAD [Blue, 1977] illustrates these points very nicely.

At the innermost level, the picture simplifies again to a set of basic small programs to carry out single self-contained computations. PORT includes subprograms for complex double-precision arithmetic and for the trigonometric functions that are not ANS Fortran. It also provides routines for initializing a vector, for moving arrays or for changing their type, for determining whether a vector is (strictly) monotone, for finding the ceiling or floor of a floating-point quantity, and for doing internal sorting. The next edition will include the basic linear algebra modules (inner products, norms, etc.) proposed by Lawson *et al.* [1979]. All the subprograms are of course implemented portably and can be called directly by the user or by other routines in the library.

The modular structure of PORT, besides conforming to the currently approved style of programming, has the advantage that modules can be extracted for local optimization, perhaps to be written in machine language or in a vectorizable form of Fortran or even as co-routines in advanced systems (although we have not actually heard of this last adaptation).

TESTING — AN AXIOMATIC APPROACH

The chore, faced by developers of mathematical software libraries, of testing a growing number, n, of programs on an increasing proliferation, p, of computers is an $n \times p$ task.

Realizing the impossibility of the task, the developers of PORT have used instead an *axiomatic* approach to testing. We take the machine-independent framework discussed above and test each part of the structure, on a new computer/compiler target: The first test to be run must be Schryer's test of floating-point arithmetic. Then the PORT linguistic assumptions, described above, are checked, and, if all is well, the consistency check is run on the values assigned to the machine-dependent constants. Finally, a thorough test of the error-handling mechanism and the storage stack allocation is carried out on the new computer. (All of these tests have been designed and written by N. L. Schryer, and all but the floating-point test are sent out on the tape with Algorithm 528 [Fox, Hall, and Schryer, 1978b].)

If the new target environment passes all the tests, we assert that PORT programs will run correctly — since each program has previously been tested on another, but conformal framework. We have reduced the testing burden from $n \times p$ to $n \times 1$.

We hope that the bravado air of our statement will challenge people to search for counterexamples.

DOCUMENTATION AND INSTALLATION

There are two documents supporting the PORT library. There is a loose-leaf binder containing reference sheets for the programs in the library, and there is a PORT installation guide.

User Reference Sheets

The manual for users of the PORT library is divided into twelve chapters covering various areas of numerical computation. Each chapter contains an introduction describing the programs available and advising the user on the appropriate applications.

In Appendix C we show, as an example of a PORT reference sheet, the usage documentation for the program ZONE, which finds the zeros of a set of nonlinear equations. The usual information is given, but some points deserve mention. The heading in the example "ZONE — zeros of nonlinear equation" attempts to include as many key words describing the program as possible, because this line is used in generating KWIC-type (Key Word In Context) or permuted indexes. Under **Usage**, input parameters are signaled by an arrow to the right, and output by a left-facing arrow. Note that in ZONE the parameter X plays both roles. If the program has, as one of its arguments, a subprogram to be written by the user, the parameters to that routine are similarly displayed. Under **Error situations** the possible errors are described and, following the PORT error philosophy discussed above, flagged "fatal" or "recoverable." For every PORT program for which it is appropriate, there is a corresponding double-precision version with a "D" prefix, e.g., DZONE. The user is warned as to which of the parameters in the calling sequence should be declared double precision.

Most of the PORT routines use the dynamic storage stack, and usually the amount used depends on some dimensions of the problem, so we give a recipe to the user to figure out stack usage and determine, for large problems, if a larger stack should be declared.

Often the most important part of the documentation is the example of use; many users find they can edit an example to obtain a program to solve their own problem. Following the example we include mention of other closely related programs (here ZONEJ, which requires the Jacobian) and give the name of the author and pointers to germane references.

The documentation for the first edition of PORT was started very soon after Bell Labs had acquired a phototypesetting machine attached to a research DEC computer, and the language to drive the device was just settling down. Fortunately B. W. Kernighan's program *eqn* [Kernighan and Cherry, 1975] to process linear input text into equations had been written, and could be used to print such things as the summation sign shown in the reference sheet. On the hardware side, things were still a bit primitive (this was 1974), and I pulled each reference sheet through the developing device. By the second edition of PORT (1976-77) the situation was more organized, and I, together with a very able person in keypunching, produced the 475-page manual on the Computer Center's phototypesetter. Also by this time we had developed a set of macro instructions which made the input of the reference sheets much simpler than before.

Installation

The portability of the library gives us the benefit of maintaining a single version of the tape. When a copy of the library is requested, this tape is copied (in the appropriate character format) and sent out together with an installation manual. The first three pages of the manual describe the tape, explaining, for example, that the programs on the tape are in sequential order. The installation procedure is then described, showing how to particularize PORT to the computer at hand by activating (i.e., removing the C's in Column 1) the appropriate machine values, and then simply compiling all the programs onto a library.

The other part of the installation manual explains to people whose computer does not allocate storage as prescribed by the Fortran Standard, how to adjust the stack allocator.

The initial part of the Fortran source on the tape consists of comment lines listing the subprograms on the tape. A second file containing the examples from the user reference sheets is also provided.

PORT SITES

The PORT Mathematical Subroutine Library is a Bell Laboratories proprietary product (©1976). An educational institution may obtain a royalty-free license to use PORT for educational and academic purposes; there is a small service charge to help defray the cost of distribution. For commercial and governmental organizations, and for educational institutions desiring to use PORT for commercial purposes, a royalty is charged, currently a one-time fee of $3500.

By the start of 1980 PORT was installed at some 20 locations within Bell Laboratories, but since that time users of DEC VAX machines within Bell Labs have tended to send each other copies of the library, and internal distribution is no longer tracked.

There are 60 PORT licenses at various non-Bell Labs sites, the majority of which are educational institutions. Many of these are abroad — Austria, Australia, Belgium, France, Germany, Israel, Taiwan, Thailand, and places in between. In the United States license agreements for the PORT library have been signed by the Department of Energy.

The computers on which PORT is installed include the Burroughs B6700, a variety of CDC CYBER systems, the CRAY-1, Data General Novas, DEC PDP 10's and 11's, and the DEC VAX, a Hewlett Packard 3000, the Honeywell 66 series, IBM 3032 and other mainframes, ICL 1904A, Interdata 8/32, PRIME P400, Telefunken TR445, and UNIVAC 70.

CURRENT PORT-RELATED ACTIVITIES

The PORT library has played a central role in two activities undertaken in the last couple of years at the Bell Labs Murray Hill Computer Center. The first of these involved running benchmarks, and the second was a study of types of computations being performed using PORT.

Benchmarking

Early in 1978 it began to appear that Bell Labs was seriously interested in evaluating the CRAY-1 computer from the point of view of user needs for large-scale "number-crunching." We undertook to collect a set of test programs from relevant fields, primarily linear algebra and differential equations, and run them with timings and costs on a variety of computers. We built these test programs over the PORT library, so when the tests were to be run on various computers, PORT was installed (if it was not already in use) and the tests run. This approach simplified our work by an order of magnitude, because the actual test programs were quite small, and of course written portably, and could easily be sent to run at the test sites. We were particularly pleased that the people at Cray Research, Inc., were able to run the benchmark programs very quickly, with no trouble at all. After setting the three functions to return the appropriate machine-dependent constants, PORT was compiled on the CRAY-1 in 10 seconds (41,000 lines) and the tests run immediately. The results we obtained from the various computers seemed particularly "fair" to us in the sense that the code had not been changed between sites.

Monitoring PORT In order to find out which programs in PORT were being used, how much, and by whom, we put "hooks" into a selected set of routines on the Honeywell 6000 computer at Murray Hill. We chose routines at the bottom level — the heart, so to speak, of the various numerical areas. For example, in quadrature there is a fundamental routine that is called by a variety of user interface top-level routines. We compiled usage by computational area and by user department, and summarized our findings in an unpublished memo. The work was completed in 1979, and in some sense we picked the last possible date to do the monitoring, because many of the users represented in the Honeywell data are now using PORT on various minicomputers or on the CRAY-1; the same study could not be performed today.

SUMMARY

As we have noted above, the development of the PORT library in a research setting has had great advantages. We have been able to find out about, and meet, many of the numerical software needs of our users, and they in turn have appreciated the ability PORT has given them to run on a variety of computers across Bell Labs. The symbiosis has been fortunate.

REFERENCES

Blue, J. L. [1977]. "Automatic numerical quadrature." *The Bell System Technical Journal* 56: 1651-1678.

Brown, W. S. [1981]. "A simple but realistic model of floating-point computation." To appear in *ACM Trans. on Math. Soft.* 7.

Fox, P. A., A. D. Hall, and N. L. Schryer [1978a]. "The PORT mathematical subroutine library." *ACM Trans. on Math. Soft.* 4: 104-126.

Fox, P. A., A. D. Hall, and N. L. Schryer [1978b]. "Algorithm 528: framework for a portable library." *ACM Trans. on Math. Soft.* 4: 177-188.

Honeywell [1976]. *Honeywell Series 60 (Level 66)//6000 Fortran Manual.* DD02B, Appendix F.

IEEE [1979]. *Programs for Digital Signal Processing.* Ed. Digital Signal Processing Committee, IEEE Acoustics, Speech, and Signal Processing Society. IEEE Press, New York.

Kernighan, B. W., and L. L. Cherry [1975]. "A system for typesetting mathematics." *Comm. ACM* 18: 151-157.

Lawson, C. L., R. J. Hanson, D. R. Kincaid, and F. T. Krogh [1979]. "Basic linear algebra subprograms for Fortran usage." *ACM Trans. on Math. Soft.* 5: 308-371.

Ryder, B. G. [1974]. "The PFORT verifier." *Software — Practice and Experience* 4: 359-377.

Appendix A

DETERMINATION OF CORRECT FLOATING-POINT MODEL PARAMETERS

N. L. Schryer
Bell Laboratories
Murray Hill, New Jersey 07974

Most mathematical software packages are built upon a model of floating-point arithmetic, even if that model is simply that each operation should "be good to within a rounding error." If the model is correctly supported on a given machine, then a software package written using the model will perform correctly (within the framework of that model) on that machine. The more realistic the model, the more realistic the term "perform correctly" will be.

The correct parameters for a model on a given host machine — the base b, the number of base-b digits t, and the exponent range — are not trivial to obtain. If the model is at all realistic, it is not sufficient simply to look in the owner's manual for the machine and conclude, for example, that it is a base 2 machine with 56 bits in the mantissa, even though those facts may be correct. The user needs to know the *dynamic* behavior of the arithmetic unit upon the stored data. The accuracy of basic operations is crucial, and may be less than expected. Similarly, the useful exponent range may be less than expected due to anomalies near the fringes. There may also be fundamental design errors. For example, the Honeywell 6080 computer once had the property that a small number (approximately -10^{-39}) when divided by roughly 2 gave a large result (approximately 10^{+38}). A subtle design feature on the CRAY-1 results in $1 \times x = \infty$ whenever x has the largest legal floating-point exponent in the machine. These are not isolated or rare cases; the computing world is a jungle of individualistic and sometimes too clever arithmetic units.

Most computers allow the representation of far too many floating-point numbers to allow exhaustive testing of the arithmetic unit. For example, the IBM 370 series, in single-precision, is a machine with 6 hexadecimal digits in the mantissa and a base-16 exponent range of $[-64, +63]$. This represents more than 10^9 floating-point numbers. Simply checking that $x + y$ gives the correct result for all pairs of x's and y's would involve 10^{18} tests or, at one test per microsecond, many thousand cpu-years.

We cannot expect to be able to **prove** the arithmetic unit of a machine correct. In general, all we can do is gain *confidence* that it is correct.

This appendix presents a small and well-motivated subset of all floating-point (FP) numbers that, when used as test operands, triggers instances of all anomalous FP behavior the author is aware of. For a base-b machine with t base-b digits in the mantissa, there are $O(b^t)$ possible mantissas. The sample subset to be presented involves only $O(t)$ mantissas. This reduces the amount of testing to a reasonable level. The careful choice of those $O(t)$ mantissas also gives a good deal of confidence that they will uncover any FP troubles in the machine being tested.

To use these selected mantissas to detect problems, we need some definition of what the "correct" result of, for example, $x + y$ is. Without some definition of "correct" we cannot tell when we have uncovered an error and have an incorrect result. We shall use and briefly describe a previously developed model of FP arithmetic that will be taken to define correct arithmetic.

SUBSET CHOICE

Since we cannot afford to test all FP numbers, we can test only a subset. The choice of that test subset is crucial to the success and effectiveness of the test. The subset should be large enough to detect all known anomalous FP behavior and, if possible, increase the list of such. Yet the subset should be sufficiently coherent that people can easily grasp it and have "confidence" in its ability to trigger any incorrect FP behavior, if it exists, in the machine being tested.

To motivate and make clear the subset choice, we need to have a simple, clear model for the *representation* of FP numbers. For this we use the tried and true representation of Wilkinson [1963] and Forsythe and Moler [1967]. This machine model is based on the assumption that FP numbers can be represented in the signed, base b, t-digit form

$$\pm b^e \left(\frac{a_1}{b} + \cdots + \frac{a_t}{b^t} \right) = \pm b^e \sum_{i=1}^{t} \frac{a_i}{b^i} , \tag{1}$$

where either $a_1 = \cdots = a_t = 0$ or $0 < a_1$, $0 \leqslant a_i < b$, $i = 1, \cdots, t$, and $e_{\min} \leqslant e \leqslant e_{\max}$.

For our purposes, the unknown parameters in (1) are the base b, the number of base-b digits t, and the exponent limits $e_{\min}$ and $e_{\max}$.

It seems a good idea to use test operands with mantissas near the smallest (b^{-1}) and largest ($\sum_{i=1}^{t} (b-1) b^{-i} \equiv 1 - b^{-t}$) normalized mantissas. It also seems desirable to use operands with strings of 0's, 1's and $(b-1)$'s in their representations, so that isolated bits and bursts of bits are used. With these thoughts in mind, there are 5 "obvious" mantissa patterns:

$$b^{-1}+b^{-i} \qquad\qquad \text{for } i = 2, \ldots, t, \qquad Type\,1,$$

$$\sum_{j=1}^{i} b^{-j} \qquad\qquad \text{for } i = 1, \ldots, t, \qquad Type\,2,$$

$$0 \qquad\qquad Type\,3, \tag{2}$$

$$(b-1) \sum_{j=1}^{i} b^{-j} \qquad\qquad \text{for } i = 1, \ldots, t, \qquad Type\,4,$$

$$(b-1)(b^{-1}+b^{-i}) \qquad\qquad \text{for } i = 2, \ldots, t, \qquad Type\,5.$$

On a base-10 machine, for $i = 5$, examples from these five mantissa classes are

$0.100010 \cdots 0$

$0.111110 \cdots 0$

0

$0.999990 \cdots 0$

$0.900090 \cdots 0$.

For $b = 2$, clearly Types 4 and 5 are redundant and are not used in the test. The mantissas of (2) are natural in that they "bunch" together near $1/b$ and 1, the endpoints of the range for normalized mantissas. The patterns in (2) also *contain* the smallest (b^{-1}) and largest ($1-b^{-t}$) mantissas. As with software, most of the errors in machine design arise in the handling of conditional statements (IF's). Re-normalization is such a conditional response to the mantissa of the computed result being less than b^{-1} or greater than $1-b^{-t}$. Thus, the FP unit is most likely to fail near these extremities, and that is where (2) samples the mantissa most densely.

Absolutely no claim is made that (2) is sufficient to detect all anomalous FP behavior. The only claim made for (2) is that it is a small, simple and well-motivated subset of all FP mantissas that can be used to detect a vast number of FP arithmetic "problems" in existing machines. In fact, the mantissas in (2) can be used to detect at least one instance of **every** FP arithmetic problem the author is aware of. However, there are specific cases of pathology not in the sample set. For example, on the Interdata 8/32, $x - 16^{-64} x \equiv 0$ for all machine x's; yet the test finds this only for sample x's.

The number of FP mantissas represented by (1) is $O(b^t)$, while (2) represents only $O(t)$ mantissas. This reduces the number of mantissas to be used in the test to a very reasonable level. For example, on an IBM 370 machine in single-precision, we have $b = 16$ and $t = 6$. Using (2) as test mantissas, rather than all of (1), reduces the number of mantissas from roughly 10^6 to a few dozen. This reduction, from $O(b^t)$ to $O(t)$, allows testing to be done in minutes instead of millennia.

We have presented a sample set of FP numbers (2) to be used as operands in testing the dynamic behavior of the FP arithmetic unit of the host computer. How do we decide if $x + y$, for example, has been computed "correctly" by the machine? We now outline a previously developed model of FP arithmetic that will be used to define correct arithmetic for the purposes of the test.

THE MODEL

Let $fl(x * y)$ be the machine-computed value of $x * y$, where op is any of $+$, $-$, $\times$ or $/$. Then one way to assess the accuracy of FP arithmetic would be to use the representation model (1). Using the relation [Forsythe and Moler, 1967],

$$fl(x * y) = (x * y) \times (1 + \delta), \qquad |\delta| \leqslant \epsilon \equiv b^{1-t} \tag{3}$$

for all x and y in the sample set, we could attempt to compute $\epsilon \equiv \underset{x,y}{Max} |\delta|$ and then, indirectly, t. However, this approach has serious problems. First, δ is very difficult or impossible to compute accurately from (3). Second, Relation (3) allows $1 + 1 = 2 + b^{1-t}$ which is not acceptable since almost everyone agrees that $1 + 1 = 2$ **must** hold on any reasonable machine. Thus, (3) is not suitable to define "correct" FP arithmetic.

Clearly, at least some FP results are going to have to be exactly right, like $1 + 1 = 2$, but which ones? A model of the dynamic FP behavior of a machine, which takes this and many other things into account, is given in Brown [1980]. That model will be used to define "correct" FP arithmetic for the purposes of the test. In the interest of completeness and conciseness, an outline of the axioms and results of that paper is presented below.

First, we need to define some terms. The numbers defined by (1) are called *model numbers*, and the parameters must be chosen so they are a subset of the machine numbers. The smallest positive model number is

$$\sigma \equiv b^{e_{min}-1},$$

and the largest model number is

$$\lambda \equiv b^{e_{max}}(1 - b^{-t}).$$

The maximum relative spacing of FP numbers is

$$\epsilon \equiv b^{1-t}.$$

For any real number x, we say that x is λ-*bounded* if $|x| \leqslant \lambda$. We say that x is *in-range* if $x = 0$ or $\sigma \leqslant |x| \leqslant \lambda$, and is *out-of-range* otherwise. If $0 < |x| < \sigma$, we say x *underflows*, while if $|x| > \lambda$, we say that it *overflows*. Since error analysis is closely akin to interval analysis [Moore, 1966], it is convenient to formulate the axioms in terms of intervals. In particular, if the endpoints of a closed interval are both model numbers, we call it a *model interval;* if they are adjacent model numbers, we call it an *atomic model interval.*

If x is a λ-bounded real number, we let x' denote the smallest model interval containing x. Thus, if x is a model number, then $x' = x$; otherwise, x' is the atomic model interval that contains x.

Since the model is presented in terms of intervals, we need similar definitions to the above for intervals. We say that a real interval X is *in-range* or λ-*bounded* if all its elements are. Similarly, a real interval underflows or overflows if one of its elements does.

If X is a λ-bounded real interval, let X' be the smallest model interval containing X.

There are machine numbers and model numbers. The former is a superset of the latter, although the two sets may be identical.

AXIOMS

We now present the essence of the axioms of Brown [1980] that define "correct" arithmetic. The actual model is somewhat more complex in that it allows division to be "weakly" supported as a composite operation. The following outline gives the basic results for the case where all operations are "strongly" supported. Let x and y be λ-bounded machine numbers.

Axiom 1

Let $*$ be one of $+$, $-$, $\times$ or $/$. Then

$$fl\,(\,x*y\,) \in (\,x'*y'\,)'$$

provided that the interval $x' * y'$ is λ-bounded.

Axiom 2

$$fl\,(-x\,)\ \in\ (-(x')\,)\,'$$

Axiom 3

In comparing λ-bounded machine numbers x and y, the computer may report any result obtainable by an exact comparison of any $\hat{x} \in x'$ and any $\hat{y} \in y'$, but it may not report any other result.

Theorems 1 and 2 of Brown [1980] show that Axioms 1-2 imply Relation (3). However, Axioms 1-3 also imply the following exactness results.

Theorem 1

Let x and y be model numbers, and let $*$ be one of $+$, $-$, $\times$ or division. If $x * y$ is also a model number, then
$$fl\,(\,x * y\,) = x * y.$$
Thus, $fl(1 + 1) = 2$ and $fl\,(1/b\,) \equiv b^{-1}$.

It is these additional exactness results that make Axioms 1-3 especially well-suited to testing FP arithmetic.

The containment assertions of Axioms 1-2 and comparison assertions of Axiom 3 are the relations to be tested. If Axioms 1-3 hold for all x and y in the sample set, we shall declare the machine to perform FP arithmetic "correctly."

To illustrate the implementation of this testing procedure, consider the result of
$$0.101_b \times 0.11_b.$$
The exact value of this is $b^{-1}.1111_b$, or
$$b^{-1}\,(1/b + 1/b^2 + 1/b^3 + 1/b^4\,). \tag{4}$$
If the number of base-b digits t carried on the host machine obeys $t \geqslant 4$, then (4) is **exactly** the result the machine must give, according to the model. On the other hand, if $t = 3$, then the model only asserts that the computed result must lie in the interval $\left[\,b^{-1}.111_b\,,\ b^{-1}.112_b\right]$, if $b > 2$, or is $\left[\dfrac{1}{2}.111_2,\ \dfrac{1}{2}\right]$ if $b = 2$. In any case, it is possible to derive the left-hand and right-hand

endpoints of the interval which should contain the computed result. These endpoints can be evaluated exactly on the host machine. For example, by Theorem 1, the value of (4) will be computed exactly if the model is supported for $t = 4$.

It is possible to derive such containment formulae for all cases of $x * y$, where x and y are sample numbers, that are **exactly** computable within the model. If the model is supported to t base-b digits, then the test will report that fact. If t digits are not correctly supported, we can expect to detect a blizzard of errors, among which will be examples of each anomaly.

Using the above testing procedure, it is possible to determine the base b, the number of base-b digits t in the mantissa, and the exponent range parameters e_{min} and e_{max} which are correctly supported by the host machine.

The test has uncovered design errors in hardware and software and/or transient (intermittent) errors in five of the ten machine architectures it has been run on.

REFERENCES

Brown, W. S. [1980]. *A Simple but Realistic Model of Floating-Point Computation.* Bell Laboratories Computing Science Technical Report 83.

Forsythe, G., and C. Moler [1967]. *Computer Solution of Linear Algebra Systems.* Prentice-Hall, Englewood Cliffs, New Jersey.

Moore, R. E. [1966]. *Interval Analysis.* Prentice-Hall, Englewood Cliffs, New Jersey.

Wilkinson, J. H. [1963]. *Rounding Errors in Algebraic Processes.* Prentice-Hall, Englewood Cliffs, New Jersey.

Appendix B*

DEFINITION OF REAL FLOATING-POINT QUANTITIES: RIMACH

```
      REAL FUNCTION R1MACH(I)
C
C  SINGLE-PRECISION MACHINE CONSTANTS
C
C  R1MACH(1) = B**(EMIN-1), THE SMALLEST POSITIVE MAGNITUDE.
C
C  R1MACH(2) = B**EMAX*(1 - B**(-T)), THE LARGEST MAGNITUDE.
C
C  R1MACH(3) = B**(-T), THE SMALLEST RELATIVE SPACING.
C
C  R1MACH(4) = B**(1-T), THE LARGEST RELATIVE SPACING.
C
C  R1MACH(5) = LOG10(B)
C
C  TO ALTER THIS FUNCTION FOR A PARTICULAR ENVIRONMENT,
C  THE DESIRED SET OF DATA STATEMENTS SHOULD BE ACTIVATED BY
C  REMOVING THE C FROM COLUMN 1.
C
C  WHERE POSSIBLE, OCTAL OR HEXADECIMAL CONSTANTS HAVE BEEN
C  USED TO SPECIFY THE CONSTANTS EXACTLY WHICH HAS IN SOME
C  CASES REQUIRED THE USE OF EQUIVALENT INTEGER ARRAYS.
C
      INTEGER SMALL(2)
      INTEGER LARGE(2)
      INTEGER RIGHT(2)
      INTEGER DIVER(2)
      INTEGER LOG10(2)
C
      REAL RMACH(5)
C
      EQUIVALENCE (RMACH(1),SMALL(1))
      EQUIVALENCE (RMACH(2),LARGE(1))
      EQUIVALENCE (RMACH(3),RIGHT(1))
      EQUIVALENCE (RMACH(4),DIVER(1))
      EQUIVALENCE (RMACH(5),LOG10(1))
C
C  MACHINE CONSTANTS FOR THE BURROUGHS 1700 SYSTEM.
C
C     DATA RMACH(1) / Z400800000 /
C     DATA RMACH(2) / Z5FFFFFFFF /
C     DATA RMACH(3) / Z4E9800000 /
C     DATA RMACH(4) / Z4EA800000 /
C     DATA RMACH(5) / Z500E730E8 /
```

----

* The program listed here is substantially the same as a program in P. A. Fox, A. D. Hall, and N. L. Schryer, "Algorithm 528, Framework for a Portable Library," *ACM Trans. on Math. Soft.,* 4 (1978) 177-188. It is republished here with the permission of the Association for Computing Machinery.

```
C
C       MACHINE CONSTANTS FOR THE BURROUGHS 5700/6700/7700
C       SYSTEMS.
C
C       DATA RMACH(1) / O1771000000000000 /
C       DATA RMACH(2) / O0777777777777777 /
C       DATA RMACH(3) / O1311000000000000 /
C       DATA RMACH(4) / O1301000000000000 /
C       DATA RMACH(5) / O1157163034761675 /
C       MACHINE CONSTANTS FOR THE CDC 6000/7000 SERIES.
C
C       DATA RMACH(1) / 00014000000000000000000B /
C       DATA RMACH(2) / 37767777777777777777777B /
C       DATA RMACH(3) / 16404000000000000000000B /
C       DATA RMACH(4) / 16414000000000000000000B /
C       DATA RMACH(5) / 17164642023241175720B /
C
C       MACHINE CONSTANTS FOR THE CRAY 1
C
C       DATA RMACH(1) / 200034000000000000000000B /
C       DATA RMACH(2) / 577767777777777777777776B /
C       DATA RMACH(3) / 377224000000000000000000B /
C       DATA RMACH(4) / 377234000000000000000000B /
C       DATA RMACH(5) / 377774642023241175720B /
C
C       MACHINE CONSTANTS FOR THE DATA GENERAL ECLIPSE S/200
C
C       NOTE - IT MAY BE APPROPRIATE TO INCLUDE THE FOLLOWING
C       CARD - STATIC RMACH(5)
C
C       DATA SMALL/20K,0/,LARGE/77777K,177777K/
C       DATA RIGHT/35420K,0/,DIVER/36020K,0/
C       DATA LOG10/40423K,42023K/
C
C       MACHINE CONSTANTS FOR THE HARRIS SLASH 6 AND SLASH 7
C
C       DATA SMALL(1),SMALL(2) / '20000000, '00000201 /
C       DATA LARGE(1),LARGE(2) / '37777777, '00000177 /
C       DATA RIGHT(1),RIGHT(2) / '20000000, '00000352 /
C       DATA DIVER(1),DIVER(2) / '20000000, '00000353 /
C       DATA LOG10(1),LOG10(2) / '23210115, '00000377 /
C
C       MACHINE CONSTANTS FOR THE HONEYWELL 600/6000 SERIES.
C
C       DATA RMACH(1) / O402400000000 /
C       DATA RMACH(2) / O376777777777 /
C       DATA RMACH(3) / O714400000000 /
C       DATA RMACH(4) / O716400000000 /
C       DATA RMACH(5) / O776464202324 /
```

```
C
C         MACHINE CONSTANTS FOR THE IBM 360/370 SERIES,
C         THE XEROX SIGMA 5/7/9 AND THE SEL SYSTEMS 85/86.
C
C         DATA RMACH(1) / Z00100000 /
C         DATA RMACH(2) / Z7FFFFFFF /
C         DATA RMACH(3) / Z3B100000 /
C         DATA RMACH(4) / Z3C100000 /
C         DATA RMACH(5) / Z41134413 /
C         MACHINE CONSTANTS FOR THE INTERDATA 8/32
C         WITH THE UNIX SYSTEM FORTRAN 77 COMPILER.
C
C         FOR THE INTERDATA FORTRAN VII COMPILER REPLACE
C         THE Z'S SPECIFYING HEX CONSTANTS WITH Y's.
C
C         DATA RMACH(1) / Z'00100000' /
C         DATA RMACH(2) / Z'7EFFFFFF' /
C         DATA RMACH(3) / Z'3B100000' /
C         DATA RMACH(4) / Z'3C100000' /
C         DATA RMACH(5) / Z'41134413' /
C
C         MACHINE CONSTANTS FOR THE PDP-10 (KA OR KI PROCESSOR).
C
C         DATA RMACH(1) / "000400000000 /
C         DATA RMACH(2) / "377777777777 /
C         DATA RMACH(3) / "146400000000 /
C         DATA RMACH(4) / "147400000000 /
C         DATA RMACH(5) / "177464202324 /
C
C         MACHINE CONSTANTS FOR PDP-11 FORTRAN'S SUPPORTING
C         32-BIT INTEGERS (EXPRESSED IN INTEGER AND OCTAL).
C
C         DATA SMALL(1) /      8388608 /
C         DATA LARGE(1) /   2147483647 /
C         DATA RIGHT(1) /    880803840 /
C         DATA DIVER(1) /    889192448 /
C         DATA LOG10(1) /   1067065499 /
C
C         DATA RMACH(1) / O00040000000 /
C         DATA RMACH(2) / O17777777777 /
C         DATA RMACH(3) / O06440000000 /
C         DATA RMACH(4) / O06500000000 /
C         DATA RMACH(5) / O07746420233 /
C
C         MACHINE CONSTANTS FOR PDP-11 FORTRAN'S SUPPORTING
C         16-BIT INTEGERS   (EXPRESSED IN INTEGER AND OCTAL).
C
C         DATA SMALL(1),SMALL(2) /   128,      0 /
C         DATA LARGE(1),LARGE(2) / 32767,     -1 /
```

```
C      DATA RIGHT(1),RIGHT(2) / 13440,      0 /
C      DATA DIVER(1),DIVER(2) / 13568,      0 /
C      DATA LOG10(1),LOG10(2) / 16282,   8347 /
C
C      DATA SMALL(1),SMALL(2) / O000200, O000000 /
C      DATA LARGE(1),LARGE(2) / O077777, O177777 /
C      DATA RIGHT(1),RIGHT(2) / O032200, O000000 /
C      DATA DIVER(1),DIVER(2) / O032400, O000000 /
C      DATA LOG10(1),LOG10(2) / O037632, O020233 /
C      MACHINE CONSTANTS FOR THE UNIVAC 1100 SERIES.
C
C      DATA RMACH(1) / O000400000000 /
C      DATA RMACH(2) / O377777777777 /
C      DATA RMACH(3) / O146400000000 /
C      DATA RMACH(4) / O147400000000 /
C      DATA RMACH(5) / O177464202324 /
C
C      MACHINE CONSTANTS FOR THE VAX-11 WITH
C      FORTRAN IV-PLUS COMPILER
C
C      DATA RMACH(1) / Z00000080 /
C      DATA RMACH(2) / ZFFFF7FFF /
C      DATA RMACH(3) / Z00003480 /
C      DATA RMACH(4) / Z00003500 /
C      DATA RMACH(5) / Z209B3F9A /
C
C
       IF (I .LT. 1 .OR. I .GT. 5)
     1    CALL SETERR(24HR1MACH - I OUT OF BOUNDS,24,1,2)
C
       R1MACH = RMACH(I)
       RETURN
C
       END
```

Appendix C

PORT USER REFERENCE SHEETS: EXAMPLE

ZONE — zeros of nonlinear equations

Purpose: ZONE finds a set, x_i, approximately satisfying the n nonlinear equations

$$f_1(\,x_1, x_2, \cdots, x_n)=0$$
$$f_2(\,x_1, x_2, \cdots, x_n)=0$$

$$\cdot$$
$$\cdot$$
$$\cdot$$

$$f_n(\,x_1, x_2, \cdots, x_n)=0$$

Usage: CALL ZONE (FSUB, N, X, EPS, JMAX, FNORM)

FSUB $\longrightarrow$ a subroutine, provided by the user, to evaluate the functions, f_i , $i=1,2,\cdots,n$.

The subroutine must be declared EXTERNAL in the program calling ZONE, and must conform to the following usage:

CALL FSUB (N, X, F)

N $\longrightarrow$ the number of equations, n

X $\longrightarrow$ the vector (of length N) of x's

F $\longleftarrow$ a vector (of length N), computed within FSUB, with F(I) containing $f_i(x_1, x_2, \cdots, x_n)$

Note: The user should know that FSUB is called by ZONE while the recovery mode is set on.

N $\longrightarrow$ the INTEGER number, n, of equations and unknowns

X $\longrightarrow$ an initial guess for the REAL x vector

$\longleftarrow$ on output X contains values for x that approximately satisfy the system of nonlinear equations

EPS $\longrightarrow$ the REAL accuracy desired:
the computation is stopped when

$$\left[\sum_i^n f_i^2 \right]^{1/2} \leqslant \text{EPS}$$

JMAX $\longrightarrow$ an INTEGER; if the above stopping criterion is not met before the subroutine FSUB has been called more than JMAX times, the computation is stopped.

FNORM $\longleftarrow$ $\left[\sum_i^n f_i^2 \right]^{1/2}$ (REAL) for the output X vector

Error situations: *(The user can elect to "recover" from those errors marked with an asterisk.)

Number	Error
1	N < 1
2*	the initial guess for X created an error state in FSUB
3*	more than JMAX calls have been made to FSUB
4*	the process is not converging
5*	ZONE is not able to form the necessary Jacobian
6*	ZONE is not able to improve the convergence with a new Jacobian

**Double-precision
version:**   DZONE, with X, EPS, and FNORM declared double pre-
                             cision.

                             FSUB must compute double-precision F's from double-
                             precision X's.

Storage: $2N^2 + 6N$ real (or double-precision for DZONE) loca-
                             tions in the dynamic storage stack are used.

Example:

                             The following example shows the use of ZONE to com-
                             pute the zeros of the pair of nonlinear equations, starting
                             from an initial guess of $x_1 = -1.2$, and $x_2 = 1$:

$$f_1 = 10 \; (x_2 - x_1^2)$$

$$f_2 = 1 - x_1$$

                             The results were obtained on the Bell Laboratories, Mur-
                             ray Hill, Honeywell 6000 computer.

```
      REAL  X(2), FNORM
      EXTERNAL  ROSEN
      IOUNIT  =  I1MACH(2)
C
      X(1)  =  -1.2
      X(2)  =  +1.0
C
      CALL ZONE( ROSEN, 2, X, 1.E-2, 100, FNORM )
C
      WRITE ( IOUNIT, 9999 ) X(1), X(2), FNORM
 9999 FORMAT ( 1P3E15.6 )
      STOP
      END
```

```
SUBROUTINE ROSEN ( N, X, F )
REAL X(2), F(2)
F(1) = 10. * ( X(2) − X(1)**2 )
F(2) = 1. − X(1)
RETURN
END
```

1.000000E 00   1.000555E 00   5.552471E−03

For this example, an EPS of 10^{-5} (instead of 10^{-2}) gives answers correct to six figures.

See also: ZONEJ

Author: J. L. Blue

Reference: J. L. Blue, *Solving Systems of Nonlinear Equations,* Computing Science Technical Report #50, October 1976.

Chapter 14

THE EVOLVING NAG LIBRARY SERVICE

Brian Ford
Numerical Algorithms Group Ltd.
Oxford, England

James C. T. Pool
Numerical Algorithms Group Inc.
Downers Grove, Illinois

INTRODUCTION

The founding members of the Numerical Algorithms Group (NAG) identified four objectives:

- To create a balanced, general-purpose numerical algorithms library to meet the mathematical and statistical requirements of computer users;

- To support the library with documentation giving advice on problem identification, algorithms selection, and routine usage;

- To provide a test program library for certification of the library; and

- To implement the library as widely as user demand required.

These objectives, which have provided a foundation for the activities of the Numerical Algorithms Group for more than a decade, will continue as the principal aims of NAG for the foreseeable future.

The cornerstone of this foundation has been, and continues to be, collaboration. The NAG community consists of contributors, validators, implementors, site representatives, and users together with the NAG staff whose combined effort produces a library service responsive to continually changing needs of the computer user.

Reflecting these changing needs, NAG increasingly emphasizes the implementation of the library on new architectures ranging from microprocessor-based systems to supercomputers. Moreover, through active participation in research and development projects for software tools, NAG is moving toward the integration of the numerical algorithms library into a comprehensive programming environment including graphics, interactive access to documentation, and intelligent Fortran editors.

HISTORY OF NAG

The NAG Project began in 1970 when six British computing centers decided to develop a library of mathematical routines for their newly acquired ICL 1906A/S computers. The efforts of these six founding centers (University of Birmingham, University of Leeds, University of Manchester, University of Nottingham, Oxford University, and Atlas Computing Laboratory) included contribution of algorithms, implementation of software, and validation of the implementation. Mark 1 of the NAG Library was released on October 1, 1971.

Recognizing the quality of this implementation, universities in the United Kingdom with other computers became interested in the activity. Implementations on other computers were initiated at Mark 2 of the NAG Library. Simultaneously, participation in the NAG Project was expanded to include additional institutions. Algorithm contribution, implementation, and validation were coordinated from the University of Nottingham until August 1973. The Central Office of the project then moved to Oxford University, and the name of the project was changed from the "Nottingham Algorithms Group" to the "Numerical Algorithms Group."

Until mid-1975, the distribution of the NAG Library was, for funding reasons, restricted to university computing centers and related institutions in the United Kingdom. However, since a broad spectrum of scientific and engineering computer users actively sought access to the NAG Library, it was decided to make the library more widely available. Therefore, a non-profit company, Numerical Algorithms Group Ltd., was formed and a Library Service based on subscriptions to Mark 5 of the NAG Library was initiated in 1976. To service the demand for the NAG Library Service in North America, a subsidiary, Numerical Algorithms Group Inc., was established in 1978.

The nature of the resources within NAG continued as before, namely, the close collaboration between a full-time coordination staff and a large number of specialists in numerical analysis, statistics, numerical software, and related areas of computer science. These specialists are primarily members of internationally recognized academic or government research institutions. The active participation and encouragement of these people have been, and continue to be, an essential factor in the development and evolution of NAG.

In 1982, the role of the North American subsidiary, Numerical Algorithms Group Inc., was expanded to provide a focal point for interactions with potential contributors in North America. Although significant portions of the NAG Library have always originated from research projects in North America, a major effort was initiated to identify potential new contributors. Simultaneously, NAG Inc. became an active participant in long-range planning to ensure that the NAG Library continues to serve the user of state-of-the-art computer systems, especially microprocessor-based personal computers and workstations and supercomputers — the two extremes of the spectrum of computer systems.

Initially, and until Mark 5, the NAG Library was developed in both Fortran and Algol 6O, in parallel versions. However, as the use of and demand for the Algol 6O Library declined, its development slowed and eventually ceased altogether after Mark 8, although a service based on the Algol 6O Library is still supported. A project to develop a library in Algol 68 was begun in 1973. The contents of the NAG Algol 68 Library are similar to those of the NAG Fortran Library, but not as extensive. Development of this library is still continuing; and the use of it, though small, is slowly increasing. Currently, a small subset of the NAG Fortran Library is being translated into Pascal. These other language versions of the NAG Library have been developed according to the same principles and in the same collaborative manner as the Fortran Library, though with some occasional differences in organization. However, the rest of this paper is focussed exclusively on the Fortran Library.

THE NAG LIBRARY "MACHINE"

Over two hundred members of the non-profit company, Numerical Algorithms Group Ltd., and its subsidiary, Numerical Algorithms Group Inc., constitute the NAG Library "Machine" [Ford et al., 1979]. The members, each of whom has a specific interest, include

- **Contributors:** researchers in algorithms for mathematical and statistical computation who contribute algorithms, software, test programs, and documentation;

- **Validators:** peers of the contributors who certify that the contributions satisfy required standards; and

- **Implementors:** members who implement the library for a specific combination of computer, operating system, and compiler.

Members of NAG's full-time staff coordinate and participate in the "Machine" activities shown in Fig. 1. Moreover, they are responsible for assembly, distribution, maintenance, and support of the NAG Library.

CONTRIBUTION   New and enhanced library algorithms, software, test programs, and documentation

VALIDATION   Algorithmic certification

ASSEMBLY   Software certification; tailoring and processing: conforming to standards

IMPLEMENTATION   Certification in specific environment

DISTRIBUTION   Delivery of product to user

MAINTENANCE   Support of library in use

SERVICE   Response to users' inquiries

Figure 1. Activities of the NAG Library "Machine"

Contribution

The primary functions of a contributor are to identify the major types of problems encountered by users in a particular area of research interest and to provide the "best" algorithm in the library for each type. The algorithms are selected following stringent performance evaluation of contending methods.

Ideally, the contributor provides software that will run, virtually without change, to prescribed efficiency and accuracy, on all combinations of computer, operating system, and compiler supported by the NAG Library Service. Adaptable algorithms [Ford and Bentley, 1976] that can be realized as transportable software [Hague and Ford, 1976] are implemented in a subset of Fortran [Ryder, 1974]. This approach eliminates the principal problems of language dialects [Bentley and Ford, 1977], yet permits the development of reliable, robust software [Cody, 1974].

As a basis for the development of all NAG software, a simple model has been developed of features of any computer that are relevant to the software. This conceptual machine is described in terms of a number of parameters (e.g., base of floating-point number representation, overflow threshold), with each parameter given a specific value to reflect a feature of a particular computer (e.g., SRADIX [Ford, 1977] is 16 for IBM 360 and is 2 for CDC 7600). Hence, the contributor writes his routines for the conceptual machine. In execution, the routines use the assigned values of the parameters to adapt their performance to the actual machine. In this way the individual demands of accuracy and efficiency for the routine are met for every machine range.

For each routine the contributor provides an implementation test which will be used by implementors to demonstrate the operational efficiency and accuracy of the routine in each environment. Contributors also supply a draft document for each routine, including a complete example program illustrating its use.

Validation

The task of the validator is to certify the algorithmic correctness and the documentation of the relevant contributor. This second stage in the development of the NAG Library seeks to ensure that the problems addressed by the library contents are relevant to user requirements, that each algorithm is selected after due consideration, and that the user documentation is clear and concise. Substandard of ill-conceived material is returned to the contributor for modification and improvement. Since these activities involve individual assessment rather than incontrovertible fact, discussion and lively debate often ensue between contributor and validator.

Assembly

Once validated, the software and draft documentation are sent to the NAG Central Office for assembling and processing for general distribution at each new library mark.

Since the library is the work of many individuals, inconsistency and confusion inevitably arise in interpretation and satisfaction of standards. Hence, it is essential that the NAG Central Office check for compliance with these standards in assembling the codes and documentation provided by contributors. Wherever possible, the standards are checked with software tools, but there are certain standards that need to be checked by hand — for example, ensuring that the relevant chapter design is being followed and that the user interface chosen for routines satisfies the demands of the general library structure.

Once the staff has checked that the draft documentation conforms to content and format standards, the material is added to a documentation data base. The input form is in a typesetting language [Hooper, 1976] to permit preparation of phototypeset masters for printing of the library manual. An on-line form of documentation is extracted by program [Hague and Nugent, 1980] to provide the user with interactive access to documentation (see Fig. 2):

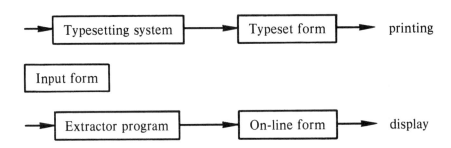

Figure 2. Processing of NAG Fortran Library Documentation

Figure 3 indicates the major steps in the Central Office procedure, established in the mid-1970's, for processing the software of the NAG Fortran Library and the principal software tools currently used. Processing of the contributed software by the Central Office, whether automatic or manual, is designed to achieve the following functions [Ford, 1983a; Du Croz, Hague, and Siemieniuch, 1977]:

- Diagnosing a programming error (algorithmic or linguistic).

- Altering a structural property of the text, e.g., imposing a particular order on non-executable statements in Fortran.

- Standardizing the appearance of the text.

- Standardizing nomenclature used, e.g., giving the same name to variables having the same function in different program units.

- Conducting dynamic analysis of the text.

- Ensuring adherence to declared language standards or subsets thereof.

- Changing an operational property, e.g., the arithmetic precision.

- Coping with arithmetic, dialect, and other differences between computing systems.

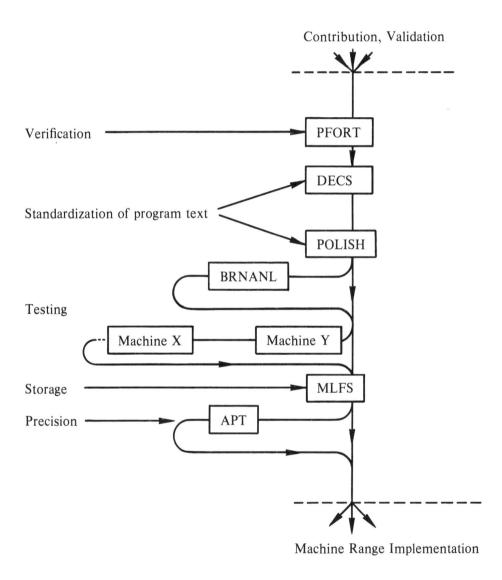

Figure 3. Processing NAG Fortran Library Software

As further certification of the efficiency, accuracy, and effectiveness of the contributed codes, Central Office staff run routines, together with implementation and example programs, on a variety of machines with different characteristics (currently ICL 2988, DEC VAX 11/78O, IBM 37O, and CDC 7600). This procedure helps gauge the extent to which test results may differ — information that is useful during the subsequent implementation stage.

Once certified, the routines are used to prepare an updated version of the "Contributed Library." This new version, known as a new mark of the library, is released approximately once a year and consists of the last generally released version of the library, supplemented by the newly certified routines and any other improvements or corrections. The implementation, example programs, and the NAG library documentation are supplemented in a similar manner.

Implementation

The number of implementations of the NAG Fortran Library depends on precisely what is counted. At the crudest level the library is available on over 3O distinct major machine ranges. If, however, we take into account minor variations in hardware, different precision versions, and different compiler versions, the number of distinct compiled versions of Mark 1O of the NAG Fortran Library is anticipated to exceed 5O.

Coordination of implementations requires the Central Office to supply the initial software, advise about any anticipated difficulties (e.g., machine code), help solve any problems that arise, and ensure that the implementation is of an acceptable standard (e.g., by examining the computed results).

Each implementation starts from its "Predicted Library" tape prepared by the Central Office. The tape holds

- The "Predicted Library" source text in the relevant precision(s);

- The example programs each with their input data and results, all in the relevant precision(s); and

- The implementation programs, each with their input data and results, all in the relevant precision(s).

The predicted source text is, of course, in the appropriate precision (single or double) and may also incorporate variants of some parts of the library that are required by a number of implementations in common. For example, double precision implementations that have no DOUBLE COMPLEX (or equivalent) data type require versions of the affected software in which this data type is simulated by DOUBLE PRECISION arrays with an extra leading dimension of 2. Machines with hexadecimal arithmetic require special modifications to some of the special function routines to deal with the problem of wobbling precision. These variants are generated by the Central Office. Subsequent modification of the source text by individual implementors is normally confined to assigning the correct values of the machine-dependent constants, editing the special function routines to select the correct code for the precision and range of the machine, and writing at most half a dozen routines in machine code (or in non-standard extended Fortran). The main function of the implementor is to read the material and then systematically to compile and to test the routines, example programs, and implementation test programs. The activity is essentially one of file handling, file management, and file comparison. Sophisticated programs have been developed for automatic comparison of results, for example.

Distribution

Each implementor prepares a certified library distribution tape. This contains a compiled version of the implemented library, the source text of the routines from which it was prepared, and the example programs, plus the input data and results computed during its certification. The structure and format of the tape are chosen for convenience in the particular computing environment. The contents and form of the tape are described in a Library Support Note, which also advises the site staff how to read the software from the tape into storage.

Maintenance/Service

The NAG Project has always emphasized the importance of basing its library service on a tested object-module library. Each site simply reads the library software into storage and may then confidently make it available immediately to its users. There is an annual update of the library mark and intermediate correction of software and documentation errors, if required. A numerical and software advisory service by telephone, telex, and letter is available from our offices in the United Kingdom and the United States.

An independent NAG Users Association was established in 1981. Its annual meetings are a valuable focus for communication between users and NAG.

LIBRARY CONTENTS

The contents of the NAG Fortran Library must evolve as research and development permit and user requirements demand. Therefore, the library is structured to enable the library contents to change with minimal inconvenience to users. The NAG Fortran Library is often used by individuals who are experts in a specific area of numerical analysis. Most users, however, are experts in a specific area of engineering or the physical, biological, or management sciences, with varying degrees of knowledge of numerical algorithms. Therefore, the library not only in contents but also in documentation must satisfy a broad spectrum of programming and problem specification requirements.

Hierarchy of Software

In general, most user requirements are satisfied by three types of library software [Ford and Bentley, 1978]:

Type	Function	Example
Problem solver	One routine to call to solve the problem	Solution of set of simultaneous real linear equations
Primary routine	One major algorithm contained in each routine	LU factorization
Basic module	Basic numerical utility designed by the chapter contributor for own use and other contributors' use	Extended precision inner product

This division into types yields an important principle for the design of the library: consistency. To ensure consistency throughout the library, a common error mechanism is used, and information is passed through calling sequences wherever possible.

At least three operational requirements can be recognized in the design of each calling sequence: 1) convenient and correct use by the programmer, 2) satisfaction of the needs of the algorithm, and 3) use of the data structures of the numerical area. These requirements underlie the preparation of interfaces for the three types of user software:

Problem solvers: minimum calling sequence

Primary routines: longer calling sequence, if necessary, to permit
                          greater flexibility and control

Basic modules: calling sequence designed to provide both flexi-
                          bility and efficiency for use within other library
                          routines.

The three types of library software will ultimately provide the three tiers
of a steady-state library structure. This, it is hoped, will prove sufficiently flexible to accommodate new development in numerical algorithms and changing
notions about software design, while preserving as much as possible of the existing structure of the library.

Organization of the Library and Documentation

The NAG Fortran Library and its documentation are organized into
chapters, each devoted to a mathematical or statistical problem area. Each
chapter has a one- or a three-character name and a title, e.g.,

S   - Approximations of Special Functions
D02 - Ordinary Differential Equations.

These names are based on the ACM Modified SHARE Classification. The names
of user-callable routines have six characters beginning with the characters of the
chapter name and ending with F, e.g., D02AJF. (When multiple-precision
implementations are available, the recommended precision ends in F and the
alternative precision ends in D if double precision and E if single precision.)

The *NAG Fortran Library Manual* is the principal documentation of the
NAG Fortran Library. It is composed of individual documents and has the same
structure as the library. For each chapter, there is an introduction designed to
help users identify the characteristics of their problem and direct them to the
specific routines included in the library to solve it.

Each chapter introduction consists of three sections: 1) scope of the
chapter, 2) background of the problems, and 3) recommendations on choice and
use of routines. The third section often includes decision trees, whose leaves
consist of the names of specific routines, and an index, whose entries point to
specific routines. Each chapter also has a contents summary, listing the routines
in the chapter.

Each routine document describes a routine of the same name and has 13 numbered sections with the following headings:

1. Purpose
2. Specification
3. Description
4. References
5. Parameters
6. Error Indicators
7. Auxiliary Routines
8. Timing
9. Storage
10. Accuracy
11. Further Comments
12. Keywords
13. Example

The sections on purpose, references, auxiliary routines, and keywords are totally implementation-independent.

The example program in Section 13 of each routine document illustrates a simple call of the routine. In addition to modification of the program text to take account of issues of precision, there arises the more fundamental question of the computed results. While ideally the sample problem will return an identical result on all machines (through selection of a well-conditioned problem with simple input data and returning the result to limited accuracy), not all numerical areas easily provide such examples; and appreciating when comparable results have been achieved may occasionally involve substantial numerical insight.

Example programs are provided with the Library Service. Many sites make the example programs available to users in machine-readable form. Users can then edit the example programs to produce a modified program that solves their problem.

For the flexibility of being able to run a program on many different machines, confident of access to the same library, the programmer must also read the appropriate *Implementation Document.* This document (revised at each release of the software) gives any necessary additional information that applies specifically to the implementation. In particular, it notes the precision of the standard library implementation and advises the user about the interpretation of italicized terms, the values of parameters of the conceptual machine, techniques for modifying the published example programs, and areas to expect significant differences in results. For each implementation of the NAG Library, the *Implementation Document* is a few (4-10) pages, comprising a manual of some 3000 pages. It is available on-line at most sites.

Algorithmic Content

At Mark 1O, the NAG Fortran Library includes 49O user-callable routines distributed across 31 chapters. The following list highlights the capabilities of the NAG Fortran Library at Mark 1O:

A02   COMPLEX ARITHMETIC. 3 routines: square root, modulus, quotient

C02   ZEROS OF POLYNOMIALS. 2 routines: zeros of real or complex polynomials

C05   ROOTS OF ONE OR MORE TRANSCENDENTAL EQUATIONS. 11 routines: zero of a real function of one variable; solution of systems of non-linear equations using function values only or first derivatives

C06   SUMMATION OF SERIES. 12 routines: discrete Fourier transform (FFT algorithm) of real, complex, or Hermitian convergence by epsilon algorithm; summation of Chebyshev series

D01   QUADRATURE. 25 routines. *One-dimensional:* adaptive integration over a finite or infinite interval allowing for various weight functions; weights and abscissae for Gaussian quadrature rules; integration of function defined by data values only. *Multi-dimensional:* adaptive integration over hyper-rectangle or hypersphere; Monte Carlo, number-theoretic or Sag-Szekeres method

D02   ORDINARY DIFFERENTIAL EQUATIONS. 34 routines. *Initial value problems:* integration over a range or until a function of the solution is zero; Runge-Kutta-Merson, Adams or Gear methods; comprehensive routines with various facilities for error-control and interrupts. *Boundary value problems:* two-point or general problems by shooting method, deferred correction or least squares collocation; second-order Sturm-Liouville problems

D03   PARTIAL DIFFERENTIAL EQUATIONS. 9 routines: parabolic problems in one space variable by method of lines; Laplace's equation for a two-dimensional domain; solution of finite-difference equations for elliptic problems in two or three dimensions

D04   NUMERICAL DIFFERENTIATION. 1 routine: derivatives of a function of a real variable

D05   INTEGRAL EQUATIONS. 2 routines: linear Fredholm equation of second kind

E01   INTERPOLATION. 7 routines: cubic spline, polynomial or rational interpolants

E02   CURVE AND SURFACE FITTING. 22 routines: least squares curve fit by cubic splines or polynomials; least squares surface fit by bicubic splines or bivariate polynomials; evaluation, differentiation and interpolation of fitted function; and l and l_∞ approximation by general linear function; Pade approximants

E04   MINIMIZING OR MAXIMIZING A FUNCTION. 35 routines: minimum of a function of one variable; minimum of a function of several variables — unconstrained, subject to simple bounds, or subject to nonlinear constraints, using function values only, first derivatives, or second derivatives; unconstrained minimum of a sum of squares; routines for checking user- supplied first or second derivatives

F01   MATRIX OPERATIONS, INCLUDING INVERSION. 60 routines: simple matrix operations; factorizations and similarity transformations of matrices (for use by routines in Chapters F02 or F04); inverses and pseudo-inverses

F02   EIGENVALUES AND EIGENVECTORS. 32 routines: standard eigenvalue problems — real, complex, real symmetric, real symmetric band, complex Hermitian; generalized eigenvalue problems — real, real symmetric, real symmetric band, complex; singular value decomposition

F03   DETERMINANTS. 9 routines: LU and LL^T factorizations

F04   SIMULTANEOUS LINEAR EQUATIONS. 26 routines: solution of systems of linear equations — real, complex, real band, real symmetric or complex Hermitian positive-definite, real sparse; least squares solutions using QR-factorization or singular value decomposition

F05   ORTHOGONALIZATION. 2 routines: orthogonalization of a set of vectors

G01   SIMPLE CALCULATIONS ON STATISTICAL DATA. 18 routines: means, standard deviations, 3rd and 4th moments; contingency tables; line printer scatterplots and histograms; statistical distribution functions and their inverses; normal scores

G02 CORRELATION AND REGRESSION ANALYSIS. 26 routines: product-moment correlation-coefficients; non-parametric rank correlation-coefficients; simple linear regression; multiple linear regression

G04 ANALYSIS OF VARIANCE. 4 routines: one-way; two-way — cross-classification or hierarchical classification; three-way — Latin square

G05 RANDOM NUMBER GENERATORS. 33 routines: random numbers from various continuous or discrete distributions; multivariate random Normal vector; random permutations or subsets; generate time series from ARMA model; initialization, saving, and restoring seeds

G08 NON-PARAMETRIC STATISTICS. 9 routines: tests of location, of dispersion, of fit, of association and correlation

G13 TIME SERIES ANALYSIS. 18 routines: differencing; auto-correlations; estimation of ARIMA models; forecasting; filtering; spectral analysis of univariate or bivariate series, with various smoothing options

H OPERATIONS RESEARCH. 7 routines: linear and quadratic programming; integer linear programming; transportation problem; M01 SORTING: real or integer vectors; rows of a real or integer matrix; character data

P01 ERROR TRAPPING. 1 routine: standard library error handling procedure

S APPROXIMATIONS OF SPECIAL FUNCTIONS. 41 routines: trigonometric and hyperbolic functions and their inverses; gamma and log gamma functions; normal distribution and error functions; exponential, sine and cosine integrals; Dawson's integral; Fresnel integrals; Bessel functions of orders O and 1; Airy functions; incomplete elliptic integrals

X01 MATHEMATICAL CONSTANTS. 2 routines: π and Euler's constant

X02 MACHINE CONSTANTS. 15 routines: fifteen constants characterizing machines

X03 INNER PRODUCTS. 2 routines: real or complex inner products using basic or additional precision

X04   INPUT/OUTPUT UTILITIES. 2 routines: unit numbers for error or advisory messages.

EVOLUTION OF THE NAG FORTRAN LIBRARY

Computer technology, both hardware and software, and user expectations are changing. Consequently, a general-purpose mathematical and statistical library must not remain static.

High-Performance Computers

The criteria for selecting algorithms for the NAG Fortran Library include user demand, robustness, numerical stability, accuracy, and speed. Ideally, these criteria should be satisfied for all combinations of computer, operating system, and compiler served by the NAG Library Service. High-performance computers, however, demand a review of algorithms relative to speed. Matrix multiplication, for example, may be programmed in six different ways by varying the order of nesting of the loops. While each variant performs the same number of arithmetic operations, the speed of execution varies significantly for some computer architectures (including both minicomputers and supercomputers with paging).

A partially tailored library for the Cray-1 has been introduced at Mark 1O of the NAG Fortran Library. Continued performance improvements for the NAG Fortran Library have the following aims:

a) To provide the user of the NAG Fortran Library with execution speeds approaching the full power of the high-performance computer;

b) To preserve generality of software where possible (including software for scalar machines), to ease management and maintenance of the NAG Fortran Library;

c) To develop limited machine-specific software, where necessary;

d) To preserve existing specification of routines to avoid changes to existing user application programs and library documentation; and

e) To introduce new routines in some areas of the library, which either by their design or by their choice of algorithms are better suited to give high performance on supercomputers, while still remaining useful on scalar machines.

The trade-offs involved in b), c) and e) imply that at some stage it may be desirable to develop special supplements to the NAG Fortran Library for specific architectures of high-performance computers.

Fortran Standards

The introduction of the Fortran 77 standard, while greatly to be welcomed, has presented problems for library distributors. NAG's policy toward the new standard has had to be governed by the availability of reliable Fortran 77 compilers and their rate of adoption by users. These factors have varied widely from one machine range to another. On some ranges, Fortran 66 and Fortran 77 compilers have coexisted for several years, and the degree of compatibility of the compiled code has varied considerably.

The Fortran 66 standards used by NAG are very largely compatible with Fortran 77, and the NAG Fortran Library has been successfully implemented using a Fortran 77 compiler on numerous machine ranges. On some ranges, however, two distinct compiled libraries have been distributed, one using the Fortran 66 compiler and one the Fortran 77 compiler [Du Croz, 1980 and 1982].

It is now proposed that the NAG Fortran Library make use of new features of Fortran 77 and, hence, cease to be compatible with Fortran 66 at Mark 12. This will allow some simplification of calling sequences, and improved software standards, especially in the areas of character handling and the generation of double-precision code. Calling sequences and documentation for existing routines will not be changed. The use of Fortran 77 features will be confined to new routines and to code within the library that is not visible to users. The previous mark will continue to be available, and will be supported, for the benefit of those sites who wished to remain compatible with Fortran 66 and those users who wished to continue working with their existing Fortran 66 programs.

This carefully planned transition from Fortran 66 to Fortran 77 is essential for the stability of the user's application programs. It also demonstrates the importance of standards in general, and Fortran standards in particular. Therefore, NAG has actively supported the participation of NAG members in the development of future standards for Fortran programming in an effort to represent the interests of the numerical software community.

Mark 11 of the NAG Fortran Library

Table I illustrates the evolution of the NAG Fortran Library. Thirty-seven routines are planned to be included in Mark 11 of the NAG Fortran Library. One routine has been scheduled for withdrawal at Mark 11; therefore, the

Table I

Number of User-Callable Routines

Mark	Added	Routines Deleted	Total
1	98	-	98
2	82	-	180
3	38	-	218
4	55	-	273
5	40	15	298
6	64	17	345
7	82	32	395
8	94	23	466
9	17	20	463
10	28	1	490

anticipated number of user-callable routines in Mark 11 is 526. Routines will be added in Chapters C06, E04, F01, F02, F04, G13, and S. For example, five additional capabilities for sparse linear algebra problems will be added:

- Incomplete Cholesky factorization of a sparse symmetric matrix;

- Eigenvalues and corresponding eigenvectors of a real sparse symmetric matrix, or a real sparse symmetric generalized eigenvalue problem, by simultaneous iteration;

- Solution of a sparse symmetric positive-definite system of linear equations by a pre-conditioned conjugate gradient method;

- Solution of a sparse symmetric indefinite system of linear equations by the SYMMLQ algorithm of Paige and Saunders; and

- Solution of a sparse linear least-squares problem by the LSQR algorithm of Paige and Saunders.

The above quantitative information about the addition and deletion of user-callable routines in past marks and this specific example of additional sparse linear algebra capabilities illustrate the planning required to maintain a stable, high-quality numerical library. Innovation prompted by new research results is demanded by the sophisticated user; however, stability is essential to preserve the investment of users in application software development.

THE FORTRAN PROGRAMMING ENVIRONMENT

In the following subsections, existing and anticipated capabilities supplementing the NAG Fortran Library are described. Since most of the relevant software has been designed to be consistent with the NAG Fortran Library, integration of current and future versions will yield a comprehensive Fortran programming environment.

Interactive Access to Documentation

Most users will not have a printed manual of 3000 pages adjacent to their terminal when programming. In any case they do not want to search through a manual to find a few essential pieces of information. Therefore, NAG has developed an on-line information system to supplement the printed documentation [Ford, 1983b].

The system provides a transportable, interactive "HELP" facility for users of the NAG Fortran Library. It has been implemented in Fortran 77 and is based on the "HELP" system developed at the University of Cambridge [Hazel and O'Donohue, 1980]. The role of the On-Line Information Supplement is to provide

• A summary of the overall facilities available in the current mark of the library (the system is, of course, updated with each mark).

• Advice on the choice of routines in a given subject area.

• An explanation of terms used (in some subject areas).

• A specification of use of chosen routines, under the following headings: purpose, (formal) specification, parameters, and error indicators and warnings.

• Details about local access to the NAG Fortran Library.

These forms of information are dispensed by the system during an interactive dialogue with the user. The interaction involves responses to topic "menus," keywords supplied by the user, or sequences of questions and answers.

The system consists of a processing program written in Fortran 77 and a data base. In installing the system, some site-dependent adjustments may be necessary; but experience suggests that typically less than two man-days are required to implement the system in its standard form, assuming a Fortran 77 compiler is available. The processing program in executable form requires about 2O kilobytes on a DEC Vax 11/78O, for instance. In fixed 8O-column card-image form, the total data base occupies around 7 megabytes.

Graphics

Recognizing the important role of graphics in numerical computation, NAG has developed an optional extension to the NAG Fortran Library, the NAG Graphical Supplement.

The Supplement has two levels of software. The top level consists of high-level graphics routines, each of which produces a substantial part or all of a plot. These routines invoke a small set of low-level primitive plotting routines, called the NAG Graphical Interface; these routines in turn call routines in locally available plotting packages. Each distinct plotting package requires a different version of the NAG Interface. For some of the widely available packages, appropriate interface versions are provided by NAG.

The Supplement serves as a bridge between the NAG Library and locally available plotting facilities. Because plotting is performed via the package-independent, high-level routines, no detailed knowledge of the local plotting system is required, beyond the calling of an appropriate plotting device nomination routine before plotting commences. Most user awareness of the low-level NAG Interface routines is limited to simple calls of routines to initialize the NAG Supplement software, set up a data-region-to-viewport mapping, advance the frame, or terminate plotting. However, further routines may be called directly by more advanced users to achieve special effects on their graphs. These users may also make direct calls to the underlying plotting package.

The Supplement as a graphical system offers NAG Fortran Library users several advantages compared with a local plotting system:

● Supplement routines are designed to interface closely with routines in the NAG Fortran Library;

● Documentation and mode of access is similar to that for the NAG Fortran Library; and

● Users' plotting programs can be portable between NAG sites, and not dependent on local plotting software.

Mark 2 of the NAG Graphical Supplement has 50 high-level routines with capabilities ranging from simple point plotting to perspective views of a single-valued bivariate surface from any viewing angle, utilizing data on either a regular or an irregular grid.

Software Tools

The software for the NAG Fortran Library, including test programs, comprises over 200,000 source text records. In the description of the activities of the members of the NAG Library "Machine," it becomes evident that the overall software management task is large, especially since the NAG Fortran Library is implemented on over 30 machine ranges. Automating repetitive processes is essential to avoid accidental introduction of errors. Automating complex processes is equally essential to gain access to both the manipulative and analytical powers of the computer.

Therefore, as previously described, NAG has continually used increasingly sophisticated software tools. Moreover, NAG has actively participated in the development of experimental software tools for use by its members. These tools have been invaluable for NAG; however, it became clear early in use of the various tools that a coordinated set of tools was highly desirable. Consequently, NAG welcomed the opportunity to participate in the Toolpack Project [Cowell and Osterweil, 1983].

The Toolpack Project is a collaborative effort initiated by six American institutions together with NAG in 1979. Its aim is to produce a systematic collection of software tools to facilitate the development and maintenance of Fortran 77 programs. These tools are designed mainly for interactive use. They are intended to be transportable between host systems with as little difficulty as possible. A strategy for using the tools in an integrated way has been developed, but the construction principles adopted also permit the use of tool components in a stand-alone mode.

NAG is contributing its experience in the implementation, validation, integration, and standardization of software and applying this experience to contributed tools. Moreover, because of its continuing effort to reduce the demands on contributors and implementors and to aid software integration, NAG is developing and will contribute, for example, a Fortran-oriented editor, an automatic precision transformer, and other tools.

The Future Scientific Workstation

A revolution in hardware technology is making available microprocessor-based workstations with substantial computing power, interactive bit-mapped graphics, large primary memories, large secondary disk-based memories, and high-capacity communication links. NAG has implemented the NAG Fortran Library, the On-Line Information Supplement, the Graphical Supplement, and various software tools including parts of Toolpack on this new generation of computer system. A systematic, step-by-step integration of this software is evolving toward a powerful programming environment for the development of application programs in Fortran. As in the development of the NAG Fortran Library, collaboration is the cornerstone of this effort. In this case, however, the collaboration will cross the boundaries of most areas of applied mathematics, statistics, and computer science.

CONCLUSION

The spectrum of activities described in the previous sections range from basic research in numerical analysis to contract negotiation. These activities are executed by over 200 members of NAG, both volunteers and full-time staff. The quality of the resulting library service is demonstrated by the continued increase in the number of subscribers worldwide. This collaborative effort through a non-profit organization is undoubtedly unique not only in high-quality results, but also in scope and scale. The reward to each member of NAG is principally the satisfaction of attacking challenging problems and achieving high-quality results.

REFERENCES

Bentley, J., and B. Ford [1977]. "On the enhancement of portability within the NAG project — a statistical survey." *Portability of Numerical Software.* Ed. W. R. Cowell. Springer-Verlag, New York, pp. 505-528.

Cody, W. J. [1974]. "The construction of numerical subroutine libraries." SIAM Review 16: 36-46.

Cowell, W. R., and L. J. Osterweil [1983]. "The Toolpack/IST programming environment." *Proceedings of SOFTFAIR, A Conference on Software Development Tools, Techniques, and Alternatives* (Arlington, Virginia, July 25-28, 1983). IEEE Computer Society Press, Silver Spring, Maryland.

Du Croz, J. J. [1980]. "The impact of Fortran 77 on the NAG Library." NAG Newsletter 1/80: 4-7.

Du Croz, J. J. [1982]. "Transition to Fortran 77." NAG Newsletter 2/82: 7-1O.

Du Croz, J. J., S. J. Hague, and J. L. Siemieniuch [1977]. "Aids to portability within the NAG project." *Portability of Numerical Software.* Ed. W. R. Cowell. Springer-Verlag, New York, pp. 389-404.

Ford, B. [1977]. "Preparing conventions for parameters for transportable numerical software." *Portability of Numerical Software.* Ed. W. R. Cowell. Springer-Verlag, New York, pp. 68-91.

Ford, B. [1983a]. "Experience in multi-machine software development: the NAG library — a case study." *Programming for Software Sharing.* Ed. D. Muxworthy. Reidel, Amsterdam.

Ford, B. [1983b]. "Transportable user documentation for numerical software." *Programming for Software Sharing.* Ed. D. Muxworthy. Reidel, Amsterdam.

Ford, B., and J. Bentley [1976]. "A library design for all parties." *Numerical Software — Needs and Availability.* Ed. D. Jacobs. Academic Press, London, pp. 3-20.

Ford, B., J. Bentley, J. J. Du Croz, and S. J. Hague [1979]. "The NAG Library 'Machine'." Software — Practice and Experience 9: 65-72.

Hague, S. J., and B. Ford [1976]. "Portability — prediction and correction." Software — Practice and Experience 6: 61-69.

Hague, S. J., and S. M. Nugent [1980]. "Computer-based documentation for a multi-machine library." *Practice in Software Adaptation and Maintenance.* Eds. Ebert, Luegger, and Goecke. North Holland, Amsterdam.

Hazel, P., and M. R. O'Donohue [1980]. "HELP Numerical: the Cambridge interactive documentation system for numerical methods." *Production and Assessment of Numerical Software.* Eds. M. A. Hennell and L. M. Delves. Academic Press, London, pp. 367-382.

Hooper, M. J. [1976]. *TSSD, A Typesetting System for Scientific Documents.* AERE-R 8574, HMSO, London.

Ryder, B. G. [1974]. "The PFORT verifier." Software — Practice and Experience 4: 359-377.